AF764490

Coal Processing and Utilization

Coal Processing and Utilization

Editor

Ombir Kumar

scitus
academics

Coal Processing and Utilization

Edited by **Ombir Kumar**

ISBN: 978-1-68117-677-2
Library of Congress Control Number: 2016958829

© 2018 by
SCITUS Academics LLC,
616, Corporate Way, Suite 2, 4766,
Valley Cottage, NY 10989

www.scitusacademics.com

This book contains information obtained from highly regarded resources. Copyright for individual articles remains with the authors as indicated. All chapters are distributed under the terms of the Creative Commons Attribution License, which permits unrestricted use, distribution, and reproduction in any medium, provided the original author and source are credited.

Notice

Reasonable efforts have been made to publish reliable data and views articulated in the chapters are those of the individual contributors, and not necessarily those of the editors or publishers. Editors or publishers are not responsible for the accuracy of the information in the published chapters or consequences of their use. The publisher believes no responsibility for any damage or grievance to the persons or property arising out of the use of any materials, instructions, methods or thoughts in the book. The editors and the publisher have attempted to trace the copyright holders of all material reproduced in this publication and apologize to copyright holders if permission has not been obtained. If any copyright holder has not been acknowledged, please write to us so we may rectify.

Printed in United States of America on Acid Free Paper

Preface

Coal preparation is the removal of undesirable material from the Run of Mine (ROM) coal using separation processes capable of distinguishing between the physical and surface characteristics of impurities and coal. It must be processed after coal has been mined. The coal burst goes to a plant where it is washed or possibly cleaned to remove contaminants. The cleaning removes impurities like rocks, ash, other substances and sulfur. Coal that is burnt in solid form can be used in a variety of preparation procedures. The easiest thing to do is to remove foreign material and to screen the size. A consistent product is achieved through the preparation of coal. The use of coal is the combustion of coal or possibly its conversion into solid, gaseous and liquid elements. The most important use of coal is by far in combustion, primarily to provide heat to the electricity plant boilers. In addition, techniques for gasifying and liquefying coal into fuels or possibly into feedstocks are well developed for the chemical industry.

This book provides comprehensive information on the characteristics of the coal industry; an overview of the methods of extraction and processing of coal and also discusses the main concerns relating to mining and processing of coal. The book will be of immense value to engineers and professionals in the fields of energy, mining, mineral, geological and metallurgical engineering, as well as to geologists and professionals in earth sciences.

Table of Contents

Chapter 1

Mineralogy, Distribution, Occurrence and Removability of Trace Elements During The Coal Preparation of No. 6 Coal From Heidaigou Mine

Xiangfei Bai[1,2] • Yue Wang[1,2] • Wenhua Li[3]

[1]*Beijing Research Institute of Coal Chemistry, China Coal Research Institute, Beijing 100013, China*

[2]*State Key Laboratory of Coal Mining and Clean Utilization,Beijing 100013, China*

[3]*National Institute of Clean-and-Low-Carbon Energy,Beijing 102209, China*

ABSTRACT

Optical microscopy and scanning electron microscopy in conjunction with energy dispersed X-ray spectrometry (SEM–EDX) were used to study the minerals and the concentrations of 33 trace elements in No. 6 coal from Heidaigou mine. The distributions, organic affinity and removability of 18 trace elements were studied by float-sink experiments. A determination of the maceral groups was also undertaken. A high mineral content, dominated by kaolinite, was found in No. 6 coal from Heidaigou mine. The bauxite content was relatively high and it was mainly present as individual particles in fusinite lumens or was intimately intergrown with carbonate minerals. The pyrite and quartz contents were low. Some marcasite with a parallel twin structure was observed by cross-polar reflected light. A small amount of bean-like goyazite was present in the calcite. The weighted trace element content in Heidaigou formations is relatively low, which is beneficial for coal processing and utilization. The concentrations of Ga, Hg, Pb, Se, Th, Ta are relatively high compared with the average values of Chinese coals. As, Hg, Mo, Ge, Ga, Ta, Ti, W, Mn are mainly present in minerals while B, Be, Th, P, Sc, Sr, V, Y, Yb are mainly found in organic matter. As, Ge, Hg, Mo are mainly present in sulfides and Be, Th, P, Sc, Sr, Y, Yb are mainly present in inertinite. B and V are mainly present in vitrinite. The high organic affinity and the low theoretical removability of most trace elements cause difficulties in removing them during coal preparation.

Keywords: Trace elements Occurrence Organic affinity Removability Heidaigou coal

1 INTRODUCTION

The Heidaigou coal mine is located in the middle of the Junger coalfield in the Inner Mongolia Autonomous Region. More than 30 million tons of raw coal is produced every year. As a preferred feedstock for power plants Heidaigou coal is primarily used for the generation of electricity. A low degree of anogenic metamorphism is the main coalification type. The main mineable coal bed, the No. 6 coal seam, was formed during the drying terrestrial delta deposition environment of the late Carboniferous period (Mao and Xu 1999).

Research into minimizing the hazards caused by trace elements is an important part of coal utilization technologies. Studies into the distribution and the occurrence of trace elements in coal are important. Bench samples from a drill core of the No. 6 coal seam were used to study the distributions and occurrence of trace elements in Heidaigou coal. It was found that Ga, Se, Sr, Zr, Hg, Pb, Th and REEs were relatively abundant while the Ga concentration was sufficient for industrial use. Ga, Th and REEs were found to be mainly distributed in boehmite in the No. 6 coal (Dai et al. 2006a). Their relative abundance in boehmite comes from the periodic drying of peat because of changes in the water table during the coal forming period (Dai et al. 2007). It has been shown that high concentrations of Ga in weathered coal come from the absorption of humic acid (Wang et al. 2011). Lead and selenium are mainly present in galena, clausthalite, and selenio-galena while Hg is mainly found in clausthalite and sulfides. High concentrations of Zr are attributed to the presence of zircon and Sr is present in goyazite (Dai et al. 2006b, c; Li and

Ren 2006). No. 6 coal from the Haerwusu surface mine that adjoins Heidaigou is enriched in Li, F, Ga, Se, Sr, Zr, REEs, Pb, and Th while Ga and F were mainly found in the clay minerals and partially in organic matter. Se and Pb are mainly present in the epigenetic clausthalite that fills the fractures. Sr and P may share the same carriers (Dai et al. 2008).

As a low-cost and mature technology coal preparation effectively reduces the amount of harmful elements in addition to improving the calorific values and lowering the ash and sulfur content of coal (Martinez-Tarazona et al. 1992). Coal preparation is thus believed to be an effective method to lower the concentrations of trace elements. This reduces the amount of environmental pollution by the trace elements (Pires et al. 1997; Klika et al. 1997, 2000; Tang et al. 2005; Song et al. 2010).

The geological setting of Heidaigou mine has been reported (Dai et al. 2006c). Production samples were used to study the coal's mineral composition. The distributions, occurrence and removability of trace elements from No. 6 coal were determined using float-sink experiments, petrological and statistical methods in this study. Correlation analysis was conducted to study the relationship among the trace elements and between the trace elements and the maceral groups, the ash content and the sulfur content. The theoretical content of trace elements in the macerals and minerals were then calculated using linear equations and the organic affinities of the trace elements are discussed. The removabilities of the trace elements were determined to provide guidance for the clean and environmentally-friendly use of Heidaigou coal.

2 EXPERIMENTS

2.1 Characteristics of the samples

The samples were obtained from the No. 6 coal seam at the Heidaigou surface mine according to GB/T 481-1993 (production coal sampling method). Proximate analyses (GB/T 212-2008), ultimate analyses (GB/T 476-2008,GB/T 19227-2008), total sulfur (GB/T 214-2007) and forms of sulfur (GB/T 215-2003), ash composition (GB/T 1574-2007), the fusibility of ash (GB/T 219-2008), the maceral groups (GB/T 8899-

2008), the microlithotype composition (GB/T 15590-2008), and the mean maximum reflectance of the vitrinite (GB/T 6948-2008) were determined following Chinese standards and the results are listed in Tables 1, 2, 3, 4, 5 and 6.

Table 1. Results of the proximate and ultimate analyses (wt%)

Proximate analysis				Ultimate analysis				
M_{ad}	A_d	V_{daf}	FC_d	C_{daf}	H_{daf}	N_{daf}	O_{daf}	S_{daf}
8.65	24.67	38.50	46.33	81.52	4.51	1.30	12.02	0.65

Table 2. Total sulfur and forms of sulfur in the coal (wt%)

$S_{t,d}$	Forms of sulfur		
	$S_{p,d}$	$S_{s,d}$	$S_{o,d}$
0.49	0.33	0.00	0.16

Table 3. Ash composition analysis (wt%)

SiO_2	Al_2O_3	Fe_2O_3	CaO	MgO	SO_3	TiO_2	P_2O_5	Na_2O	K_2O
23.84	57.18	3.58	5.12	0.43	0.72	2.48	0.18	0.24	0.81

Table 4. Fusibility of the coal ash (°C)

DT	ST	HT	FT
1,380	>1,500	>1,500	>1,500

Table 5. Maceral groups and the mean maximum reflectance of vitrinite

Maceral group (vol%)				R^o_{max} (%)
Vitrinite	Inertinite	Liptinite	Mineral	
31.87	44.88	8.78	14.48	0.52

Table 6. Microlithotypes in Heidaigou coal (vol%)

Vitrite	Clarite	Inertrite	Vitrinertite	Trimacerite	Liptite	Durite	Minerite
11.33	4.30	17.38	12.11	13.48	0.00	29.88	11.52

Heidaigou coal is a low-rank bituminous coal with a high ash content. The sulfur content is low and dominated by pyritic sulfur. The Al_2O_3 content is high while SiO_2 is relatively low in the ash. This results a high ash fusion temperature. The mine is located at the north edge of the carboniferous-permian coal accumulation basin in Northern China. The vitrinite content is lower than the average of the area while the inertinite and liptinite contents are higher (Han1996). The dominant vitrinite macerals are telocollinite and desmocollinite. A small amount of telinite is also present. The liptinite group is mainly represented by sporinite, which disperses in the desmocollinite and in the matrix of the inertinite in the bedding direction. Fragmental fusinite usually mixes with semifusinite. The main microlithotypes are vitrite, trimacerite and durite while a small amount of clarite, inertrite and vitrinertite are present. Vitrite is usually uniform while trimacerite is fragmental or of lineation shape. Durite is composed of an inertinite matrix and a clastic texture.

2.2 Mineral analysis

Minerals are the main carriers of most trace elements in coal. The distributions, occurrence and removability of the trace elements are largely affected by the presence of minerals. They are closely related to each other in terms of their genesis. Samples were prepared for microscopic analysis (ZEISS AXIOSKOP 40 optical microscope) according to the Chinese standard (GB/T 16773-2008). The optical characteristics of the minerals are described below and photographs were taken. A scanning electron microscope in conjunction with energy-dispersive X-ray spectrometry (SEM–EDS) was used to study the characteristics of the minerals and also to determine the distributions of elements in the macerals and in the minerals by spot and area scanning. Backscattered electron images (BSE) of the minerals were then obtained. The mineral compositions were determined by EDS both qualitatively and quantitatively. The acquisition time for the EDS analysis was 50 s and the scanned areas were determined by the size and morphology of the mineral particles.

2.3 Distributions and occurrence of the trace elements

2.3.1 Determination of the trace elements

The concentrations of Ag, As, B, Ba, Be, Cd, Co, Cr, Cs, Cu, F, Ga, Ge, Hg, Mn, Mo, Ni, P, Pb, Sb, Sc, Se, Sn, Sr, Ta, Th, Ti, U, V, W, Y, Yb, Zn in the No. 6 coal seam were determined.

The trace elements were determined by inductively coupled-plasma mass spectrometry (ICP-MS) except for As, F, Ge, Hg, Sb and Se. For ICP-MS analysis the samples were digested using a microwave high pressure reactor. F was determined by pyrohydrolysis in conjunction with a fluoride ion-selective electrode according to GB/T3558-1996. Hydride generation atomic fluorescence spectrometry analysis (HGAFS) was used to determine the concentrations of As, Ge, Sb and Se while cold-vapor atomic fluorescence spectrometry analysis (CVAFS) was used to determine the concentration of Hg.

2.3.2 Occurrence of the trace elements

The coal samples were crushed to less than 3 mm and fine particles of less than 0.5 mm were collected. Float-sink experiments were conducted using particles between 3 and 0.5 mm following the Chinese standard GB/T478-2008. The ash and sulfur content, the maceral groups and minerals (divided into clay minerals, pyrite, quartz and carbonates) and the contents of As, B, Be, Ga, Ge, Hg, Mn, Mo, P, Sc, Sr, Ta, Th, Ti, V, W, Y, Yb were determined in products of different densities and in the fines of particle size less than 0.5 mm. The precision of the float-sink experiment and the determination of the trace elements were evaluated by balancing the trace element content between the raw coal and the weighted mean values in the products of different densities in the float-sink experiments and in the fines of particle size less than 0.5 mm. The calculation was conducted according to Eq. (1):

$$R = \sum_{j=1}^{n} C_j W_j / C \tag{1}$$

where, R is the ratio of the weight mean trace element content in the products of different densities and of that in raw coal (%), C is the content of trace elements in the raw coal (μg/g), C_j is the content of trace elements in the products of a certain density including the fines of particle size less than 0.5 mm (μg/g), W_j is the yield of products of certain density including the fines of particle size less than 0.5 mm (%).

The dominant mineral components are clay minerals and the pyrite content is low resulting in a simple mineral composition for the Heidaigou coal. The coal can be divided into vitrinite, inertinite and various minerals in terms of coal petrology while liptinite is ignored. The mass fraction of organic matter that consists of vitrinite and inertinite, the minerals in the raw coal and the products of different densities can be calculated according to Eqs. (2) and (3).

$$MM = 1.1A_{\mathrm{d}} + 0.5S_{\mathrm{p.d}} \tag{2}$$

$$ORG = 100 - MM \tag{3}$$

where, MM is the mineral content of the coal (mass fraction), ORG is the organic matter in the coal (mass fraction), A_d is the yield of ash in the coal (%), $S_{p,d}$ is the pyritic sulfur content of the coal (%).

Because the macerals and the minerals in the coal petrological analysis are expressed as volume fractions the densities of vitrinite, inertinite and the minerals were 1.33, 1.40 and 2.60 g/cm^3, respectively, for the calculation of mass fraction in this paper.

A linear relationship was found for the trace element content of the raw coal (C_m) and for the different macerals and minerals (C_{mi}). The theoretical trace element (C_{mi}) content of the macerals and the minerals was calculated by multi-factor linear regression of the maceral and mineral compositions (W_{mi}). The trace element (C_m) content of the products of different densities were determined using Eq. (4):

$$C_{\mathrm{m}} = \sum_{i=1}^{n} C_{\mathrm{mi}} W_{\mathrm{mi}} \tag{4}$$

where, C_m is the content of a specific element in the coal (μg/g), C_{mi} is the theoretical content of a specific element in vitrinite, inertinite and in the minerals (μg/g), and W_{mi} is the maceral composition of the products of different densities (%).

The amount of trace elements in the organic matter of the coal can be expressed by organic affinity. The organic affinity index (A_o) of different trace elements in the coal was calculated using Eq. (5):

$$A_o = C_o W_o / C \tag{5}$$

where, C_o is the average content of a specific trace element in the organic matter (μg/g), C is the content of trace elements in the raw coal (μg/g), and W_o is the content of organic matter in the coal (mass fraction).

2.4 Removability of the trace elements

The theoretical degree of trace element removal depends on the organic affinity and on fine coal recovery. It is the ratio of trace element content in waste coal to that in raw coal. Suitable separation density and trace element washability was obtained by float-sink experiments. The theoretical degree of trace element removal from the coal preparation can be calculated using Eq. (6).

$$\lambda = (C - C_{\mathrm{r}} \times FCR)/C \tag{6}$$

where, λ is the theoretical degree of trace element removal (%), *C* is the trace element content of the raw coal (μg/g), C_r is the trace element content of the clean coal (μg/g), *FCR* is the fine coal recovery (%).

3 RESULTS AND DISCUSSIONS

3.1 Mineralogical composition of No. 6 coal from Heidaigou mine

A high mineral content, dominated by kaolinite, was found in No. 6 coal from Heidaigou mine. Its bauxite content is relatively high while pyrite and quartz are low.

(1)*Clay minerals* Kaolinite is the main clay mineral in Heidaigou coal with some illite also present. Kaolinite is mainly dispersed in thin-layers or as individual particles (Fig. 1) and it is also found in the fusinite lumens. It appears uneven in reflected light with a gray–brown color and the kaolinite has a dim gray to yellow interference color under cross-polarized reflected light. The interference color changes slightly when the stage is rotated. The Si/Al ratio of Kaolinite is between 0.9 and 1.1 (Table 7) indicating a relatively high Al content and a comparatively low Si content. A low impurity content of zircon (Fig. 2), rutile (Fig. 3), chalcopyrite (Fig. 4) and Pb-bearing minerals (Fig. 5) was found in the Kaolinite by SEM–EDX analyses.

Table 7. Si/Al ratio of the clay minerals

Sample No.	Scanning area (μm × μm)	Acquisition time (s)	Si/Al ratio
1	250 × 250	50	1.04
2	250 × 250	50	1.06
3	100 × 80	50	0.90
4	250 × 250	50	1.01

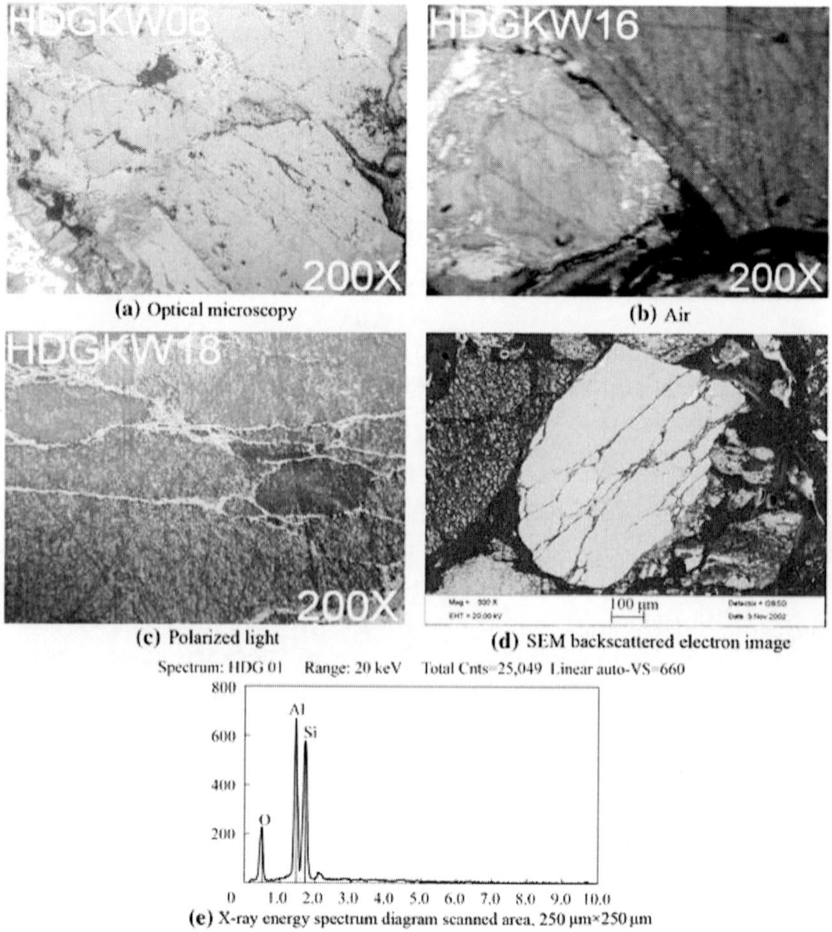

Figure 1. Kaolinite in Heidaigou coal

(2)*Bauxite* The presence of bauxite in Heidaigou coal results in a relatively high Al_2O_3content for the coal ash (57.18 %) and a high coal ash fusion temperature (temperature higher than 1,500 °C). A gray reflection color and an uneven surface were found for bauxite while a dim gray–yellow or no interference color was found when using cross-polarized reflected light. The interference color does not change when the stage is rotated. Therefore, it is difficult to distinguish bauxite from kaolinite using optical microscopy. Individual particles (Fig. 6) are the main form of bauxite, which is also found in the fusinite lumens (Fig. 7) and it is intimately intergrown with carbonate minerals (Fig. 8).

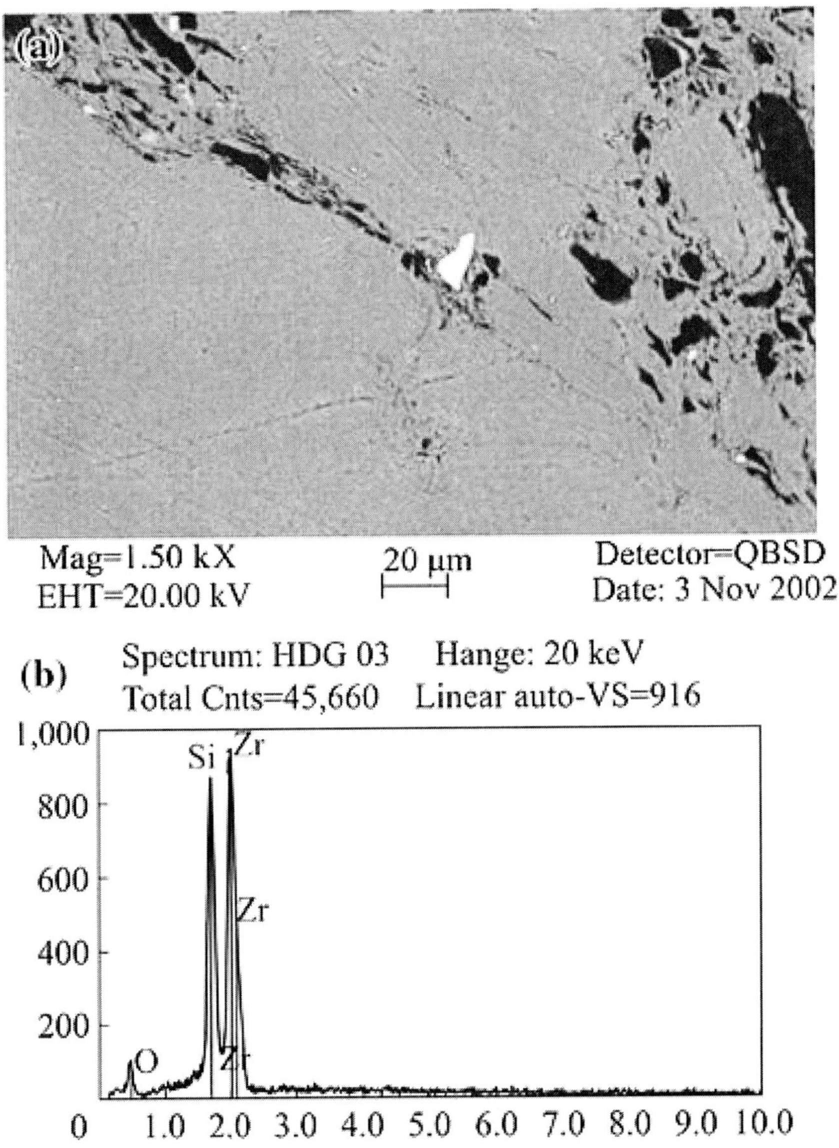

Figure 2. Fine-grain zircon in the kaolinite

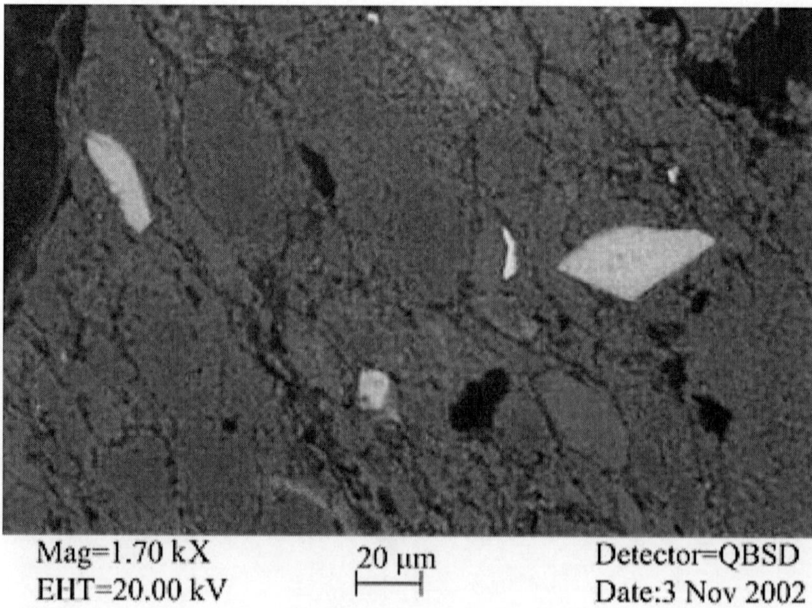

Figure 3. Fine-grain rutile in the kaolinite (SEM backscattered electron image)

(3)*Carbonates* The carbonate mineral content is relatively high and the dominant mineral is calcite. Calcite mainly fills fractures (Fig. 9a) and cell cavities (Figs. 10,11a, c). Some calcite was intimately intergrown with pyrite (Fig. 9b) and bauxite (Fig. 8). A significant amount of Fe and Mg that probably displaced Ca as isomorphs were found at the edge of the calcite (Fig. 11b, d). Small amounts of impure minerals such as irregular Al-bearing minerals and bean-like goyazite were found in the calcite particles (Fig. 11b, e).

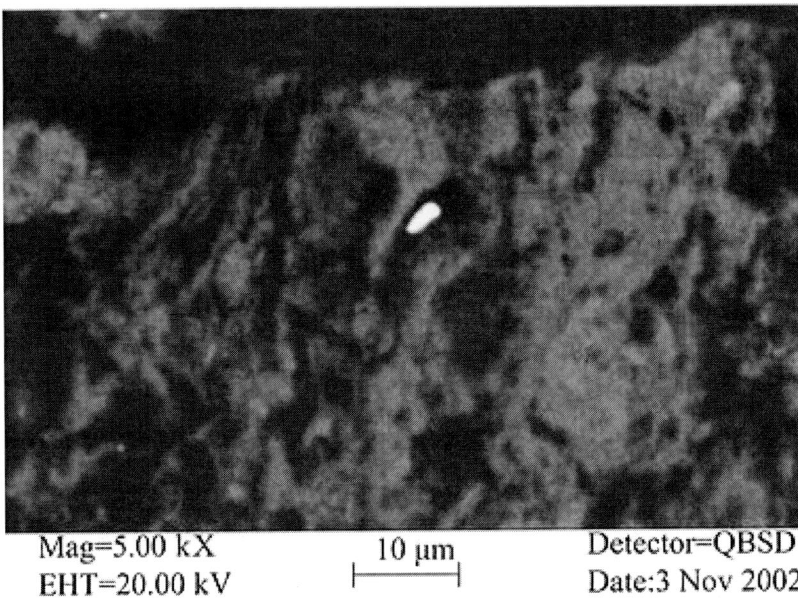

(a) SEM backscattered eelctron image

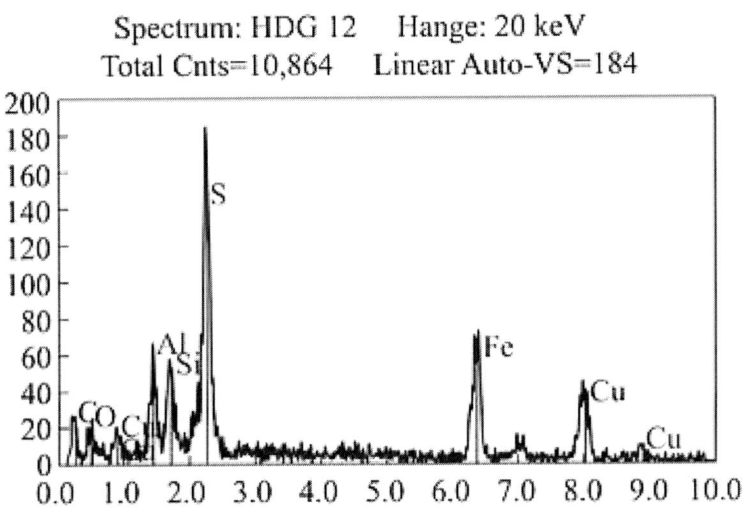

(b) X-ray spectrum diagram, spot scan

Figure 4. Fine grain chalcopyrite in the kaolinite

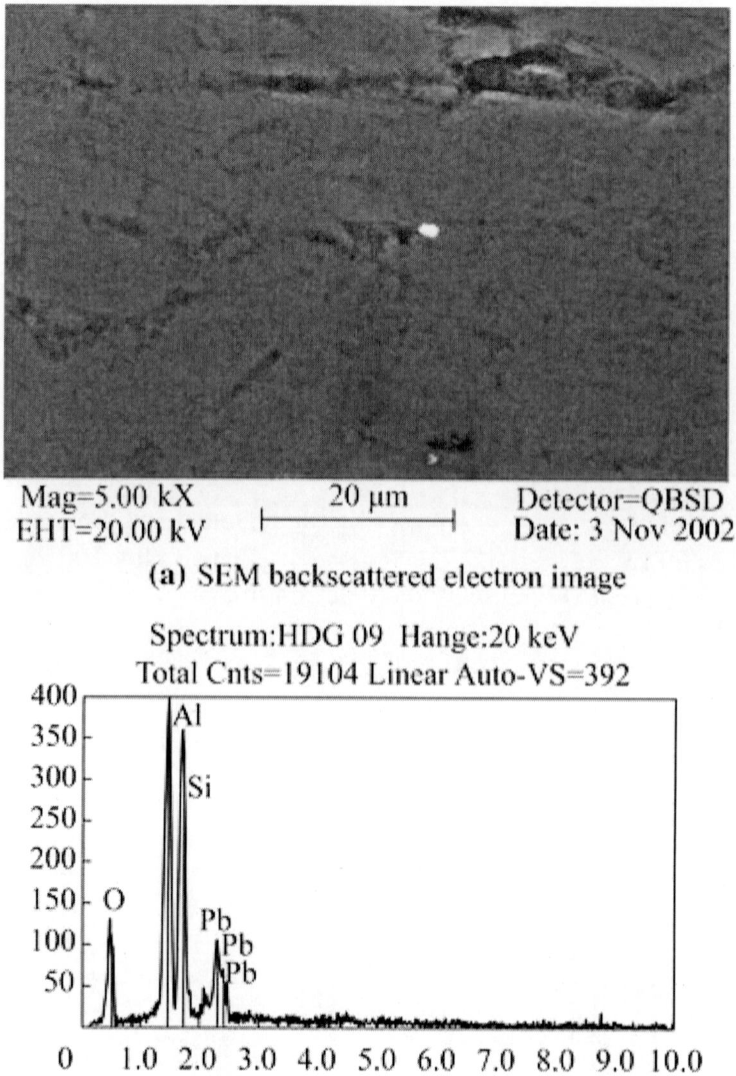

Figure 5. Pb-bearing minerals in the kaolinite

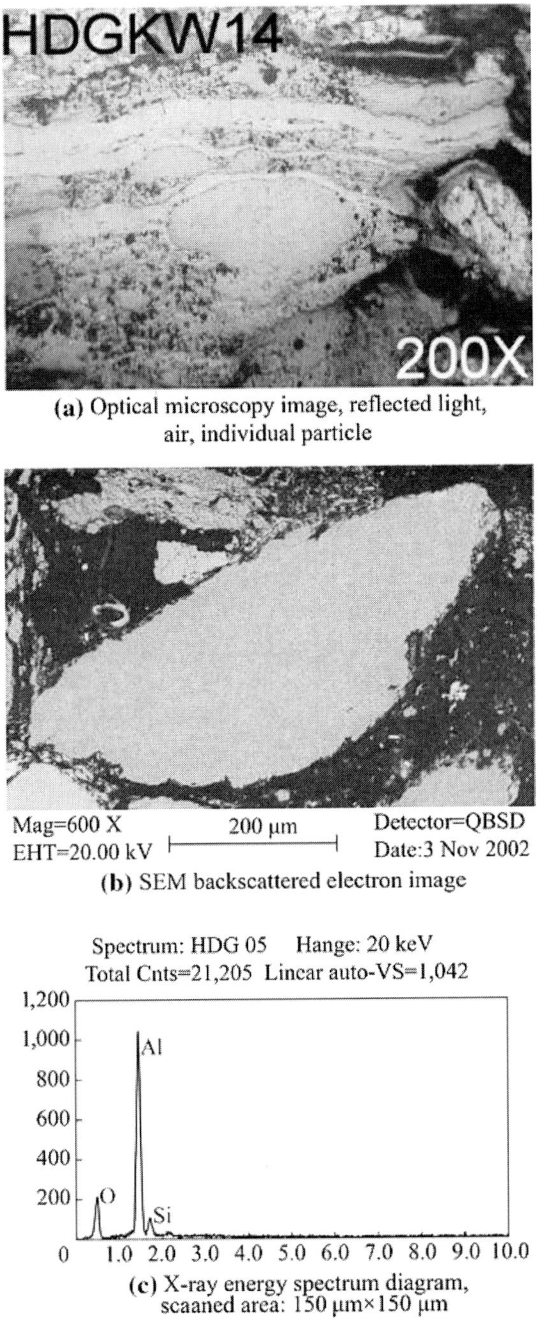

Figure 6. Bauxite in Heidaigou coal

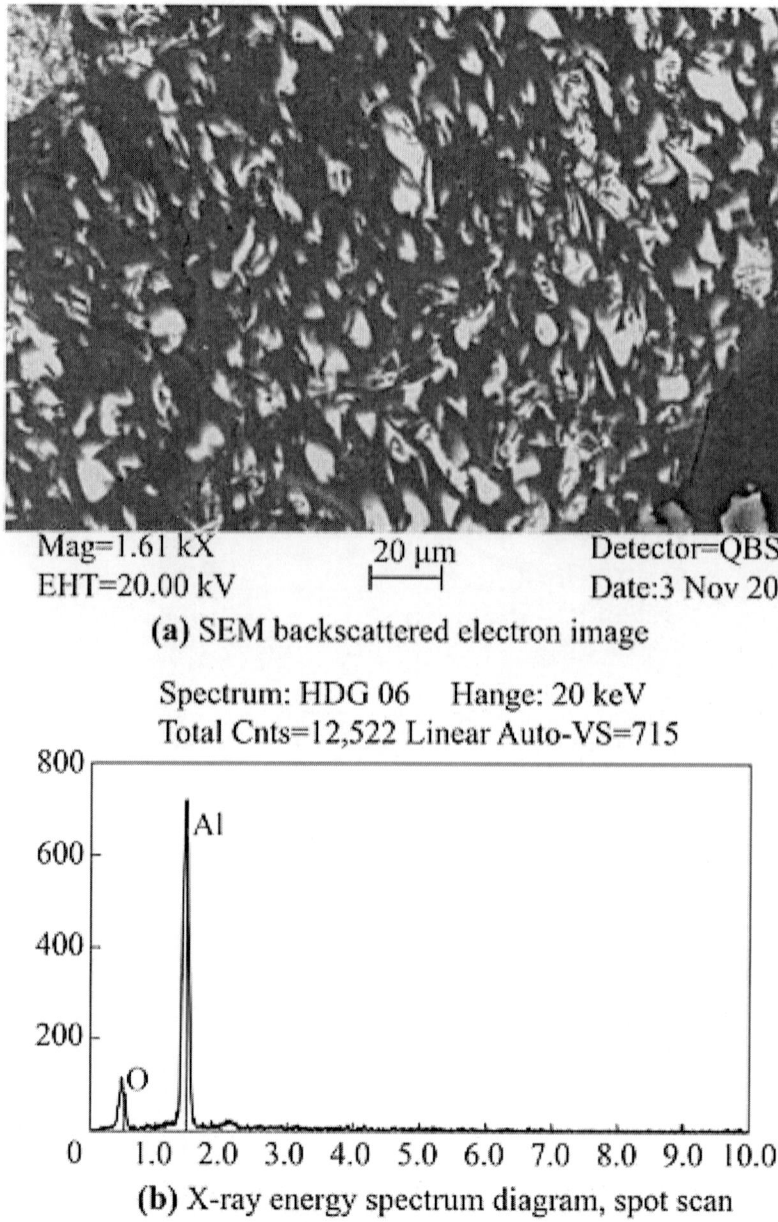

Figure 7. Bauxite in the fusinite lumens

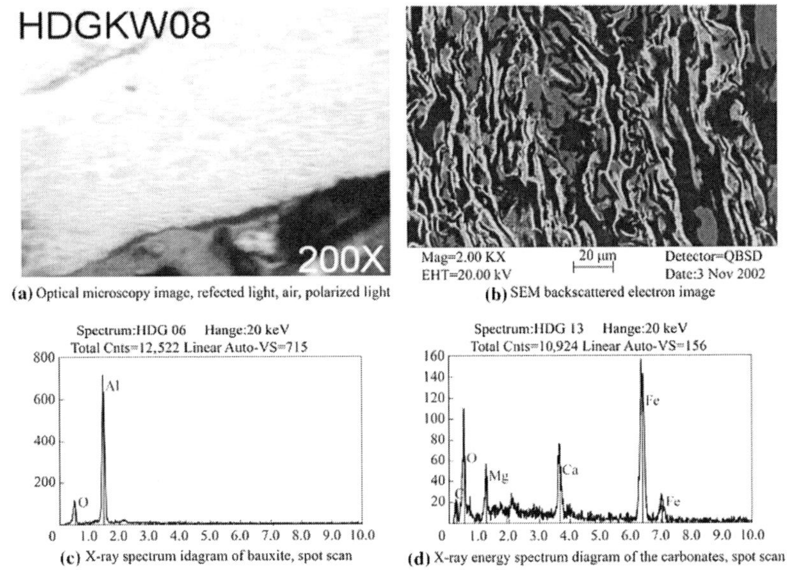

Figure 8. Bauxite intergrown with carbonates

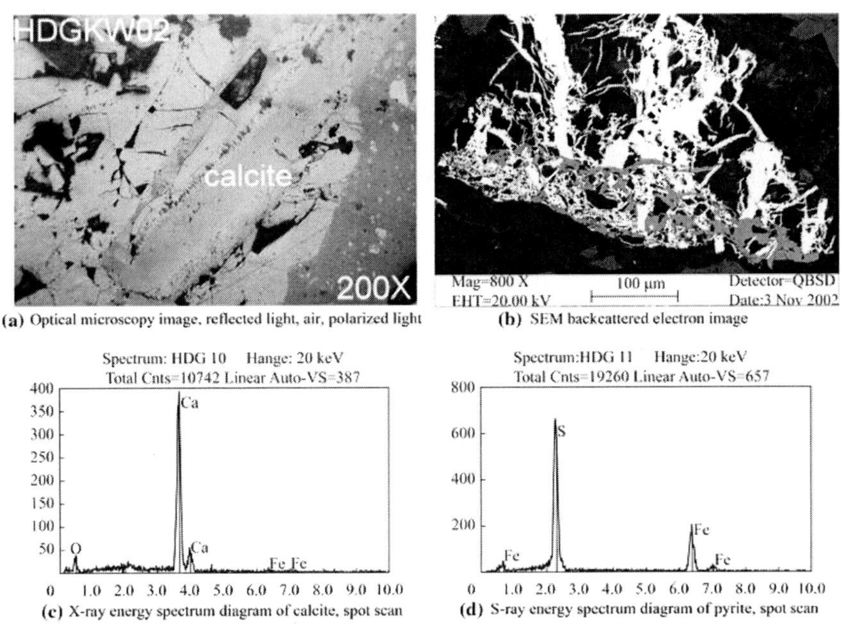

Figure 9. Pyrite and calcite that fill fractures

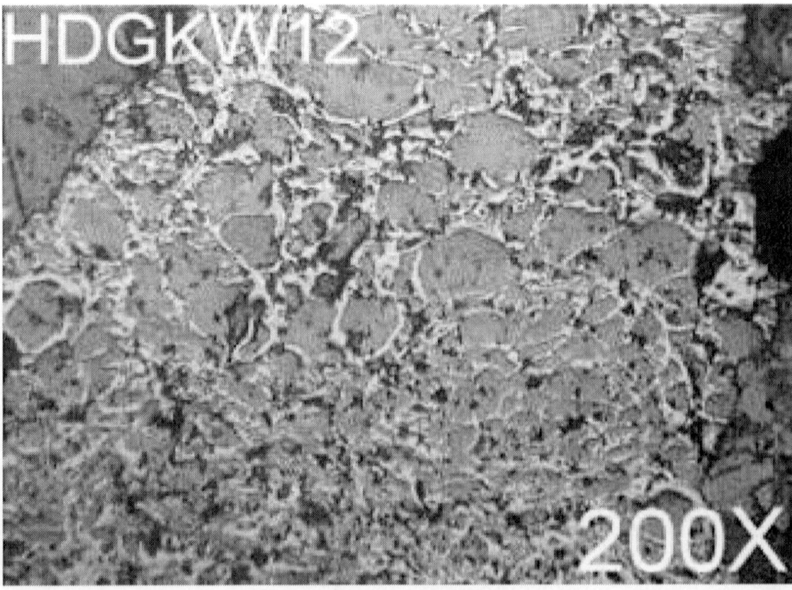

Figure 10. Calcite in the fusinite lumens (reflected light; air; polarized light; ×200)

(4)*Sulfides* The sulfide content mainly comes from sulfides such as pyrite and it is low in coal. The pyrite mainly fills fractures (Fig. 9b) or disseminates in the fine grains. Some marcasite is present in the coal (Fig. 12). A similar yellowish-white reflected color was obtained for marcasite and pyrite when subjected to polarized light and it is thus difficult to distinguish marcasite from pyrite. Marcasite presents obvious anisotropy and it has a parallel twin structure and a dim gray-shallow celadon interference color in cross-polarized reflected light, which is significantly different to that of pyrite (Fig. 13).

3.2 Distributions and occurrence of trace elements in No. 6 coal from Heidaigou mine

3.2.1 Trace element concentrations in the coal samples

The concentrations of 33 trace elements in No. 6 coal from Heidaigou mine were compared to average values of these trace elements in Chinese and international coals and the data are listed in Table 8.

Compared to the average values of trace elements in Chinese coals, the weighted value of the trace elements in Heidaigou coal is relatively low. The concentrations of Ta, Ga, Hg, Th, Se, Pb, U and Ti are 1.5–3 times higher than the average of these elements in Chinese coals while the W, Y, P, Sr contents are almost the same as the average concentrations in Chinese coals. The concentrations of the other trace elements are lower than those in Chinese coals. Compared to the average values of trace elements in international coals the

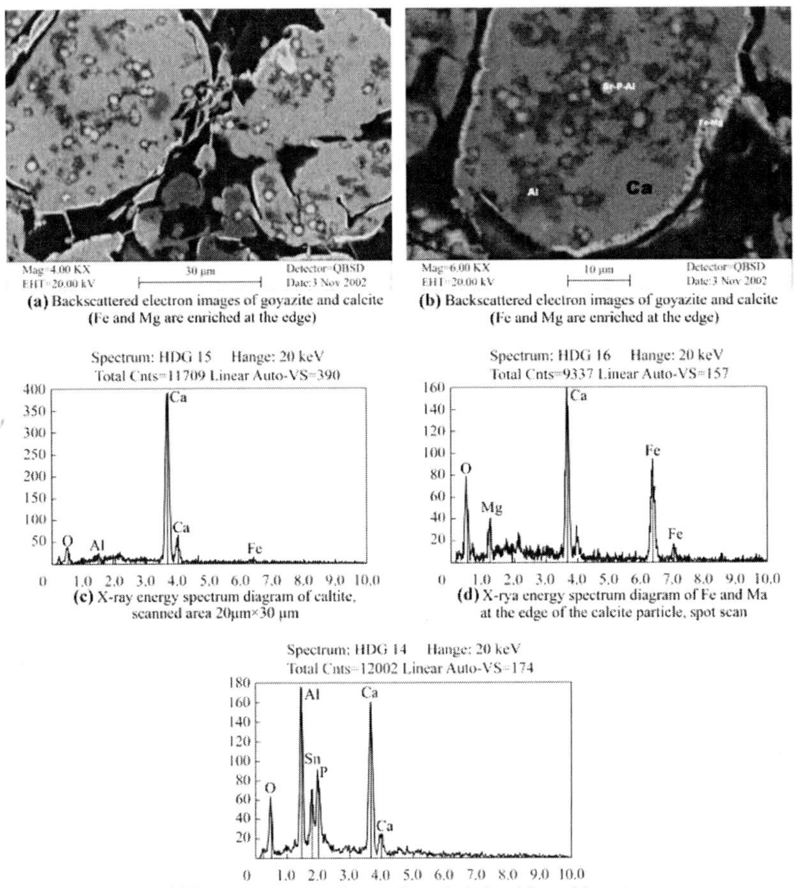

(a) Backscattered electron images of goyazite and calcite (Fe and Mg are enriched at the edge)

(b) Backscattered electron images of goyazite and calcite (Fe and Mg are enriched at the edge)

(c) X-ray energy spectrum diagram of caltite, scanned area 20μm×30 μm

(d) X-rya energy spectrum diagram of Fe and Ma at the edge of the calcite particle, spot scan

(e) X-ray energy spectrum diagram of goyazite in the calcite particle, spot scan

Figure 11. Calcite and goyazite (bean-like) in the fusinite lumens

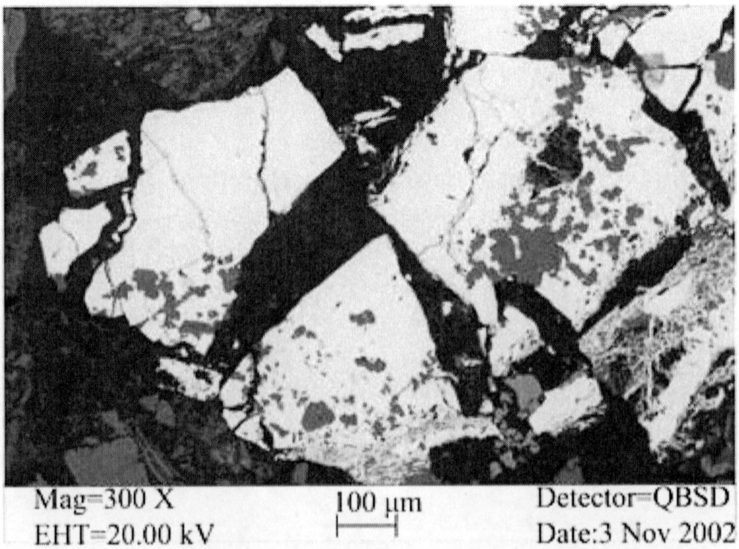

(a) Backscattered electron image of marcasite

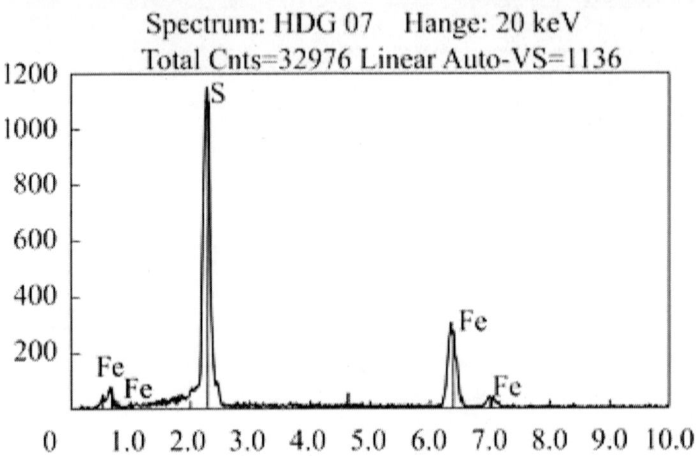

(b) X-ray energy spectrum diagram of marcasite

Figure 12. Marcasite in Heidaigou coal

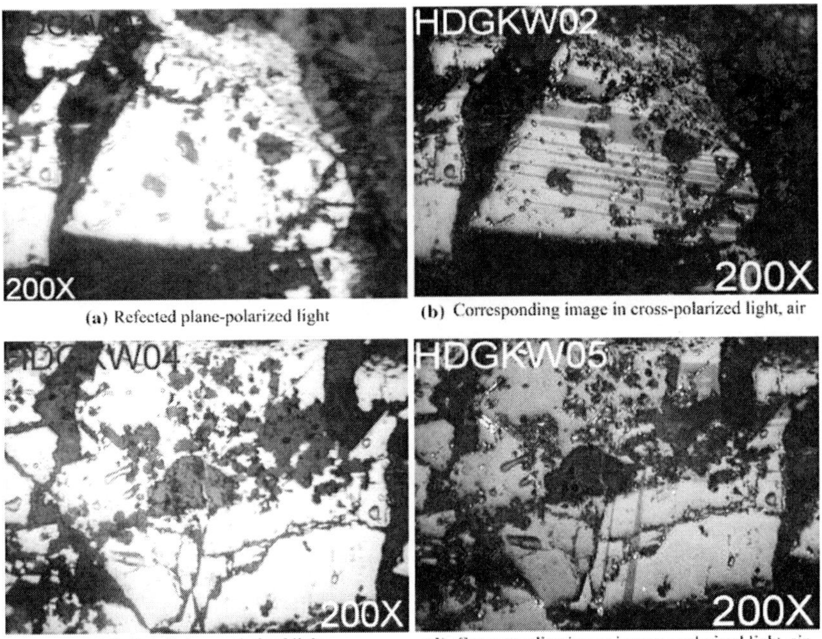

(a) Refected plane-polarized light (b) Corresponding image in cross-polarized light, air

(c) Refected plane-polarized light (d) Corresponding image in cross-polarized light, air

Figure 13. Comparison between marcasite upon receiving reflected plane-polarized light and cross-polarized light

concentrations of Se, Ta, Ti, Th, Hg, Pb, Ga are 3–5 times higher while Sr, Yb, and U are only slightly higher. Concentrations of Y, W, F, V, Sc, P, Be are almost the same as the average content of these elements in international coals. The concentrations of the other trace elements are lower than those in international coals. The ratios of these elemental concentrations in Heidaigou coal to Chinese coals and international coals are shown in Figs. 14 and 15.

Table 8. Element concentrations in No. 6 coal from Heidaigou mine (μg/g)

Element	Concentrations	China coal*	World coal**
Ag	0.02	0.04	0.095
As	1.34	4.09	8.3
B	32.21	53****	52
Ba	18.1	270	150
Be	1.51	1.75	1.6
Cd	0.06	0.81	0.22
Co	1.24	10.62	5.1
Cr	3.00	16.94	16
Cs	0.10	1.51	1.0
Cu	12.53	17.87	16
F	100	157	88
Ga	19.06	6.84	5.8
Ge	0.09	2.43	2.2
Hg	0.413	0.154	0.10
Mn	40.9	271.22****	50
Mo	1.31	2.70	2.2
Ni	3.13	14.44	13
P	219.90	216	230
Pb	31.63	16.64	7.8
Sb	0.09	0.71	0.92
Sc	3.90	4.40	3.9
Se	7.09	2.82	1.3
Sn	0.51	2.11***	1.1
Sr	178.0	195	110

Ta	1.42	0.40	0.28
Th	15.31	5.88	3.3
Ti	2,515	1,685****	500
U	3.51	2.33	2.4
V	27.59	51.18	25
W	1.28	1.05	1.1
Y	10.68	9.07	8.4
Yb	1.58	1.76	1.0
Zn	18.0	41.4***	23

* Bai et al. (2007)

** Ketris and Yudovich (2009)

*** Dai et al. (2012)

**** Ren et al. (1999)

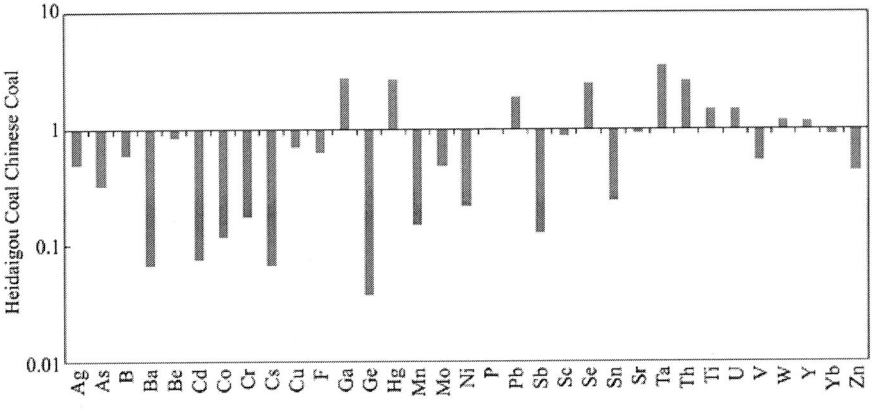

Figure 14. Comparisons between the concentrations of trace elements in Heidaigou coal and those in other Chinese coals

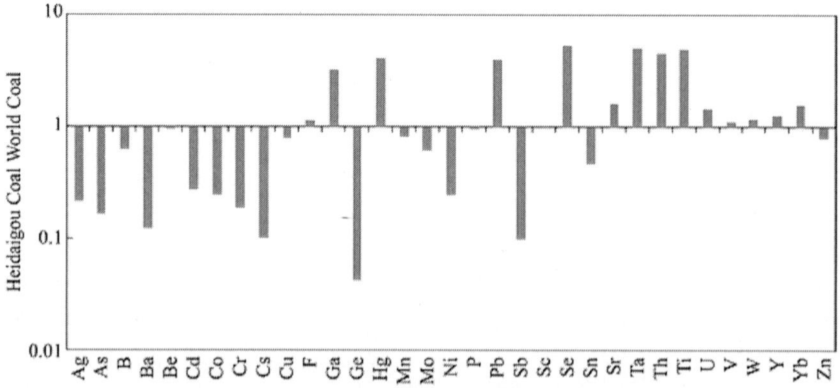

Figure 15. Comparisons between the concentrations of trace elements in Heidaigou coal and those in international coals

3.2.2 Occurrence of trace elements in No. 6 coal from Heidaigou mine

The maceral composition and the content of 18 trace elements in the products from the float-sink experiments are listed in Table 9. The maceral compositions in the products of different densities are shown in Fig. 16. The precision of the float-sink experiments was in accordance with GB/T 478-2008, as shown by the balanced ash and sulfur content calculations. The *R*values of most trace elements were between 50 % and 150 %, which are acceptable for trace elements in float-sink experiments (Querol et al. 2001; Reed et al. 2001). The dark reflectance colors of clay minerals and quartz make a quantitative determination in oil immersed reflected light difficult, which results in deviations in the results of the balance calculations.

As shown in Table 9: As, Ga, Hg, Mn, Mo, Ta, Ti, W concentrate in the heavy density products and, therefore, these trace elements are probably present in the minerals. B, P, Sr, V, Y, Yb are mainly present in the organic matter. B and V are mainly present in vitrinite while P, Sc, Sr, Y, Yb are present in inertinite. Be and Th are mainly present in inertinite with small amounts found in the minerals. Ge is evenly distributed in the minerals and in the organic matter.

Table 9. Maceral compositions and concentrations of 18 trace elements in the products from the float-sink experiments

Size grade (mm)	3–0.5						<0.5	Raw coal	*R* (%)
Yield of production (%)	54.44						45.56		
Density level (kg/L)	−1.30	1.3–1.4	1.4–1.5	1.5–1.6	1.6–1.8	+1.80			
Yield of products of different density (%)	6.29	31.12	19.82	13.34	8.46	20.97			
Coal quality									
A_d (%)	2.62	7.13	14.47	23.89	34.67	68.11	23.15	24.67	99.36
$S_{t,d}$ (%)	0.53	0.46	0.34	0.27	0.33	0.85	0.44	0.49	94.91
Maceral group (vov%)									
Vitrinite	79.09	40.54	14.21	10.56	8.53	5.99	27.62	31.87	80.13
Inertinite	12.55	40.54	65.94	63.34	56.81	22.66	50.90	44.88	105.63
Liptinite	7.27	13.21	5.10	3.84	0.91	0.75	4.69	8.78	63.56
Clay mineral	0.55	4.82	12.75	17.66	28.13	59.93	15.16	11.87	156.17
Pyrite	0.18	0.18	0.18	0.38	0.36	5.81	0.36	0.65	142.70
Quartz	0.18	0.18	1.28	1.54	2.54	2.25	0.36	0.33	249.84
Carbonates	0.18	0.54	0.55	2.69	2.72	2.62	0.90	1.63	72.81
Concentrations of trace element (μg/g)									
As	0.95	0.95	0.95	0.95	0.95	5.97	0.95	1.34	113.97
B	51.77	31.63	23.94	28.53	27.74	26.60	31.80	32.21	94.96
Be	1.11	1.63	2.10	2.03	1.94	0.83	1.63	1.51	106.88
Ga	14.73	14.76	15.27	19.86	25.05	27.10	18.99	19.06	99.65
Ge	0.19	0.13	0.06	0.06	0.06	0.24	0.08	0.09	117.71
Hg	0.063	0.091	0.113	0.155	0.175	0.971	0.342	0.412	76.62
Mn	10.00	13.10	24.60	42.30	48.60	41.70	45.80	40.90	88.39
Mo	1.32	1.05	0.98	1.10	1.37	2.02	1.21	1.31	95.72
P	145.50	253.90	315.00	262.70	200.10	97.57	222.80	219.90	101.38
Sc	2.19	4.28	5.40	5.11	3.75	0.95	5.48	3.90	116.20
Sr	152.40	248.40	258.60	204.00	111.10	41.73	203.60	178.00	108.24
Ta	0.32	0.58	1.02	1.29	1.91	2.51	1.39	1.42	93.24
Th	4.90	14.10	21.33	17.57	16.07	8.21	15.73	15.31	97.83
Ti	1,325.00	1,661.00	2,292.00	2,632.00	3,221.00	3,298.00	2,571.00	2,515.00	97.87
V	48.11	30.81	27.60	28.05	29.86	16.51	29.75	27.59	104.01
W	0.18	0.46	0.89	1.33	1.98	2.19	1.19	1.28	90.70
Y	12.93	15.99	17.08	13.72	9.49	3.53	14.21	10.68	124.58
Yb	1.51	1.80	2.06	1.79	1.45	0.53	1.74	1.58	103.09

The balance calculations for the maceral groups are based on volume fraction

The trace element content of the products of different densities of the No. 6 coal from Heidaigou mine is shown in Fig. 17. The contents of As and Hg in the products of different densities have the same trend as the change in density. This is also the case for Ga, Ta, W, Ti and Mn. The trend for these elements is the same as that of the minerals upon a change in density. Therefore, these trace elements are probably present in the minerals. As the density changes the B and V content of the products of different densities have the same trend as that of vitrinite. Therefore, these trace elements are probably present in the vitrinite. The Be, Th, P, Sc, Yb, Y and Sr content of the products of different densities have the same trend as the change in density. This is the same as the behavior of inertinite. Therefore, these trace elements are probably present in the inertinite.

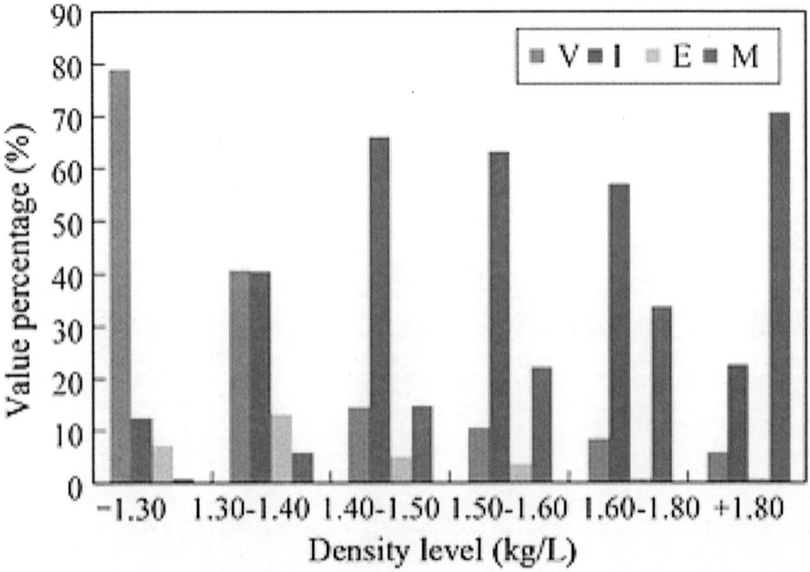

Figure 16. Histogram of the maceral groups in the products of different densities

(1)*Correlation analyses* The relationship between the trace element content and the maceral groups, the ash content, and the sulfur content are listed in Table 10respectively. The content of trace elements, vitrinite, inertinite, clay minerals, pyrite, quartz, carbonates, ash and sulfur in the 6 products of different densities that were obtained from the float-sink experiments, raw coal and the fines of less than 0.5 mm were used to calculate the correlation coefficients. A significant positive correlation exists between As and the ash content, the sulfur content, the clay mineral content, and the pyrite at a confidence level of 95 % (critical correlation coefficient$R = 0.7068$). This is also the case for Hg and Mo. A significant positive correlation was found between Ga and the ash content, the clay mineral content, the quartz content and the carbonates content. It had a negative correlation with virtrinte and liptinite. This was also the case for Ta, Ti and W. A positive correlation was found between Mn and the carbonates content while a negative correlation exists between Mn and virtrinte. Significant negative correlations exist between Sr and the ash content, the sulfur content, the clay mineral content and the

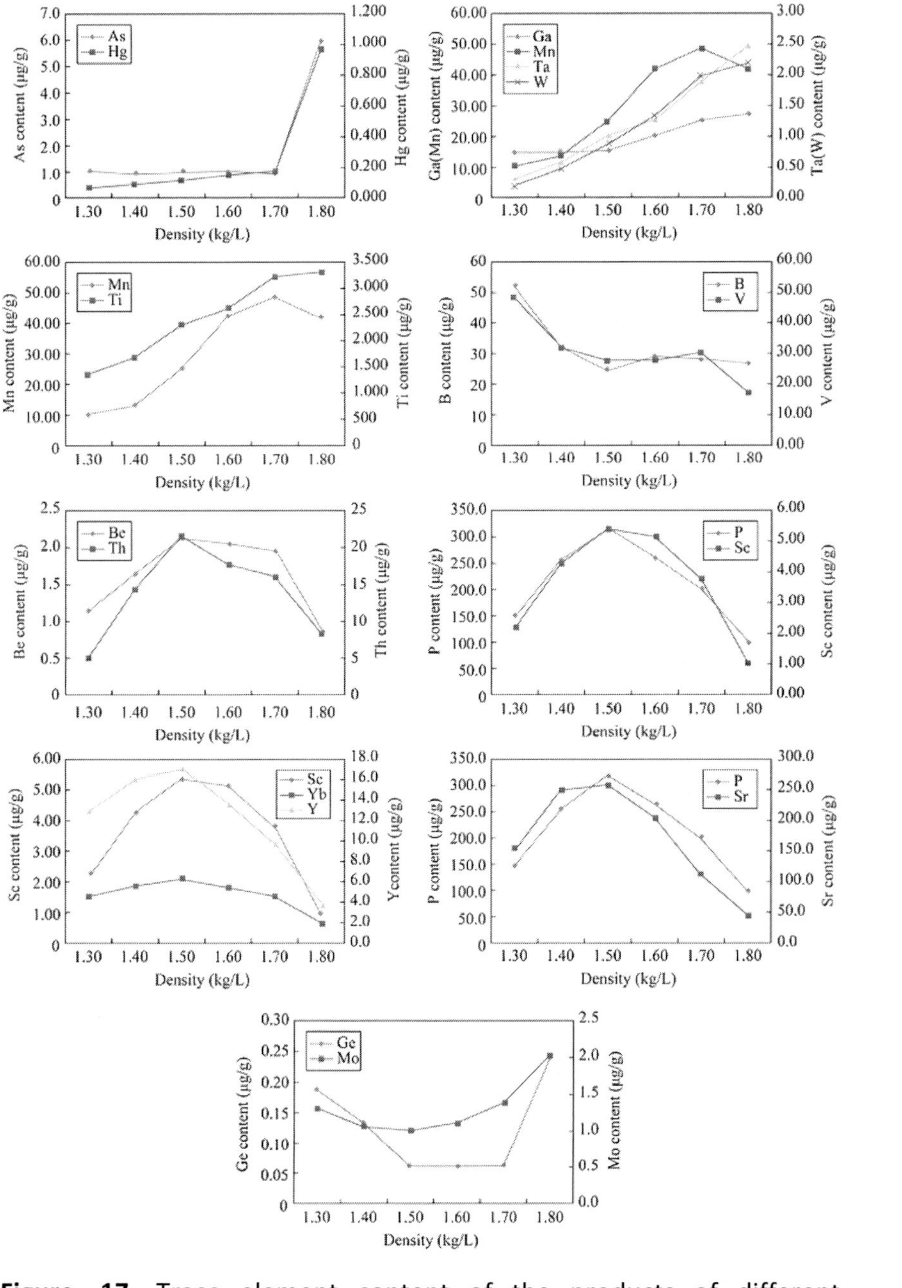

Figure 17. Trace element content of the products of different densities from No. 6 coal from Heidaigou mine

pyrite content. This was also the case for Y and Yb. A significant positive correlation existed between Be and inertinite and a negative correlation with the sulfur content. This was also the case for P, Sc and Th. A significant positive correlation existed between B and virtrinte and a negative correlation exists with inertinite. A significant positive correlation exists between V and virtrinte and a negative correlation exists with the minerals. A significant positive correlation exists between Ge and the sulfur content. The correlation coefficients of the trace elements are listed in Table 11. A significant positive correlation exists among As, Hg, Mo and Ge. This is also the case among Ga, Ta, Ti, W, Mn, among Be, Th, P, Sc, Sr, Y, Yb and between B and V. High correlation coefficients indicate that these grouped elements have a similar occurrence. As, Hg, Mo, Ge, Ga, Ta, Ti, W, Mn are mainly concentrated in the minerals while Be, Th, P, Sc, Sr, Y, Yb, B, V are mainly found in the organic matter according to the correlation analyses of the trace elements and between the trace elements and the maceral groups.

Table 10. Relationship between trace element content and the maceral group content, the mineral content, the ash content and the sulfur content

Element	A_d	$S_{t,d}$	Vitrinite	Inertinite	Liptinite	Clay mineral	Pyrite	Quartz	Carbonates
As	0.8676	0.8810	−0.3512	−0.4,729	−0.4488	0.8905	0.9990	0.4708	0.4366
B	−0.5092	0.1370	0.9445	−0.7056	0.3505	−0.5078	−0.2594	−0.5762	−0.5163
Be	−0.3976	−0.9342	−0.3833	0.9369	0.0176	−0.3967	−0.6935	0.1181	0.0530
Ga	0.9323	0.4011	−0.6666	−0.0220	−0.7861	0.9056	0.7019	0.8163	0.8776
Ge	0.3744	0.9143	0.3432	−0.9200	−0.0179	0.4243	0.7266	−0.0155	−0.0705
Hg	0.9047	0.8304	−0.4204	−0.3560	−0.4554	0.8706	0.9392	0.3875	0.4869
Mn	0.6685	−0.0540	−0.7474	0.4421	−0.6863	0.5763	0.2755	0.5761	0.7903
Mo	0.8581	0.8630	−0.2093	−0.5941	−0.5389	0.8525	0.9184	0.4830	0.5038
P	−0.5626	−0.8124	−0.2050	0.8307	0.3495	−0.5597	−0.6967	−0.2154	−0.2674
Sc	−0.5142	−0.8395	−0.2097	0.8488	0.2303	−0.5431	−0.7325	−0.2555	−0.2096
Sr	−0.7936	−0.6761	0.1861	0.5303	0.6482	−0.7859	−0.7616	−0.5937	−0.6070
Ta	0.9636	0.4143	−0.7827	0.1020	−0.7567	0.9269	0.7338	0.7698	0.8166
Th	−0.1508	−0.7180	−0.6006	0.9756	−0.0224	−0.1736	−0.4434	0.1487	0.1118
Ti	0.8685	0.1581	−0.8751	0.3399	−0.7930	0.8275	0.5445	0.8169	0.8605
V	−0.7908	−0.3102	0.8760	−0.3295	0.3959	−0.7678	−0.6451	−0.5563	−0.6161
W	0.9155	0.2631	−0.8241	0.2135	−0.7847	0.8774	0.6260	0.8311	0.8875
Y	−0.8823	−0.7275	0.3110	0.4600	0.5839	−0.8518	−0.8419	−0.5803	−0.6737
Yb	−0.8354	−0.8647	0.1817	0.6264	0.4864	−0.8381	−0.9148	−0.4778	−0.5081

Table 11. Correlation coefficients for the trace elements in No. 6 coal from Heidaigou mine

	As	B	Be	Ga	Ge	Hg	Mn	Mo	P
As	1.0000								
B	−0.2441	1.0000							
Be	−0.7076	−0.4683	−1.0000						
Ga	0.6751	−0.4453	−0.2534	1.0000					
Ge	0.7448	0.4102	−0.9481	0.2056	1.0000				
Hg	0.9333	−0.3136	−0.6535	0.7304	0.5841	1.0000			
Mn	0.2380	−0.5743	0.1855	0.7763	−0.3424	0.4713	1.0000		
Mo	0.9135	−0.0241	−0.7859	0.7817	0.7459	0.8911	0.3393	1.0000	
P	−0.6954	−0.4054	0.8931	−0.5561	−0.8303	−0.6481	−0.0941	−0.8931	1.0000
Sc	−0.7446	−0.3534	0.8758	−0.4517	−0.9009	−0.5996	0.1326	−0.8568	0.9151
Sr	−0.7499	−0.0606	0.6605	−0.8360	−0.6126	−0.7218	−0.4080	−0.9413	0.8986
Ta	0.7099	−0.6181	−0.1930	0.9544	0.1326	0.8090	0.8206	0.7341	−0.4150
Th	−0.4555	−0.7285	0.9033	−0.1204	−0.8875	−0.3329	0.3456	−0.6225	0.8822
Ti	0.5146	−0.7107	0.−637	0.9164	−0.1225	0.6360	0.9053	0.5445	−0.2121
V	−0.6319	0.8771	−0.0125	−0.6477	0.0007	−0.7142	−0.6184	−0.4314	0.0028
W	0.5981	−0.6440	−0.0556	0.9620	0.0024	0.6957	0.8654	0.6463	−0.3345
Y	−0.8313	0.1007	0.6532	−0.8739	−0.5889	−0.8436	−0.4854	−0.9636	0.8363
Yb	−0.9107	0.0012	0.7972	−0.7671	−0.7791	−0.8612	−0.2830	−0.9921	0.9039
	Sc	Sr	Ta	Th	Ti	V	W	Y	Yb

	Sc	Sr	Ta	Th	Ti	V	W	Y	Yb
As									
B									
Be									
Ga									
Ge									
Hg									
Mn									
Mo									
P									
Sc	1.0000								
Sr	0.8338	1.0000							
Ta	−0.3262	−0.7140	1.0000						
Th	0.8587	0.6142	0.0500	1.0000					
Ti	−0.1068	−0.5763	0.9640	0.2653	1.0000				
V	0.0285	0.2709	−0.8117	−0.3619	−0.7888	1.0000			
W	−0.2455	−0.6787	0.9829	0.1363	0.9885	−0.7744	1.0000		
Y	0.7964	0.9637	−0.8063	0.5133	−0.6593	0.4673	−0.7581	1.0000	
Yb	0.8845	0.9376	−0.6972	0.6595	−0.5025	0.4100	−0.6146	0.9572	1.0000

(2)*Theoretical trace element content in the maceral groups and the minerals in No. 6 coal from Heidaigou mine* The trace element content of the maceral groups and of the minerals can be calculated using float-sink experiments and statistical methods, which is of great importance in determining the trace element occurrence and their behavior during coal processing and conversion. The theoretical content of trace elements in the maceral groups and in the minerals of No. 6 coal from Heidaigou mine are listed in Table 12. A ternary diagram showing the distribution of trace elements is given in Fig. 18. As, Ga, Hg, Mn, Mo, Ta, Ti, W are obviously concentrated in the minerals with low concentrations found in organic matter. B, Be, P, Sc, Sr, Th, V, Y, Yb are concentrated in the organic matter. B and V are mainly present in virtrinte while Be, P, Sc, Sr, Th, Y, Yb are present in inertinite. The Ge content of the minerals and of the organic matter is similar, which indicates an even distribution of Ge in the minerals and in the organic matter (no significant correlation from an F test).

Table 12. Theoretical trace element content in the maceral groups and in the minerals of No. 6 coal from Heidaigou mine, calculated values (in μg/g)

Element	Component			Sample amount	Significance testing			
	Vitrinite	Inertinite	Minerals		R	F	$F_{0.05}$ (3, $n-4$)	Significance
As	1.82	−3.01	7.01	8	0.93	10.73	6.59	Significant
B	54.45	16.34	29.29	8	0.94	12.61	6.59	Significant
Be	0.69	3.19	0.52	8	0.96	18.23	6.59	Significant
Ga	14.37	12.8	31.83	8	0.94	12.42	6.59	Significant
Ge	0.04	0.04	0.04	8	0.97	24.12	6.59	Significant
Hg	0.128	0.114	0.413	7	0.32	0.15	9.28	Non-significant
Mn	11.58	2.78	109.33	7	0.89	5.09	9.28	Non-significant
Mo	1.47	0.28	2.35	8	0.99	72.63	6.59	Significant
P	105.8	460.94	20.28	8	0.56	0.77	6.59	Non-significant
Sc	1.23	9.63	−0.56	8	0.96	22.45	6.59	Significant
Sr	138.63	388.46	−48.11	8	0.94	12.29	6.59	Significant
Ta	0.27	0.66	3.04	8	0.97	24.06	6.59	Significant
Th	1.17	32.64	4.01	8	0.99	82.17	6.59	Significant
Ti	1,082.59	2,404.66	3,738.02	8	0.96	17.89	6.59	Significant
V	46.93	27.65	16.46	8	0.88	5.86	6.59	Significant
W	0.03	0.81	2.71	8	0.96	17.64	6.59	Significant
Y	11.43	24.04	−1.27	8	0.95	15.28	6.59	Significant
Yb	1.26	3.03	0.06	8	0.99	137.42	6.59	Significant

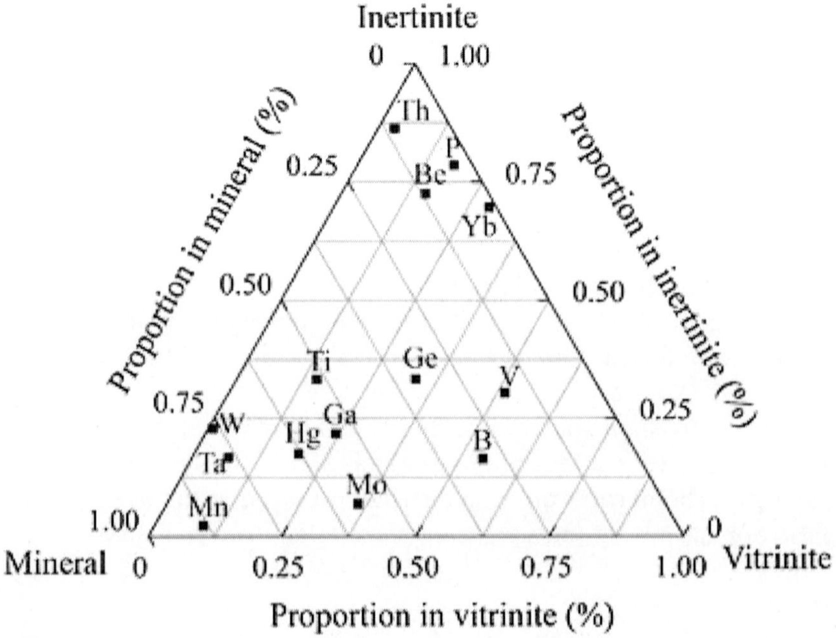

Figure 18. Ternary diagram showing the distribution of trace elements in No.6 coal

(3)*Organic affinities of the trace elements in No. 6 coal from Heidaigou mine* The organic affinities of the trace elements in No. 6 coal from Heidaigou mine are listed in Table 13. The value for Ta in the raw coal was obtained from a weighted mean value of Ta in the different density products. As shown in Table 13, most of the trace elements are associated with the organics except for As, Hg, Mn, Ta, and W. Their low organic affinity indicates that these elements are mainly integrated into an inorganic phase. It is worth noting that trace elements with high organic affinity are not necessary organically integrated because some trace elements were present in the fine grained minerals in the organic matter that filled the lumens.

Table 13. Organic affinity of the trace elements in No. 6 coal from Heidaigou mine

Element	As	B	Be	Ga	Ge	Hg	Mn	Mo	P
A_0	26.66	77.75	92.11	56.05	–	–	–	51.61	–
Element	Sc	Sr	Ta	Th	Ti	V	W	Y	Yb
A_0	100	100	32.05	93.07	58.93	86.98	33.05	100	99.10

3.3 Removal of trace elements during the coal preparation of No. 6 coal from Heidaigou mine

The washability curves of the trace elements were obtained during coal preparation of No. 6 coal from Heidaigou mine and are shown in Fig. 19. The theoretical removal of the trace elements is listed in Table 14. The value for Ta in the raw coal was obtained as a weighted mean value of Ta in the different density products.

The washability curves of As and Hg are similar, which is also the case for Ga and W, Mn, Ta and Ti. High removal rates were obtained for these elements during coal preparation because sharp washability curves were obtained. The elements with flat washability curves are evenly distributed in the organic matter and in the minerals and they cannot be effectively removed during coal preparation.

The ash and sulfur content of the clean coal were 9.18 % and 0.43 %, respectively, and poor washability was obtained for No. 6 coal. The recovery of trace elements is closely related to their washability, as shown in Table 14. Only Hg, Mn, W, Ta, As, Ti, Ga, Mo have high washability. The relationship between organic affinity and the theoretical removal rate of trace elements in No. 6 coal from Heidaigou mine are shown in Fig. 20. The theoretical removal rates of the trace elements decrease as the organic affinity increases. Most trace elements are closely related to the organic matter in No. 6 coal from Heidaigou mine. This is also shown by the correlation analyses and the theoretical concentration calculations of the trace elements in the maceral groups and in the minerals. Therefore, it is difficult to remove these trace elements during coal preparation.

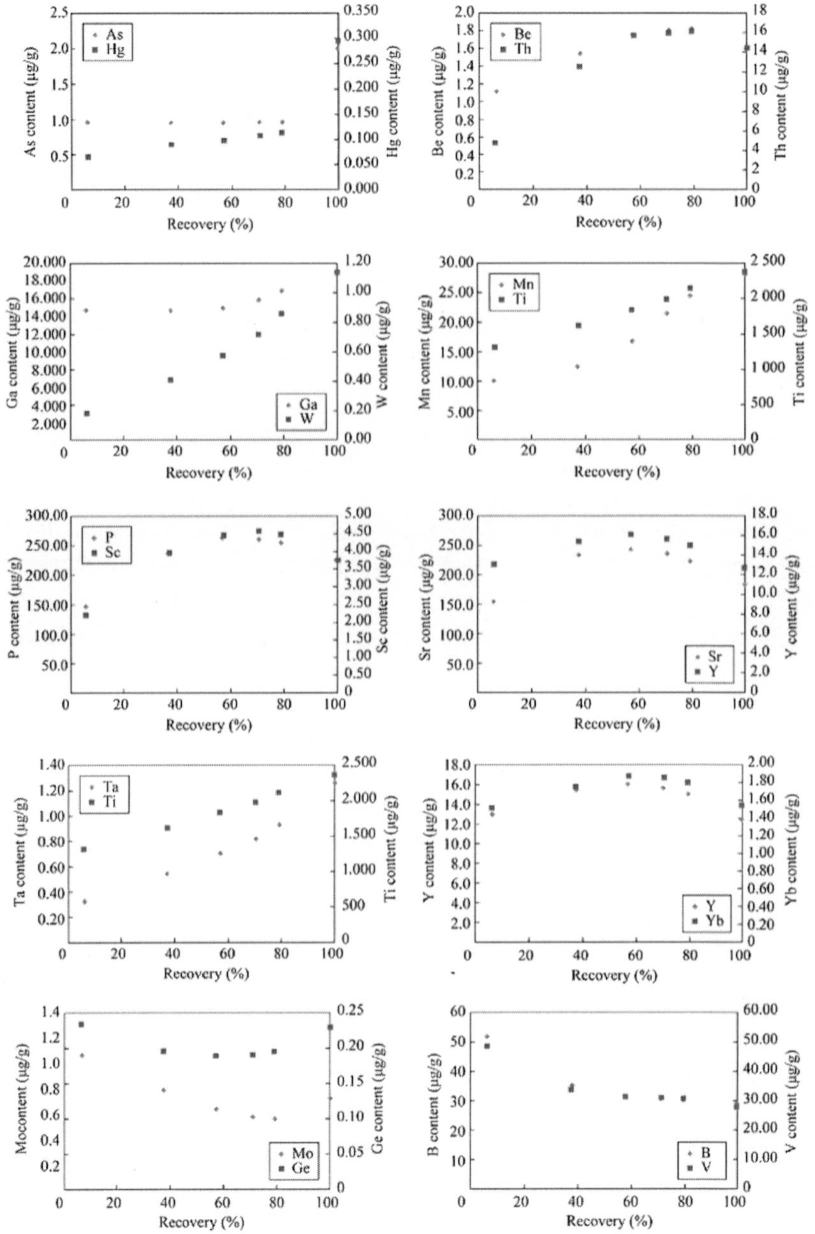

Figure 19. Washability curves for the trace elements in No. 6 coal from Heidaigou mine

Table 14. Theoretical removal of trace elements in No. 6 coal from Heidaigou mine (Recovery of clean coal was 57.23 % at a separation density of 1.50 kg/L)

Element	As	B	Be	Ga	Ge	Hg	Mn	Mo	P
Theoretical removal rate (%)	59.29	44.60	34.28	55.16	28.64	86.70	76.57	53.92	31.51
Element	Sc	Sr	Ta	Th	Ti	V	W	Y	Yb
Theoretical removal rate (%)	34.87	22.39	71.55	41.71	58.07	34.45	74.05	14.10	32.72

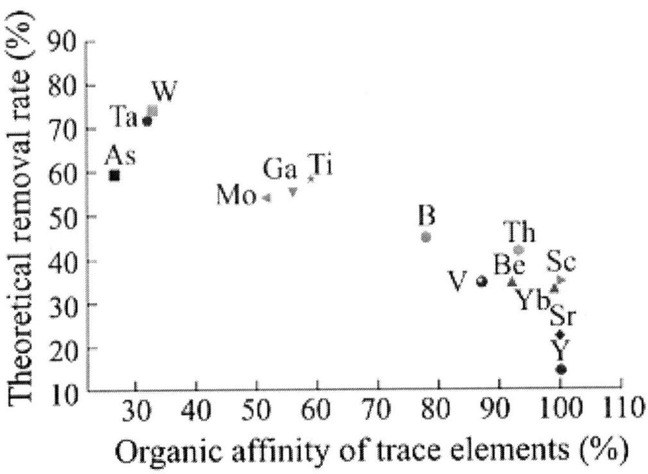

Figure 20. Relationship between organic affinity and theoretical trace element removal rate in No. 6 coal from Heidaigou mine

4 CONCLUSIONS

Optical microscopy and SEM were used to study the characteristics of minerals and the concentrations of 33 trace elements in No. 6 coal from Heidaigou mine. The distribution of trace elements in the macerals and in the minerals as well as the organic affinity and removability of the trace elements are also discussed.

1. A high mineral content dominated by kaolinite clay minerals was found in No. 6 coal from Heidaigou mine. Bauxite was mainly present as individual particles in the fusinite lumens or was intimately intergrown with carbonate minerals. Its content was relatively high while that of pyrite and quartz was low. Some marcasite with a parallel twin structure, as

observed by cross-polarized reflected light, was also found. A small proportion of bean-like goyazite was found in the calcite.

2. The weighted content of trace elements in Heidaigou formations is relatively low, which is beneficial for coal processing and utilization. The concentrations of Ga, Hg, Pb, Se, Th, Ta are relatively high compared to the average values of Chinese coals.

3. As, Hg, Mo, Ge, Ga, Ta, Ti, W, Mn are mainly concentrated in the minerals while B, Be, Th, P, Sc, Sr, V, Y, Yb are mainly found in the organic matter in No. 6 coal from Heidaigou mine. As, Ge, Hg, Mo are mainly present in sulfides, Be, Th, P, Sc, Sr, Y, Yb are mainly present in inertinite while B and V are mainly present in vitrinite.

4. The high organic affinity and low theoretical removability of most trace elements causes difficulties in their removal during coal preparation.

ACKNOWLEDGEMENTS

Our research was funded by the National Key Basic Research Development Plan (Grant Agreement number 2014CB744302).

REFERENCES

1. Bai XF, Li WH, Chen YF, Jiang Y (2007) The general distributions of trace elements in Chinese coals. Coal Qual Technol 1:1–4

2. Dai SF, Ren DY, Li SS, Chou CL (2006a) A discovery of extremely-enriched Boehmite from coal in the Junger coalfield, the Northeastern Ordos Basin. Acta Geol Sin 80(2):294–300

3. Dai SF, Ren DY, Li SS (2006b) Discovery of ultra-large gallium deposit in the Junger Coalfield Inner Mongolia. Chin Sci Bull 51(2):177–185

4. Dai SF, Ren DY, Chou CL, Li SS, Jiang YF (2006c) Mineralogy and geochemistry of the No. 6 Coal (Pennsylvanian) in the Junger Coalfield, Ordos Basin, China. Int J Coal Geol 66:253–270C

5. Dai SF, Ren DY, Li SS, Zhao L, Zhang Y (2007) Succession characteristics of coal facies of the main coal seam in Heidaigou mine from the Jungar Coalfield, Inner Mongolia. China Earth Sci D 37(supp.):119–126

6. Dai SF, Li D, Chou CL, Zhao L, Zhang Y, Ren DY, Ma YW, Sun YY (2008) Mineralogy and geochemistry of boehmite-rich coals: new insights from the Haerwusu Surface Mine, Jungar Coalfield, Inner Mongolia, China. Int J Coal Geol 74:185–202

7. Dai SF, Ren DY, Chou CL, Finkelman RB, Seredin VV, Zhou YP (2012) Geochemistry of trace elements in Chinese coals: a review of abundances, genetic types, impacts on human health, and industrial utilization. Int J Coal Geol 94:3–21

8. Han DX (editor in chief) (1996). Coal petrology of China. China University of Mining and Technology Press, Xuzhou

9. Ketris MP, Yudovich YE (2009) Estimations of Clarkes for carbonaceous biolithes: world averages for trace element contents in black shales and coals. Int J Coal Geol 78:135–148

10. Klika Z, Kolonaznik I (2000) New concept for the calculation of the trace element affinity in coal. Fuel 79:659–670

11. Klika Z, Weiss Z, Roubicek V (1997) Calculation of element distributions between inorganic and organic parts of coal. Fuel 76(14/15):1537–1544

12. Li SS, Ren DY (2006) Analysis of anomalous high concentration of lead and selenium and their origin in the main minable coal seam in the Junger Coalfield. J China Univ Min Technol 35(5):612–616

13. Mao JH, Xu HL (1999) Prediction and evaluation of coal resources in China. Science Press, Beijing

14. Martinez-Tarazona MR, Spears DA, Tascon JMD (1992) Organic affinity of trace elements in Asturian bituminous coals. Fuel 71:909–917

15. Pires M, Fiedler H, Teixeira EC (1997) Geochemical distribution of trace elements in coal: modeling and environmental aspects. Fuel 76(14/15):1425–1437

16. Querol X, Klika Z, Weiss Z, Finkelman RB, Alastueya A, Juana R, López-Solera A, Plana F, Kolker A, Chenery SRN (2001) Determination of element affinities by density fractionation of bulk coal samples. Fuel 80:83–96

17. Reed GP, Ergüdenler A, Grace JR, Watkinson AP, Herod AA, Dugwell D, Kandiyoti R (2001) Control of gasifier mercury emissions in a hot gas filter: the effect of temperature. Fuel 80(5):623–634

18. Ren DY, Zhao FH, Wang YQ, Yang SJ (1999) Distributions of minor and trace elements in Chinese coals. Int J Coal Geol 40:109–118

19. Song DY, Zhang XK, Zhang JY, Zheng CG (2010) Migration characteristics of hazardous trace elements in coal in the process of flotation. J China Coal Soc 35(7):1170–1176

20. Tang YG, Chang CX, Zhang YZ (2005) Migration and distribution of fifteen toxic trace elements during the coal washing of the Kailuan Coalfield Hebei Province. Geochimica 34(4):366–372

21. Wang WF, Qin Y, Liu XH, Zhao JL, Wang JY, Wu DG, Liu JT (2011) Distribution, occurrence and enrichment causes of gallium in coals from the Jungar Coalfield Inner Mongolia. Sci China Earth Sci 41(2):181–196

Chapter 2

Geodetic Monitoring of the Surface within the Trial Operation of the Room and Pillar Mining Method in the Ostrava-Karvina Coal District

Eva Jiránková[1*], Jitka Mučková[1], Petr Jadviščok[1], Robert Vochta[2], Vlastimil Molčák[1], Martina Havlicová[1]

[1]VŠB-Technical University, Ostrava, Czech Republic

[2]Survey and Geology Manager, OKD, A.S., Mine Plant 2, Stonava, Czech Republic

ABSTRACT

Mine Plant 2 is a part of the Ostrava-Karvina Coal District (OKD) that is located in the Czech part of the Upper Silesian Basin. The first coal was exhausted from Mine Plant 2 in 1968. The most used method of mining so far in this area has been strike longwall mining with controlled caving. Due to extensive changes in the surface, which occur as a consequence of deep mining by the method of longwall mining with controlled caving, it is not possible to use this method in densely populated areas. At the present time, therefore, the trial operation of a new mining method called room and pillar is carried out. The method was chosen with the aim to minimize subsidence and deformations of the surface. The room and pillar mining method has never been used before in the conditions of the OKR, therefore it is necessary to prove the real effect of mining by this method on the surface. For this purpose, a surface observation station was designed, consisting of 36 surface points. The position and height of all points of the observation station is determined in stages three times a year.

Keywords: Surface Subsidence, Room and Pillar, Protective Pillar of the Shaft

1. INTRODUCTION

Due to deep mining, movements and deformations of the surface occur. Even though the extraction of coal deposits involves the remedy of damages incurred and the reclamation of the surrounding landscape, changes in the surface caused by mining are essential and irreversible.

The essential factor that limits mining activity is the degree to which the objects concerned are affected by it on the surface, i.e. the magnitude of movements and deformations of surface objects. With respect to the relatively demanding legislation that regulates consent to mining, the extension of a mining area in a densely populated area is very difficult. One of the solutions of the conflict of interest of the mining company and the owners of real estate is to find a suitable alternative method of mining, which will prove the least effect on the surface. For this purpose, the trial operation of the new room and pillar mining method was commenced. The task of the trial operation is to prove the usability of this method of mining in the conditions of deep bituminous coal mines of the Ostrava-Karvina Coal District. The principle of this method of mining is to leave stable pillars of coal that will be the permanent support of the roof, thus avoiding the layering of the roof, and the manifestation of mining on the surface will be minimal [1] - [4] . Geodetic monitoring of the surface is carried out simultaneously with geo-mechanical monitoring of coal pillars left in place. Within geo-mechanical monitoring, deformometry of the overlying rocks of tunnels is carried out, as well as monitoring and measurement of the stress state in the rock mass in the conditions in situ, deformometry of coal pillars, loading of installed anchor elements, measurement of the convergence of tunnels and seismological monitoring [5] .

This paper is concerned with geodetic measurements, the purpose of which is the collection of data required for the evaluation of surface subsidence that will occur as a result of the effect of the room and pillar mining method. The trial operation of this new, not yet approved mining method for coal mining in the area of the OKR is carried out in the protective pillar of the shafts Sever (North) of Mining Plant 2. The tunnels of the stope are 5.2 m wide and 3.5 m high on average. The extent of the stope using the room and pillar method for the year 2014 is graphically depicted in Figure 1.

2. DESCRIPTION OF THE SITE

From the mining-technical point of view, the selected area is not quite ideal for the trial operation of the new mining method. Equally as in the other parts of the coal district, the rock mass is relatively much tectonically fractured there, with the layers dipping up to 20 degrees and the depth of mining being 750 to 900 metres below the ground surface.

In the protective pillar of the shafts Sever, there is a downcast and an upcast shaft (Figure 1). Deformations of the shafts are measured once a year; the methods of measurement are still being developed [6] . Even though it is the first mining in the protective pillar of the shafts Sever, this area suffers subsidence of the surface being about 15 mm a year; the to talsubsidence from the beginning of mining activity is about 630 mm. Therefore, it is necessary to gather a required amount of geodetic data so that it can be possible to prove in the future, what part of the value of surface subsidence the effect of mining using the room and pillar method is accounted for, under the current effect of longwall mining with controlled caving of coal faces at the efficient distance from the protective pillar of the shafts Sever [7] - [10] .

For the purposes of geodetic measurements, an observation station was stabilized on the surface, consisting of three aligning bases and surface points in a reclaimed area along the state road 475 and the bridge construction of a floating pipeline. The stabilization of surface points was carried out in a manner corresponding to the technical requirements for the stabilization of points of the vertical point network. In all, twenty seven new surface points were stabilized; the observation station also includes points stabilized earlier—six points of

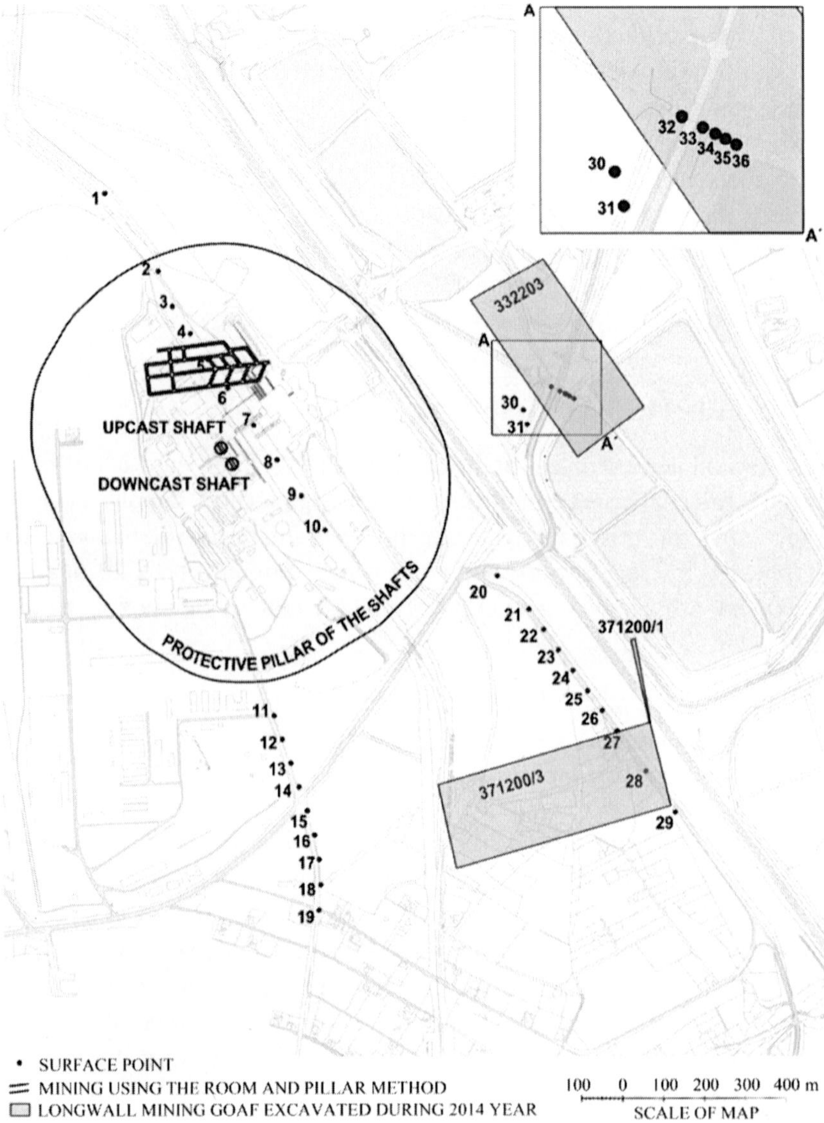

Figure 1. Surface pointsof aligning bases and the extent of mining for the year 2014.

the base point network of Mine Plant 2 and three points earlier stabilized by the workers of Mine Plant 2 to measure surface movements.

Surface line no. 1 is formed by ten surface points situated in the premises of Mine Plant 2 Sever. Surface line no. 2 is formed by nine surface points situated along the access road to Mine Plant 2 Sever around the crossroads with the state road no. 475. Surface line no. 3 is formed by ten surface points stabilized in the road shoulder of a closed road running along a railroad siding of Advanced World Transport a.s. The observation station also includes seven surface points stabilized in a reclaimed area-two points were stabilized along the road 475 and five points were stabilized into the abutments of the bridge construction of the floating pipeline. The position of all points of the observation station is graphically depicted in Figure 1 and Figure 2.

3. THE METHOD OF MEASUREMENT

In all, three stages of geodetic measurements of the height and position of surface points were carried out in 2014. The first measurement was made on 2 April 2014, the second one on 29 July 2014 and the third one on 7 November 2014.

The measurement of the heights of surface points was carried out using geometric levelling from the centre between the staffs in the category of precise measurements. A teach stage of measurement, the points of the observation station were vertically connected from a non-undermined area. The measurements were carried out using an electronic levelling instrument DNA 03 Leica with invar staffs 3 metres long. The mean kilometre error given by the manufacturer of this instrument is 0.3 mm. The length of the levelling line, including connection from the non-undermined area, is 8.1 km.

The heights of points of the linear observation station were determined from two independent measurements. For each stage of measurement, the characteristics of accuracy were calculated; the mean kilometre error of the first stage of measurement carried out on 2 April 2014 is 1.8 mm, the mean kilometre error of the second stage of measurement carried out on 29 July 2014 is 1.3 mm and the mean kilometre error of the third stage of measurement carried out on 7 November 2014 is 1.7 mm.

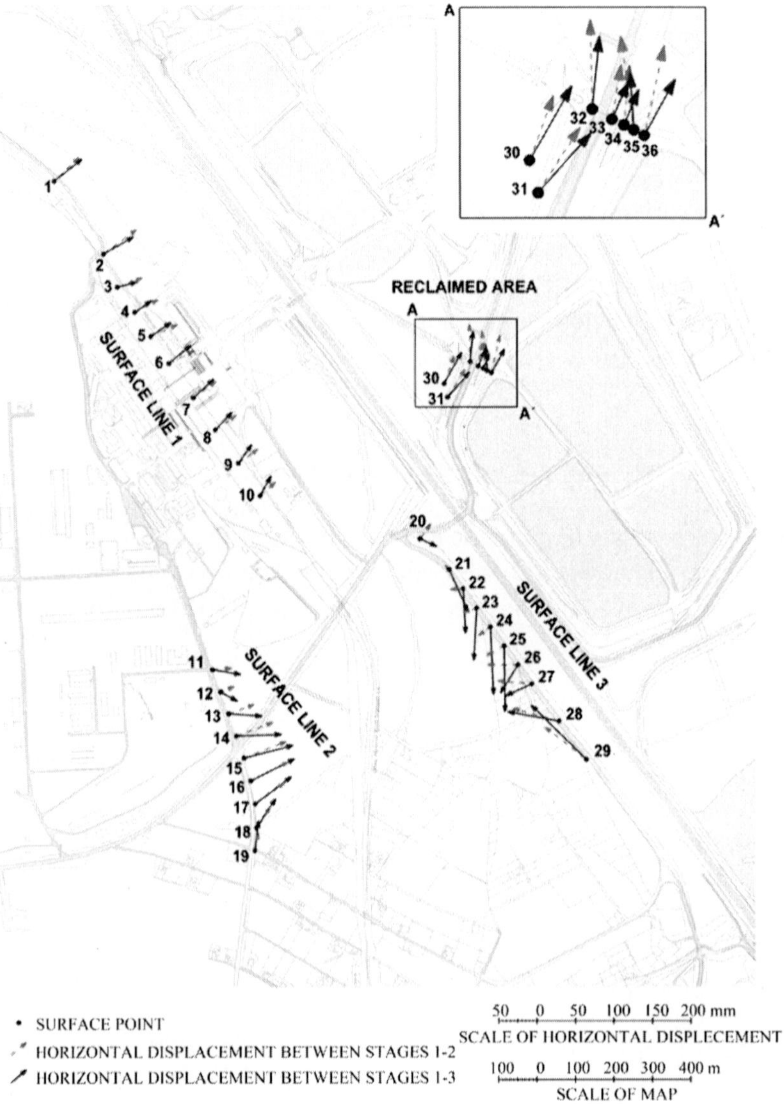

Figure 2. Graphical depiction of horizontal displacements of surface points.

The measurement of the position of surface points was carried out on aligning bases using traverses oriented and connected on both sides. The accuracy of the measurement of traverses was in the category of precise measurements. For precise measurement, an instrument TCR

1201 Leica with accessories was used. The parameters of accuracy given by the manufacturer of this instrument are 1 mm for length measurement with a prismatic reflection system, 0.3 mg on for measuring horizontal angles in one position of the telescope and 0.3 mg on as well for measuring vertical angles. Connection to S-JTSK was carried out using the points of the end sides of traverses, which were surveyed by the GNSS method. The position of all seven surface points stabilized in the reclaimed area was also measured by the GNSS method.

Of the GNSS methods, a rapid static method was used, with the time of observance being approximately 20 minutes at one point. At each point it was necessary to measure at least one repeated independent measurement. An independent measurement is considered to be a new surveying of the position of points by the same method with the lapse of time of at least 1 hour. The accuracy of this method depends on instantaneous atmospheric conditions and disturbing effects of the environment, if any. The accuracy of this method ranges in values of the order of around 3 mm in the position of the determined point. Measurement was carried out using an instrument RX1250XC Leica with an antenna ATX 1230 and accessories.

4. THE RESULTS OF MEASUREMENT

The results of vertical and horizontal measurement were used to determine subsidence and displacements of surface points. Subsidence is the vertical component of the spatial movement of the surface point and is determined as the difference between the heights of the same point over a certain period of time. The mean error of subsidence was calculated as the root of the quadratic sum of mean kilometre errors determined from the individual stages of measurement. Displacement is the horizontal component of the spatial movement of the surface point and is determined over a certain period of time from the difference of the position of the point measured at the beginning and at the end of the period of time. The vector of displacement is governed by its magnitude and the bearing in S-JTSK. The mean error of horizontal displacements was calculated as the root of the quadratic sum of the mean horizontal errors determined from the individual stages of measurement.

Subsidence and displacements determined by the values exceeding the confidential interval are designated as proved with an appropriate risk. The confidential interval with a 5% risk was determined as the double the mean error of subsidence and displacements. Both the subsidence and the displacements of aligning bases 2 and 3 and the points of the reclaimed area were with the 5% risk; only displacements with the 5% risk were proved for the surface points of aligning base 1. A graphical depiction of subsidence is given in Figures 3-6 and displacements are shown in Figure 2.

5. INTERPRETATION OF THE RESULTS

The surface observation station was stabilized and surveyed at the beginning of mining in the protective pillar of the shafts using the room and pillar method. An uplift of the points of aligning base 1, with the values ranging from 1.9 to 3.2 mm, has been identified from the second measurement of heights of surface points (Figure 3). The results of the third measurement of heights show evident subsidence of surface points of aligning base 1 as compared to the second stage, yet points 1 to 7 show uplift, as compared to the first stage of measurement. In relation to the length of the traverse connected to the non-undermined area, it was not possible to prove these low values of uplifts by the method of measurement. Subsidence of the surface points of aligning bases 2 and 3 and the surface points stabilized in the reclaimed area has been proved by the method of measurement and their values correspond to the intensity and extent of strike longwall mining with controlled caving (Figures 4-6).

The protective pillar of the shafts is constructed using the pillar angle, which is steeper relative to the critical angle of influence. Therefore, the effect of surrounding mining is partially manifested in the protective pillar of the shafts. The surface in the area of the protective pillar is thus affected both by surrounding mining and by mining in the protective pillar using the room and pillar method. In order to separate the effect of room and pillar mining from the overall effect of mining on the surface in the area of the protective pillar, it was necessary to design the surface observation station so that it could be possible to evaluate

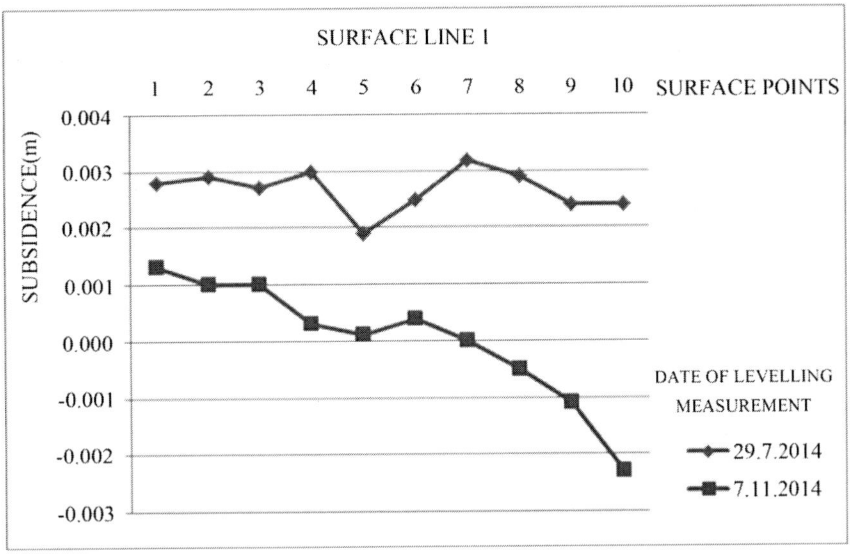

Figure 3. Graphical depiction of subsidence of surface points of aligning base 1.

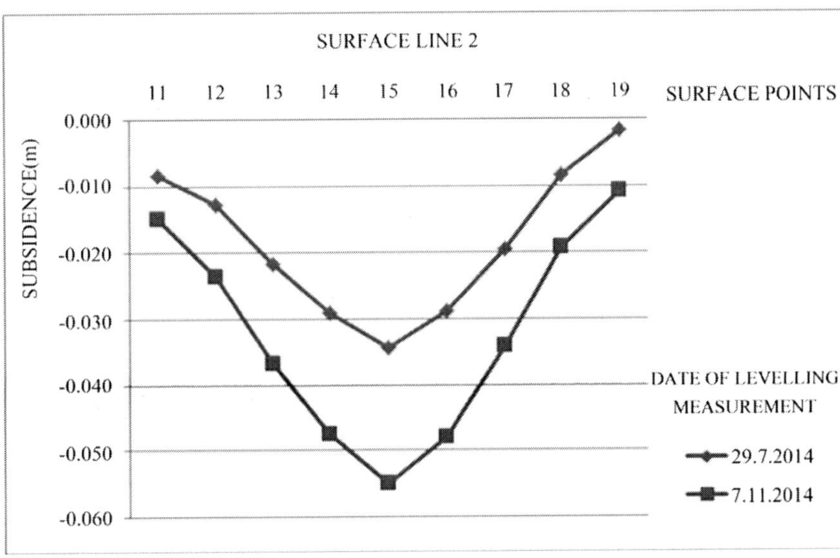

Figure 4. Graphical depiction of subsidence of surface points of aligning base 2.

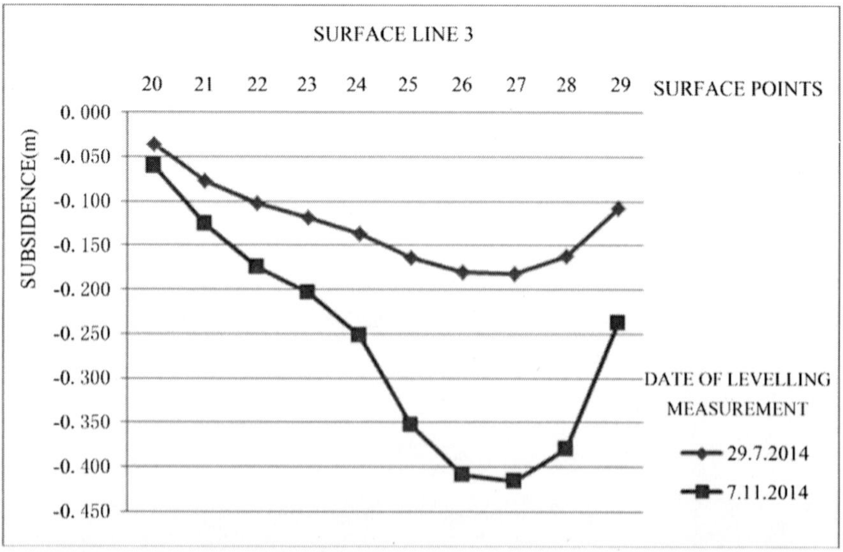

Figure 5. Graphical depiction of subsidence of surfacepoints of aligning base 3.

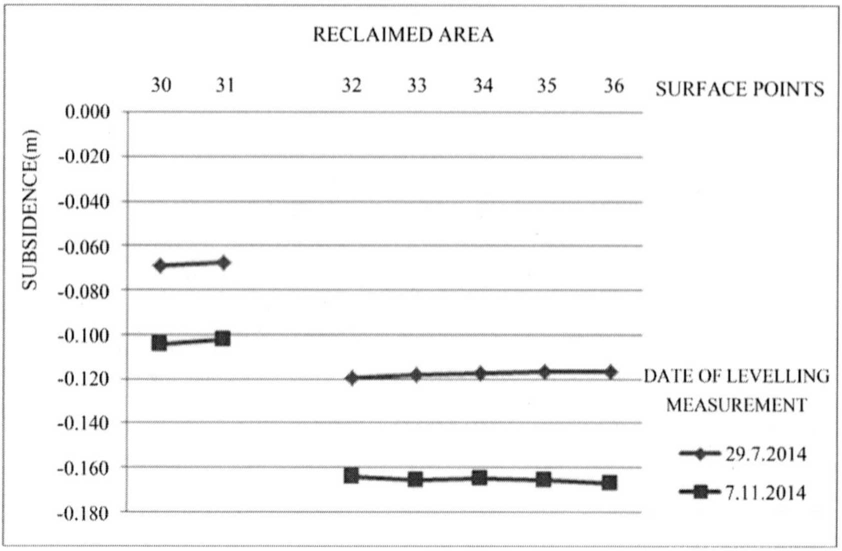

Figure 6. Graphical depiction of subsidence of surface points stabilized in therec- laimed area.

also the effect of mining that takes place at the effective distance from the protective pillar. At the same time, it was necessary to measure not only the subsidence of surface points, but the whole vector of movement, i.e. also the horizontal displacement. Based on the results of measurement, a refined theoretical model of subsidence troughs formed around the protective pillar can be developed and the effect of surrounding mining on the surface in the area of the protective pillar can be determined. Then it will be possible to separate the effect of the room and pillar mining method itself on the surface.

The results obtained clearly show that in the first year of trial operation of the new mining method, the body of the protective pillar was drawn towards the centre of gravity of the stope of active strike longwall mining with controlled caving in coal seam 33 (Figure 2). The stope in coal seam 33 includes coal face 332,203, which is located at the effective distance from the protective pillar and was mined during the first year of trial operation. This fact is shown by the directions of the vectors of displacements of surface points of aligning base 1 and surface points stabilized in the reclaimed area, which generally aim towards the centre of gravity of the stope. Another active strike longwall mining with controlled caving takes place in coal seam 37. The stope in coal seam 37 also includes coal face 371,200/3, which was mined during the first year of trial operation of the new mining method. The vectors of displacements of surface points of aligning bases 2 and 3 generally aim towards the centre of gravity of the stope in coal seam 37 (Figure 2). Based on these findings it can be presumed that a subsidence trough is being formed above the stope in coal seam 37 at the time of mining coal face 371200/3, the slope of which has a minimum effect on the movement of surface points of aligning base 1 and thus also on the area of the surface of the protective pillar of the shafts.

6. CONCLUSIONS

The room and pillar method of mining has never been used in the Ostrava-Karvina Coal District. The aim of trial operation is to prove the possibility of using the technology of room and pillar mining in the conditions of the OKR and to achieve consent of the State Mining Administration to the implementation of the new mining method on the basis of results. A great asset would then be the use of this method of mining in heavily developed areas.

The purpose of geodetic measurement is to collect data required for the evaluation of surface subsidence that will occur as a consequence of the effect of the room and pillar mining method in the protective pillar of the shafts Sever of the Mine Plant 2. With respect to the expected values of movements of surface points, the method of measurement of both the position and the height of the observed points was chosen in the category of precise measurements. The objective of geodetic measurements is to prove the real effect of the room and pillar mining method on the surface.

REFERENCES

1. Peng, S.S. (2008) Coal Mine Ground Control. 3rd Edition, Department of Mining Engineering/College of Engineering and Mineral Resources, Morgantown, 750 p.

2. Feng, Q.B. (2008) Numerical Simulation of Coal Pillar Stability in Room-Pillar Mining. Coal Mining Technology, 6, 10.

3. Yang, J.X., Liu, C.Y., Yang, Y. and Li, J.W. (2013) Study of the Bearing Mechanism of the Coal Roof and the Dimension Selection of the Room and Pillar in the Shallow and Close Distance Coal Seam. Journal of China University of Mining and Technology, 2, 161-168.

4. Ghasemi, E., Ataei, M., Shahriar, K., Sereshki, F., Jalali, S.E. and Ramazanzadeh, A. (2012) Assessment of Roof Fall Risk during Retreat Mining in Room and Pillar Coal Mines. International Journal of Rock Mechanics and Mining Sciences, 54, 80-89.

5. Waclawik, P., Ptacek, J., Konicek, P., Kukutsch, R. and Nemcik, J. (2015) Stress-State Monitoring of Coal Pillars during Room and Pillar Extraction. Gornicze Zagrozenia Naturalne, Targanice, 3-5 November 2015.

6. Cernota, P., Staňková, H., Gasinec, J. and Labant, S. (2014) New Method of Swing Observation to Determine Accurate Plummet Position at the Plumb Line for Both Connecting Surveys and

Orientation Measurements. Acta Montanistica Slovaca, 19, 221-227.

7. Sheorey, P.R., et al. (2000) Ground Subsidence Observations and a Modified Influence Function Method for Complete Subsidence Prediction. International Journal of Rock Mechanics and Mining Sciences, 37, 801-818.

8. Jiránková, E. (2012) Utilisation of Surface Subsidence Measurements in Assessing Failures of Rigid Strata Overlying Extracted Coal Seams. International Journal of Rock Mechanics and Mining Sciences, 53, 111-119. Reddish, D.J. and Whittaker, B.N. (2012) Subsidence: Occurrence, Prediction and Control. Elsevier.

9. Woo, K.S., Eberhardt, E., Elmo, D. and Stead, D. (2013) Empirical Investigation and Characterization of Surface Subsidence Related to Block Cave Mining. International Journal of Rock Mechanics and Mining Sciences, 61, 31-42.

Chapter 3

The Static Breaking Technique for Sustainable and Eco-Environmental Coal Mining

Hao Bing-yuan, Huang Hui, Feng Zi-jun, and Wang Kai

Taiyuan University of Technology, Taiyuan, Shanxi 030024, China

ABSTRACT

The initiating explosive devices are prohibited in rock breaking near the goaf of the highly gassy mine. It is effective and applicable to cracking the hard roof with static cracking agent. By testing the static cracking of cubic limestone (size: mm) with true triaxial rock mechanics testing machine under the effect of bidirectional stress and by monitoring the evolution process of the cracks generated during the acoustic emission experiment of static cracking, we conclude the following: the experiment results of the acoustic emission show that the cracks start from the lower part of the hole wall until they spread all over the sample. The crack growth rate follows a trend of "from rapidness to slowness." The expansion time is different for the two bunches of cracks. The growth rates can be divided into the rapid increasing period and the rapid declining period, of which the growth rate in declining period is less than that in the increasing period. Also, the growth rate along the vertical direction is greater than that of the horizontal direction. Then the extended model for the static cracking is built according to the theories of elastic mechanics and fracture mechanics. Thus the relation formula between the applied forces of cracks and crack expansion radius is obtained. By comparison with the test results, the model proves to be applicable. In accordance with the actual geological situation of Yangquan No. 3 Mine, the basic parameters of manpower manipulated caving breaking with static crushing are settled, which reaps bumper industrial effects.

1. INTRODUCTION

Coal energy is the most important source of energy in China. Safe coal mining under the condition of protecting eco-environment is more necessary [1–9]. Taiyuan, China, is one of the most resource-consuming coal cities, facing serious eco-environmental issues [10–19]. How to find sustainable and eco-environmental coal mining methodology is the challenge. Currently, the fully mechanized coal mining face, when exploited, gradually enters the mining roadway of the goaf. However, due to the influence of the dynamic pressure and some other factors generated during the service term, it is difficult to haul off and dismantle the anchor; therefore it is still in suspension and is being a part of reinforcement [1, 2]. Meanwhile, it is positioned at triangle area of the stope roof arch, thus the tunnel roof cannot inbreak along with the mining face roof caving, and also a hanging arch is formed on the back of the work face, which becomes universally encountered hidden danger across the sector of exploitation [5–9].

By far, the main method of tackling the issue the hanging arch are on other than hydraulic fracturing and pre splitting blasting [3]. The first method requires high-quality top tray, coupled with long processing period and does not render satisfying result, which is basically at the testing stage [4–7]. The second method, being time efficient, convenient, viable, and cost effective and having good effect, is the best way to tackle the issue of top tray and is widely utilized all over the world [6–9]. However, due to the prohibition of utilizing the explosive device for rock breaking in the goaf of highly gassy mine, it is inappropriate to tackle the issue in this way.

In this paper, the static crushing technique is first introduced to provide solution to the issue of top tray. It is a way to place the gob side roof of the roadway with manpower and shorten the spacing in between the goaf roadway, thereby solving the problems of methane accumulation in the upper corner, air leak on the work face, and great pressure in the supporting roadway. Therefore it promotes the safety management of mines.

As a well-developed technology, static breaking has been applied in several sectors and has yielded bumper harvest. Ruiping and Yongqi have studied on expansion mechanism and controllability of the static cracking agent [10]. Yujie researched into the expansion mechanism and rock breaking mechanism under the effect of static breaking techniques [11]. Furthermore, Tang et al., by carrying out physical and data analog tests on the destruction process of concrete model under the influence of static cracking agent, conclude that the destruction pattern of the crack is formed before the unstable expansion process occurs [12]. However, the above illustrated documents have never extended to the evolution principles of cracks, which also do not appear in other documents that are available to me.

This paper will report the crack evolution pattern in the process of crushing limestone with static cracking agent by monitoring the crack evolution details in the process of static destruction with acoustic emission instrument, thereby giving you a clear depiction of the three-dimensional coordinates of the location of the crack events, which is an eco-environmental sustainable method for coal mining in heavy-metal polluted region. By combining the theoretical analysis and experimental data, we acquire the calculation formula of crack expansion radius. As per this, the relationship between crack expansion radius and the expansive pressure of the static cracking agent is revealed. Also, according to the practical geological conditions of K8113 Yangquan No. 3 Mine, we determine the basic parameters of manpower forced static breaking technique. Based on the practical application, the industrial result is excellent.

2. THE EXPERIMENT ACOUSTIC EMISSION

2.1. Experimental Equipment and the Sample

This experiment employs the multichannel sound emission instrument with the AEwin data processing software produced by The America Physical Acoustics Co (PAC).

The breaking material used is K2 limestone, hexahedron limestone with side length of 200 mm, as is shown in Figure 1. Along the center position on the side perpendicular to the stretch of rock bedding, the rock core with 50 mm diameter and 150 mm length shall be chosen to be processed into standard test sample with height of 100 mm. The basic parameters needed for dynamic experiment are provided as in Table 1.

Table 1: The main mechanical parameters of limestone needed for the experiment.

Number	Specification			Max. compressing force/KN	Compressive strength/MPa	Poisson's ratio	Modulus of elasticity *E*/GPa
	Diameter/mm	Basal area/mm^2	Height/mm				
1	50.90	2034.76	104.60	98.70	48.51	0.23	28.4
2	51.00	2042.76	106.60	113.05	55.34	0.26	30.2
3	50.94	2037.96	102.46	102.73	50.41	0.22	25.9
		Average		104.83	51.42	0.237	28.12

Figure 1: Limestone sample.

Mix the static cracking agent with water at the mass ratio of 7 : 3. After stirring evenly, drip it into the hole. To simulate a practical working condition of the mine roadway, the temperature during acoustic emission experiment shall be controlled at about 10°C.

2.2. Calibration of the Detecting Points

Select two fine parallel planes. On the two ends of the diagonal line of one plane calibrate points 1 and 2, the other points 3 and 4. Points 1 and 3 shall be upward and points 2 and 4 shall be downward, and the four points calibrated shall be 70 mm away from their nearest corresponding endpoint, as shown in Figure 2.

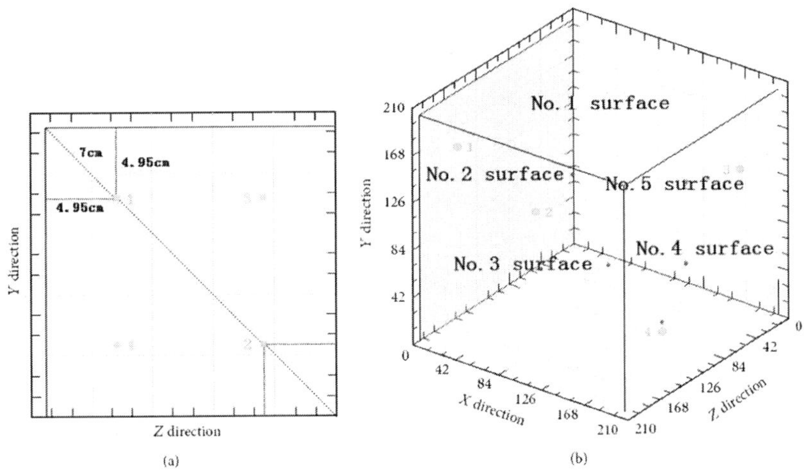

Figure 2: The position of the calibrated points.

2.3. Pressurization and Filling

After the debug of the acoustic emission instrument, check to make sure the press machine works properly. In the beginning of the experiment, prepare the agent first and then apply pressure to axial and lateral side, respectively, till 5 MPa (Figure 3).

Figure 3: Pressurization.

When pressurized operation is completed, click on the “reacquisition” button on the interface of AEwinTM, filter off the acoustic emission data collected during the pressurizing process to initiate the data collection process, and record the start time of the data acquisition.

2.4. Experimental Results

Figure 4 is a reflection of the relationship between the three-dimensional coordinates of the acoustic emission events and time, as can be seen from the figure.(1)Within 0 to 15 min, the number of acoustic emission events remains approximately unchanged; from 15 min to 20 min, the number of events significantly increases and the static cracking agent begins to react; at 30 min, the number of events continues to rise; at the same time in the lower part of the hole wall, the events are more dense; therefore, it can be inferred that the first occurring crack begins there and then expands outward.(2)At 40 min, the cracks continue to expand all around, and the expansion velocity at the hole bottom is greater than that at the orifice; at about 50 min, the first two cracks have basically expanded through the whole limestone sample, and meanwhile, the two latter occurring cracks begin to take shape; after 60 min, the first two cracks continue to expand as ever and the latter two gradually begin to grow clear; from 60 min to 300 min, the first two and the latter ones continue to expand. The comparison between the graphs at 240 min and at 300 min reveals no obvious differences. As per this, it can be concluded that, from 240 min, the formation process of the cracks basically completed for the reaction velocity of the static cracking agent is minimal.

Seen from the overall status of the three-dimensional coordinates of the acoustic emission events, from 0 min to 300 min, the number of events increases along with the elapsing of time and the growth rate is greater at first than later on and finally becomes stable; the swelling pressure of the static cracking agent breeds cracks and the cracks formed at first are approximately vertical to the latter occurring ones.

Figure 5 describes the expansion status of the first occurring cracks A and B. It reveals that the cracks start from the hole wall and extend vertically and horizontally till they go through the whole limestone sample. Select the coordinates of the main areas where events concentrate from the acoustic emission data recorded during the process to calculate the range of the occurring events and regard the result as the growth length of the cracks. According to the data distributed along directions and in Figure 5, the curve reflected in Figure 6 can be obtained.

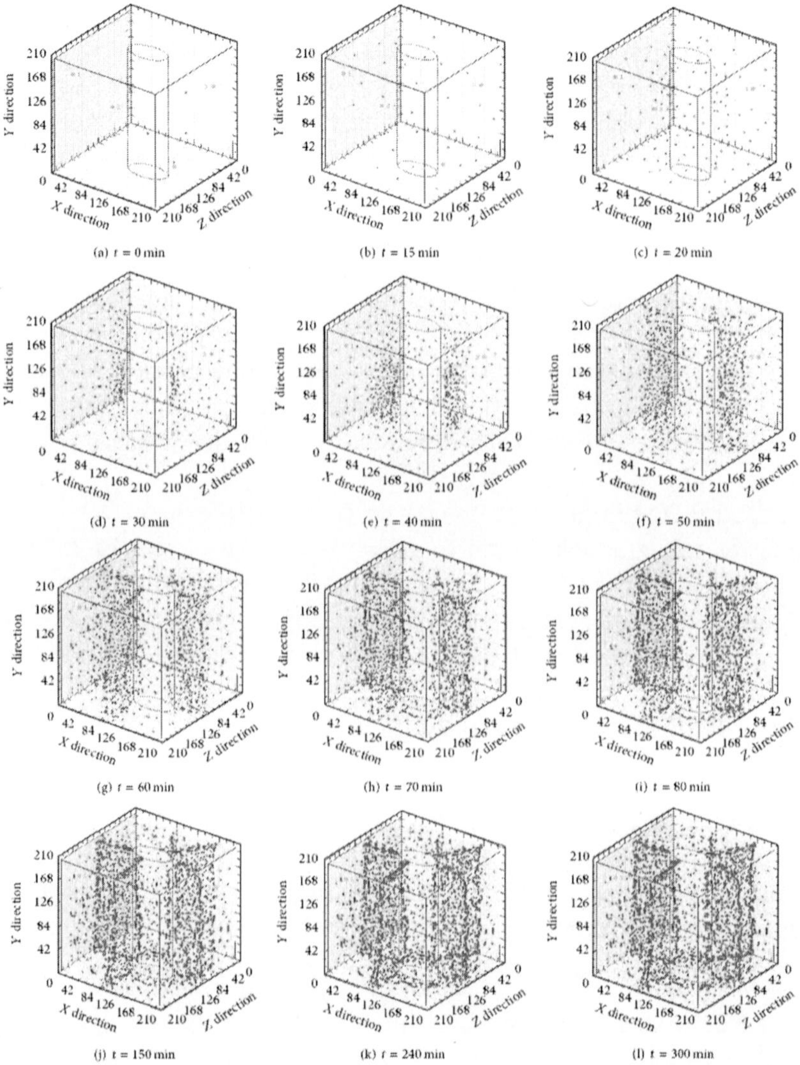

Figure 4: The three-dimensional coordinates graph of the acoustic emission events.

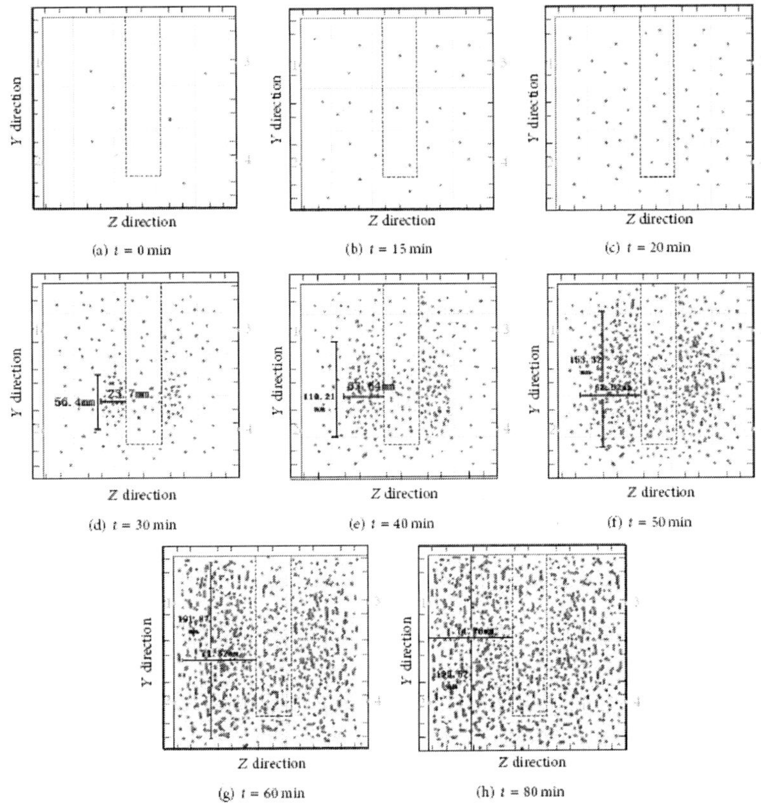

Figure 5: The expansion status of the first occurring cracks A and B.

On the basis of the expansion lengths' changes over time reflected in Figure 6, the curves in Figure 7 can be obtained. A combination analysis of Figures 6 and 7 demonstrates the following.(1)The swelling agent begins to react after about 20 min, and within the 20 min prereaction period, no cracks occur.(2)After the swelling agent takes chemical reaction, the reaction rate rises over time gradually till it reaches the maximum level at 40 min. At 40 min, the growth rate of the cracks also reaches its peak.(3)The growth rate of the cracks presents a trend of "from rapidness to slowness, and to be steady finally"; from 20 min to 30 min, the growth rate of the vertical cracks increases rapidly. After 30 min, the growth rate drops radically, and the growth rate during the declining period is less than that of the growth period. From 20 min to 40 min, the growth rate of the cracks along the horizontal bedding direction rises greatly until 40 min, when it begins to decline rapidly. Also, the growth rate during the declining period is less than that of the growth period.(4)The growth rate along the horizontal direction is greater than that of the vertical direction.(5)At 80 min, the cracks almost spread all over the whole limestone sample.

Figure 8 reveals the expansion status of the latter occurring cracks C and D. It shows that after 40 min from the occurrence of cracks A and B, cracks C and D appear. Based on the data on and from Figure 8, the curves in Figures 9 and 10 can be obtained.

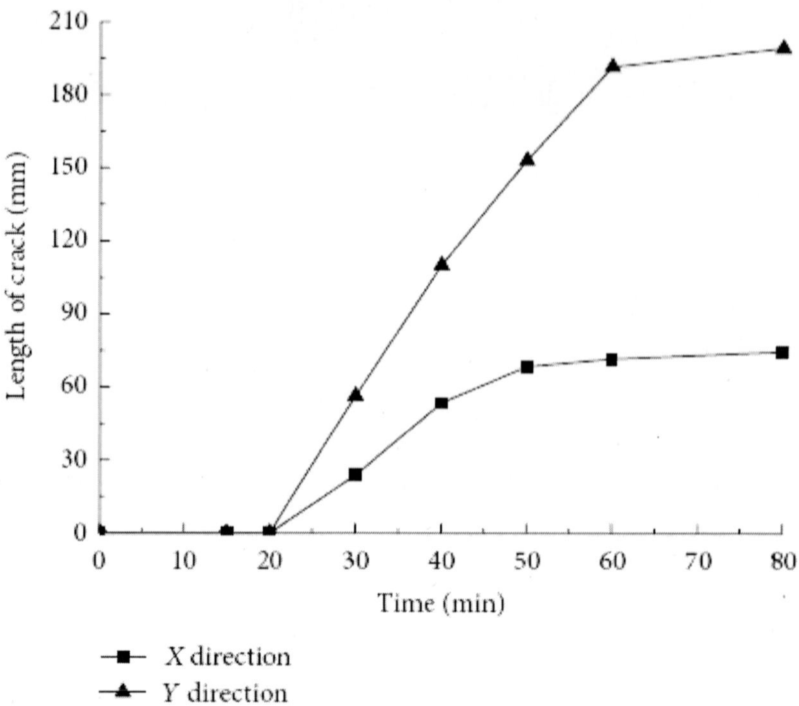

Figure 6: The relation curves between the expansion lengths of cracks A and B and time.

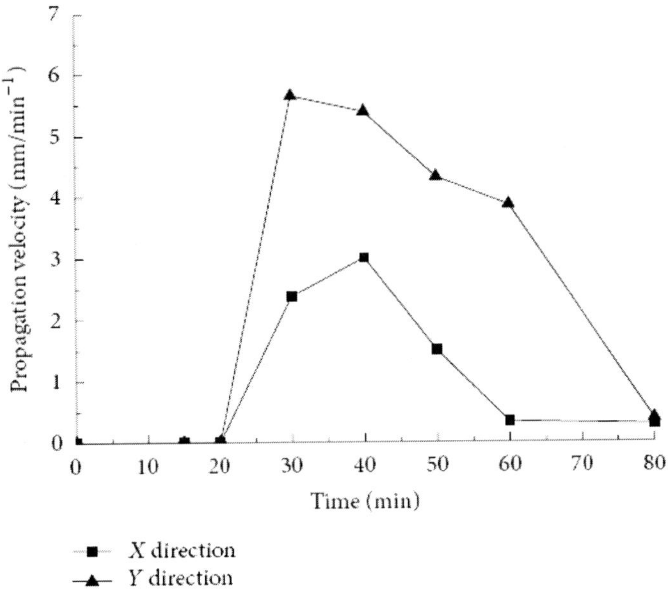

Figure 7: The relation curves between the average expansion velocities of cracks A and B and time.

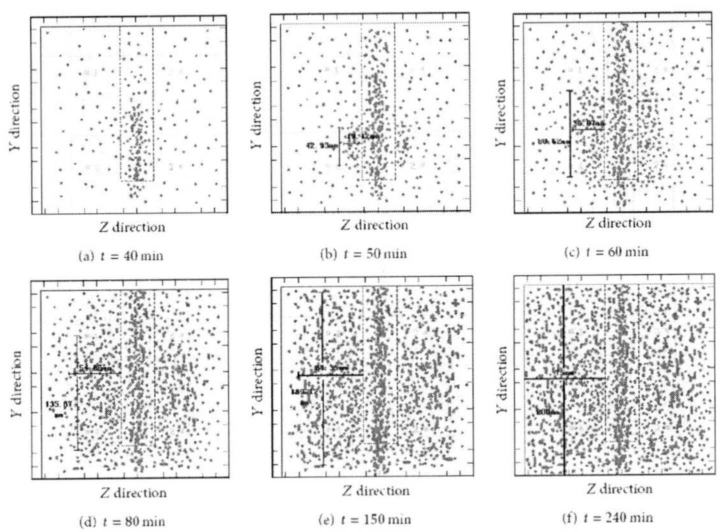

Figure 8: The expansion status of the latter occurring cracks C and D over time.

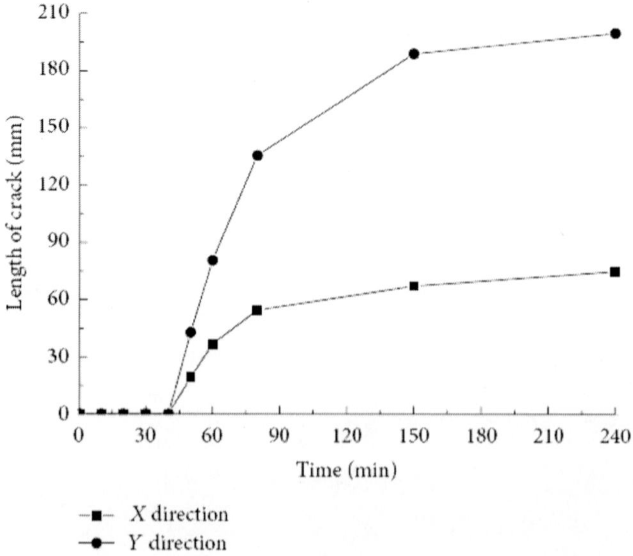

Figure 9: The relation curves between the growth lengths of C and D and time.

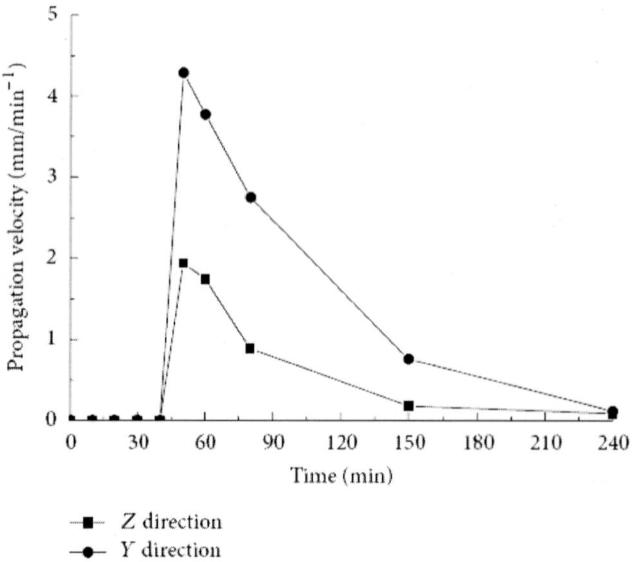

Figure 10: The relation curves between the average growth rate of C and D and time.

From the curves in Figures 9 and 10, we can see that the following occur.(1)Cracks C and D have no obvious changes from 0 to 40 min.(2)The growth rate of cracks C and D presents a trend of "from rapidness to slowness"; from 40 min to 60 min, the growth rate of the vertical cracks increases rapidly. After 60 min, the growth rate drops radically, and the growth rate during the declining period is less than that of the previous period. From 40 min to 50 min, the growth rate of the cracks along the horizontal direction rises greatly until 50 min, when it begins to decline rapidly. Again, the growth rate during the declining period is less than that of the growth period.(3)The growth rate along the vertical direction is greater than that of the horizontal direction; at 240 min, the cracks almost spread all over the whole limestone sample.

3. ESTABLISHMENT OF CRACK EXPANSION MODEL

According to the acoustic emission experiment and based on existing data, the crack expansion model is obtained as Figure 11 shows.

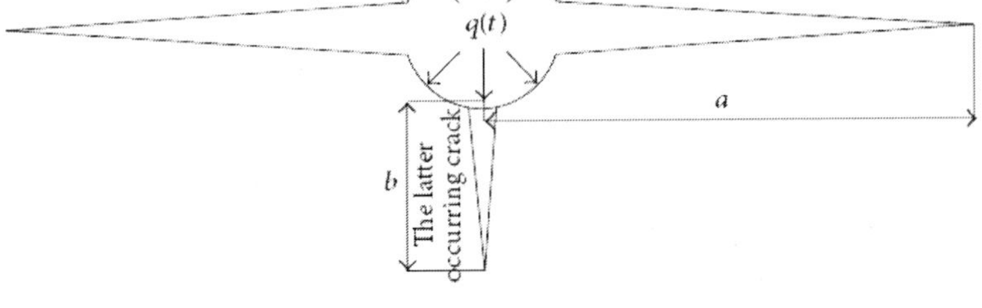

Figure 11: The crack expansion model.

When the first occurring crack length is minimal enough, the drill hole can be regarded as part of the cracks. Then if the whole model is regarded as an infinite medium, the stress intensity factor of the swelling pressure which changes over time should be [13]

$$K^{*} = 2\sqrt{\frac{a}{\pi}}q(t)\arcsin\left(\frac{d}{a}\right).$$
(1)

In the formula: is the stress intensity factor of first occurring cracks, ; is the diameter of the drill hole; is the crack length, mm.

When is greater than , the model can be regarded as the cracks generated by a couple of concentrated stress , which acts on the central, upper, and lower surfaces. ; in this formula, stands for the distance from any point to the center of the drill hole. Therefore, the stress intensity factor can be expressed as

$$K_{a}^{*} = \frac{P}{\sqrt{\pi a}},$$
(2)

$$K_{b}^{*} = \frac{P}{\sqrt{\pi b}}.$$
(3)

4. CALCULATING THE EXPANSION RADIUS OF THE CRACKS

When the static cracking agent expands its volume due to hydration, the swelling pressure exerted on the drill hole rises correspondingly. Thus the stress intensity factor around the hole increases accordingly. When it rises to the fracture toughness of the rock, the cracks begin to expand all around. From formulas (2) and (3), we can see that when stress remains unchanged, the less length is, the greater the stress intensity factor is. When reaches fracture toughness , the cracks begin to expand and along with the increasing of the crack length, the stress intensity factor drops rapidly until it is less than , when the crack expansion stops. When rises constantly, the above illustrated crack expansion will not take place. Through the measurement of the uniaxial compressive strength , the corresponding fracture toughness can be obtained [14] by

$$K_{\mathrm{IC}} = 0.0265\sigma_c + 0.0014. \tag{4}$$

Because , therefore . The formula of minimum crack expansion should be [15]

$$K^* = K_{\mathrm{IC}}. \tag{5}$$

Bring the formulas (2) and (4) into formula (5), we can obtain the following formula (6):

$$a = \frac{q^2(t)}{\pi\left(0.0256\sigma_c + 0.0014\right)^2}. \tag{6}$$

Theoretically speaking, formula (6) should be the expansion radius of the cracks. On the basis of it, the appropriate spacing between the drill holes during the process of utilizing static breaking method to tackle the issue of hanging arch should be .

5. ANALYSIS AND DISCUSSION

The following results can be concluded from the acoustic emission experiment.(1)The crack expands outward from down the middle part of the hole wall till it spreads all over the whole sample.(2)After certain time when the first occurring cracks A and B show up, cracks C and D appear. It can be inferred that the latter occurring cracks always appear later than the first cracks.(3)The expansion process of crack A is in conformity to that of crack B. It is also true for cracks C and D.(4)The growth rate of the cracks presents a trend of “from rapidness to slowness”; the growth rate can be divided into two parts, namely, period of rapid increasing and period of rapid declining. The growth rate during the declining period is less than that of the growth period.(5)The growth rate along the vertical direction is greater than that of the horizontal direction.

The result of crack expansion during the acoustic emission experiment is shown in Figure 12.

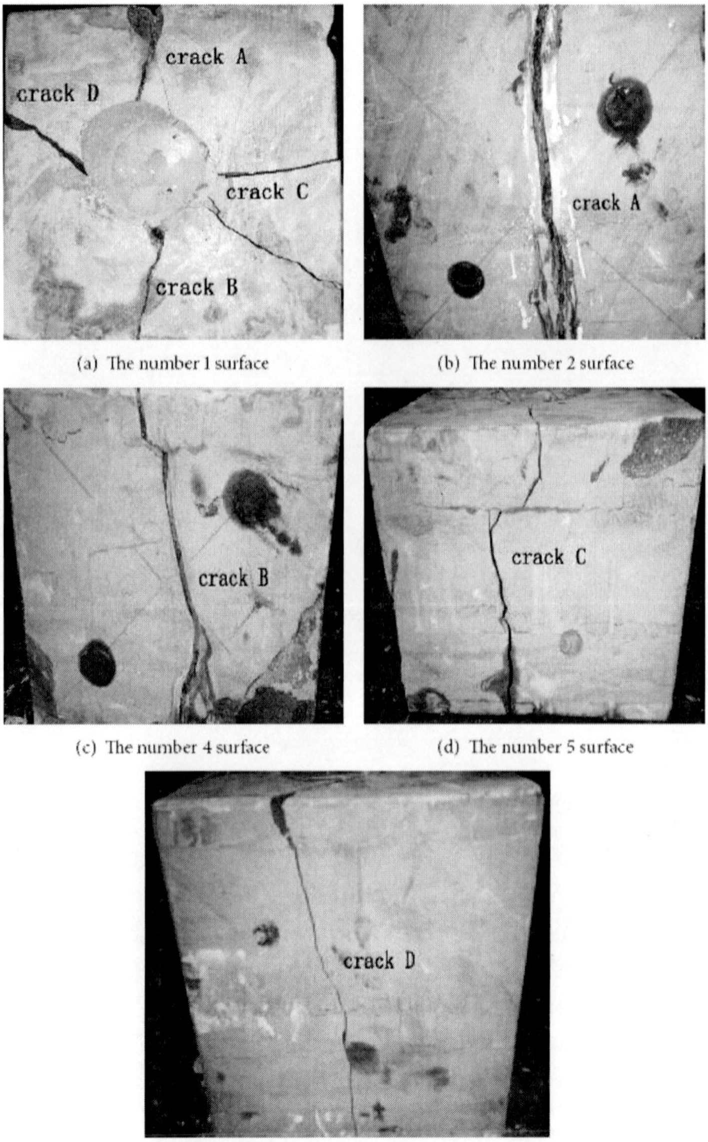

(a) The number 1 surface

(b) The number 2 surface

(c) The number 4 surface

(d) The number 5 surface

(e) The number 3 surface

Figure 12: Crack expansion actual results.

By comparing Figure 12 with Figures 4, 5, 8, and 11, we can see that the result of the crack expansion monitored from the acoustic emission experiment is in conformity with the actual situation and the crack expansion model is in accordance with the practical situation. Thus the model proves to be reasonable.

6. APPLICATION IN PROJECT

6.1. General Situation of the Project

Yangquan No. 3 Mine is a highly gassy mine. Number 15 coal seam to be mined is in K8113 working face. The average thickness is 7.18 m and the uniaxial comprehensive strength is about 50 MPa.

Static crushing material used in this study is "high efficiency coal mine static expanding agent"; the maximum swelling pressure is 80 MPa.

6.2. Scheme Design

In static crushing technology, the spacing between holes remains a problem. At present, there are only some certain empirical methods, which have not been studied scientifically in a systematical fashion. On the basis of the crack expansion model built according to result from the acoustic emission experiment, this paper offers the formula for calculating the spacing in between drill holes so as to provide theoretical support for properly designing the hole spacing.

By putting the maximum swelling pressure and the uniaxial compressive strength on the immediate roof into formula (6), we can obtain such formula as mm; then mm; namely, the most reasonable spacing between drill holes should be 800 mm. Other parameters needed are mainly the hole depth 4.5 m; the horizontal contained angle with the worked out section 45°; specification of the hole sealing machine ZF-A22; the grout amount can be calculated as ; in the formula, is grouting coefficient, and put it between 1.4 and 1.5; stands for material volume-weight, unit: kg/m3; is diameter of the drill hole, unit: m; stands for hole depth, unit: m; is hole sealing length, unit: m; is mass of the swelling agent, unit: kg.

6.3. Test Results

Before the industrial test was taken, when the hauling-off anchor and net cut are completed on working face K8113, the hanging arch is still as long as 20 m. Several hours after the grouting is completed, an obvious sound of fracture can be heard in the rock of the top tray. When working face and nose move 4 coal cutters, namely, 3.2 m forward, the coals on top inside the setting roadway begin to inbreak. Meanwhile, the limestone on the upper roof gradually starts to collapse. With further moving forward of 12 coal cutters, namely, 9.6 m, the whole test area will have already collapsed thoroughly. By being processed with the static crushing method, the length of the hanging arch lessens by 7.2 m, which improves the safety production environment of the working face effectively.

7. CONCLUSION

(1)The experiment results of the acoustic emission show that the cracks start from the lower part of the hole wall. The processes of expansion for the first occurring cracks are in conformity with each other, which is also true of the latter occurring cracks. Expansion time for the two bunches of cracks is distinct. The crack growth rate follows a trend of "from rapidness to slowness." The growth rates can be divided into the rapid increasing period and the rapid declining period, of which the growth rate in declining period is less than that of the increasing period. Also, the growth rate along the vertical direction is greater than that of the horizontal direction.(2)Based on the acoustic emission test results, the extended model of the static breaking crack, by comparison, is in conformity with the actual situation, which indicates that the model is reasonable; the fracture radius in static crushing can be calculated through the formula(3)Research on the crack expansion principles and the static technique, combined with application in the project of Yangquan Coal Group, demonstrates that the static crushing technique satisfies the demand for tackling the issue of hanging arch on highly gassy mine working face.

CONFLICT OF INTERESTS

The authors declare that there is no conflict of interests regarding the publication of this paper.

ACKNOWLEDGMENTS

The study has been supported by National Key Technology R&D Program of China, (Grant no. 2009BAB48B03) and Key Technology R&D Program of Shanxi Province, China (Grant no. 20121101009-03).

REFERENCES

1. K. Hongpu and W. Jinhua, Theories of Anchor Bolt Support and Technology, China Coal Industry Publishing House, Beijing, China, 2007.

2. C. Li, J. H. Xu, R. Wu, and J. T. Dong, "Mechanism and practice of support release of mining roadway about fully-mechanized top-coal caving mining stope," Journal of the China Coal Society, vol. 36, no. 12, pp. 2018–2023, 2011.

3. K. Wang, T. H. Kang, H. T. Li, and W. Han, "Study of control caving methods and reasonable hanging roof length on hard roof," Chinese Journal of Rock Mechanics and Engineering, vol. 28, no. 11, pp. 2320–2327, 2009.

4. S. H. Yan, Y. Ning, L. J. Kang, Y. W. Shi, Y. G. Wang, and Y. F. Li, "The mechanism of hydrobreakage to control hard roof and its test study," Journal of China Coal Society, vol. 25, no. 1, pp. 32–35, 2000.

5. Y. Feng and H. Kang, "Test on hard and stable roof control by means of directional hydraulic fracturing in coal mine," Chinese Journal of Rock Mechanics and Engineering, vol. 31, no. 6, pp. 1148–1155, 2012.

6. X. J. Yang and Z. P. Shen, "Application of softening hard roof with deep bore explosion in full-mechanized mining face with

large mining height," Coal Mining Technology, vol. 12, no. 6, pp. 30–32, 2007.

7. C. R. Li, L. J. Kang, Q. X. Qi, D. B. Mao, and G. Xu, "Numerical simulation of deep-hole blasting and its application in mine roof weaken," Journal of the China Coal Society, vol. 34, no. 12, pp. 1632–1636, 2009.

8. Y. P. Wu, K. F. Li, and Y. L. Zhang, "Simulation of hard roof weakening ahead of working face with fully mechanized caving," Journal of Mining and Safety Engineering, vol. 26, no. 3, pp. 273–277, 2009.

9. D. Y. Guo, D. Y. Shang, P. F. Lv, S. Y. Wang, and J. M. Wang, "Experimental research of deep-hole cumulative blasting in hard roof weakening," Journal of the China Coal Society, vol. 38, no. 7, pp. 1149–1153, 2013.

10. G. Ruiping and Y. Yongqi, "Swelling mechanism and controllability of SCA," Journal of China Coal Society, vol. 19, no. 5, pp. 478–485, 1994.

11. W. Yujie, Research on Mechanism and Technology of Non-Explosive Demolition, Wuhan University of Technology, Wuhan, China, 2009.

12. L. Tang, C. Tang, S. Tang, Y. Cui, and L. Song, "Physical experiment and numerical simulation on effect of soundless cracking agent," Chinese Journal of Geotechnical Engineering, vol. 27, no. 4, pp. 437–441, 2005.

13. C. Jin and Z. Shushan, Fracture Mechanics, Science Press, Beijing, China, 2012.

14. L. Jiangteng, G. Desheng, C. Ping, et al., "Interrelated law between mode-I fracture toughness and compression strength of rock," Journal of Central South University (Science and Technology), vol. 40, no. 6, pp. 1695–1699, 2009. L. Shiyu and Y. Xiangchu, Introduction of Rock Fracture Mechanics, Press of University of Science and Technology of China, Beijing, China, 2010.

15. G. Xu, Y. C. Lv, J. N. Sun, H. B. Shao, and L. L. Wei, "Recent advances in biochar applications in agricultural soils: benefits and environmental implications," Clean—Soil, Air, Water, vol. 40, no. 10, pp. 1093–1098, 2012.

16. G. Wu, H. B. Kang, X. Y. Zhang, H. B. Shao, L. Y. Chu, and C. J. Ruan, "A critical review on the bio-removal of hazardous heavy metals from contaminated soils: issues, progress, eco-environmental concerns and opportunities," Journal of Hazardous Materials, vol. 174, no. 1–3, pp. 1–8, 2010.

17. H. B. Shao, Ed., Metal Contamination: Sources, Detection and Environmental Impact, NOVA Science Publishers, New York, NY, USA, 2012.

18. H. B. Shao and L. Y. Chu, "Some progress in the study of plant-soil interactions in China," Plant Biosystems, vol. 147, no. 4, pp. 1163–1165, 2013.

Chapter 4

Grinding Analysis of Indian Coal Using Response Surface Methodology

Twinkle Singh[1], Aishwarya Awasthi[1], Pranjal Tripathi[1], Shina Gautam[2], Alok Gautam[2]

[1] *Jaypee University of Engineering & Technology, A-B Road, Raghogarh, Guna, India*

[2] *Shroff S. R. Rotary Institute of Chemical Technology, Ankleshwar, Gujarat, India*

ABSTRACT

The present work discusses a systematic approach to model grinding parameters of coal in a ball mill. A three level Box–Behnken design combined with response surface methodology using second order model was applied to the experiments done according to the model requirement. Three parameters ball charge (numbers 10–20), coal content (100–200 g) and the grinding time (4–8 min) were chosen for the experiments as well as for the modeling work. Coal fineness is defined as the d_{80} (80 % passing size). A quadratic model was developed to show the effect of parameters and their interaction with fineness of the product. Three different sizes (4, 1 and 0.65 mm) of Indian coal were used. The model equations for each fraction were developed and different sets of experiments were performed. The predicted values of the fineness of coal were in good agreement with the experimental results (R^2 values of d_{80} varies between 0.97 and 0.99). Fine size of three different coal sizes were obtained with larger ball charge with less grinding time and less solid content. This work represents the efficient use of response surface methodology and the Box–Behnken design use for grinding of Indian coal.

Keywords: Coal grinding Ball mill Grinding time Response surface methodology

1 INTRODUCTION

The consumption of solid fossil fuel is gaining importance in world energy production. There has been an increase in studies on coal since it is an alternative to petrol and has a high reserve. The pulverized coal injection is used to obtain high heat capacity and high burning efficiency in thermal reactors and boilers based on coal. For this purpose, coals are ground at a micron size and then sent to boilers to achieve maximum calorific value. Comminution of coal is essential for effective utilization through combustion, carbonization, gasification and/or beneficiation (Prasher 1987). Preheating a coal in an oven has been shown to reduce grind strength, although it is unlikely that this would be an economically viable method, at least not through associated energy requirements (Harrison and Rowson 1997). The grindability of coal is an important practical and economical property to coal handling and utilization aspects. Thus engineering experiments wish to find the grinding conditions under which a certain process reaches its response optimum. The optimum could be either a maximum or a minimum of a function of the design parameters. One of the methodologies is the surface response methodology (Kwak 2005). Grinding media charge ratio in a conventional grinding system is approximately 40 %. This leads to a better delivery of the grinding energy (Kincl et al. 2005; Lester et al. 2005; Barranco et al. 2006). Grinding in a ball mill has shown that the ball charge has a significant effect on the product fineness for same specific energy consumption and increases power draws of the mill linearly (Tuzun 1994 and Jankovic 2001). The ball size should be just heavy and large so that the charge should be grinded efficiently. In the present work the ball size is kept constant and these are considered to be efficient for the grinding media. In the grinding of Turkish coal and ore, the charge ratio, Bond work index and the ball diameter were taken as the variables in the experiments however, time of grinding is also a significant parameter in the grinding experiments (Aslan and Cebeci 2007; Edadnejad et al. 2013).

The objective of the present work is to establish the optimum conditions for the grinding of an Indian coal. The grinding of coal may depend on several parameters however, in the present study the grinding time, charge content and number of balls in a ball mill are considered as parameters. These three parameters are most important parameters in grinding of coal which were ignored in previous works. According to the requirement of the response surface methodology, experiments were done for three different size fractions of coal. Box-Behnken design to model the experimental results was used.

2 MATERIALS AND METHODS

Initially 7–10 cm size of Indian coal was crushed in a jaw crusher and roll crusher successively to prepare the feed for the ball mill. Three different sizes, 80 % passing of specified screen (d_{80}) for grinding test of coal particles were chosen before milling in a ball mill. The d_{80} sizes of coal particles of feed were 4, 1, and 0.65 mm. After grinding d_{80} of these sizes were 1.65, 0.7, 0.25 mm. Grinding tests were performed in a laboratory scale ball mill of 37.5 cm long and 27 cm diameter. The size of the balls used were of 2.5 cm diameter. The rotation speed of the ball mill was kept adequate for the charge to mix and impact together. A proximate analysis of coal is also performed before the actual experiments, Table 1. The density of the coal used was 1.38 gm/cm^3.

Table 1. Proximate analysis of a typical Indian coal

Parameter	Dry basis (%)
Moisture	6.01
Ash	39.48
Volatile matter	18.65
Fixed carbon	35.6

3 RESPONSE SURFACE METHODOLOGY

Response surface methodology (RSM) is a collection of statistical and mathematical system for the analysis of complex engineering and scientific problems. Using this technique one can evaluate the effect of different important parameters on response.

The design procedure is as follows (Gunaraj and Murugan 1999; Aslan and Cebeci 2007; Edadnejad et al. 2013):

1) Designing of a series of experiments for adequate and reliable measurement of the surface response of interest.

2) Developing a mathematical model of the second-order response surface with the best fittings.

3) Finding the optimal set of experimental parameters that produce a maximum or minimum value of response.

4) Representing the direct and interactive effects of process parameters through two and three dimensional (3D) plots.

If all the independent variables are measurable, the response of the surface methodology can be expressed as

$$y = f(x_1, x_2, x_3, \ldots, x_n) \tag{1}$$

where *y* is response of the system.

A Box–Behnken design with three levels was utilized to evaluate the effect of parameters on fineness of coal. So these three independent parameters were coded at three levels that is +1, 0, and −1 with same step size, where +1 represents the maximum value, 0 represents to the center, and −1 represents the minimum value of each parameter, which are considered for analysis. Within the present research framework, the discussion is focused on the effect of number of balls (A), amount of feed charged (B), and grinding time (C) on the fineness of coal using a Box-Behnken design. Table 2 shows the experimental design matrix with coded and actual values of parameters A, B, and C.

Table 2. The level of variables in Box-Behnken design

Parameter	Low (−1)	Center (0)	High (+1)
A (Grinding time)	4	6	8
B (Number of balls)	10	15	20
C (Amount of feed)	100	150	200

To find the optimum conditions, a quadratic model is used to relate fineness of coal to independent parameters. This model can be written as:

$$y = \alpha_0 + \sum_{j=1}^{n} \alpha_j x_j + \sum_{j=1}^{n} \alpha_{jj} x_j^2 + \sum_{j=1}^{n-1} \sum_{k=2}^{n} \alpha_{jk} x_j x_k + \delta \qquad (2)$$

The set of regression coefficients αα's are unknown and estimated by least squares. In vector–matrix, the equation for least square fit is given by:

$$Y = X\alpha + \delta \qquad (3)$$

where *Y* is defined as the measured values and X to be a matrix of independent variables. Regression coefficients αα is solved by premultiplying by the non singular regression matrix transpose X′X′

$$\beta = (X'X)^{-1}X'Y \tag{4}$$

where X′X′ is the transpose of the matrix XX and (X′X)−1(X′X)−1 is the inverse of the matrix X′XX′X. Applying the present variables in Eq. (2), the equation will expand in the form as follows:

$$y_{d_{80}} = \alpha_0 + \alpha_1 A + \alpha_2 B + \alpha_3 C + \alpha_4 AB + \alpha_5 AC + \alpha_6 BC + \alpha_7 A^2 + \alpha_8 B^2 + \alpha_9 C^2 + \delta_i \tag{5}$$

where yy is response (fineness of coal of 80 % passing), and AA, BB, and CC are the independent parameters. α0α0 is the constant coefficient, α1α1, α2α2, and α3α3, are the coefficients for linear effect, α4α4, α5α5, and α6α6, are the coefficients for interaction effect, α7α7, α8α8, and α9α9, are the coefficients for quadratic effect, and δiδi is the error.

Response surface methodology uses the experimental data points of the design matrix to a proposed model and the unknown coefficients. To find the minimum number of experimental run required for Box-Benhken design one can use equation given below:

$$N = 2N_f(N_f - 1) + C_p \tag{6}$$

where NfNf is the number of parameters used to fit the model, and CpCp is the number of the central points. To evaluate the coefficient values 15 experiments were carried out for each particle size and total 45 experiments were done for three different particle size according to the Box–Behnken design. Figure 1 shows the distribution of the experimental data points in the graphical form. Accordingly experiments were designed and Table 3 shows the design matrix of the experiments for three different particle sizes.

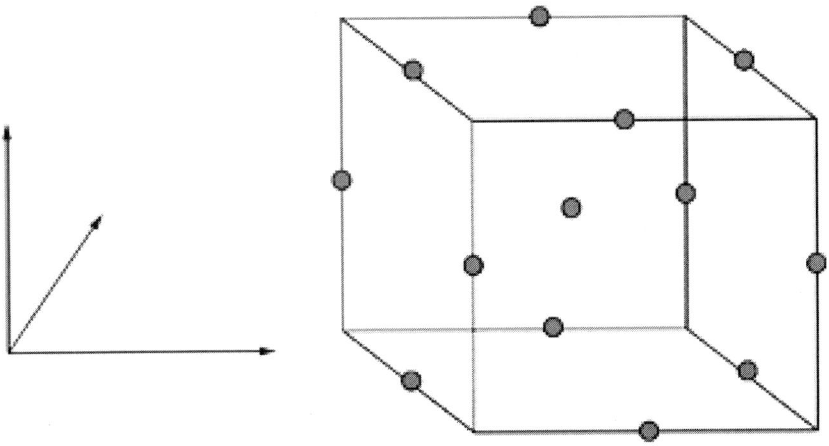

Figure 1. The graphical representation of a three parameters Box-Behnken design

Table 3. Design matrix with coded and actual design parameters

Run no.	Coded values			Actual values		
	A	B	C	A	B	C
1	+1	0	−1	8	15	100
2	0	+1	+1	6	20	200
3	+1	−1	0	8	10	150
4	−1	0	−1	4	15	100
5	−1	+1	0	4	20	150
6	0	0	0	6	15	150
7	+1	+1	0	8	20	150
8	0	−1	−1	6	10	100
9	−1	0	+1	4	15	200
10	0	0	0	6	15	150
11	−1	−1	0	4	10	150
12	+1	0	+1	8	15	200
13	0	−1	+1	6	10	200
14	0	0	0	6	15	150
15	0	+1	−1	6	20	100

4 RESULT AND DISCUSSION

There were 15 experiments which were done for each coal particle size as listed in Table 4. Model equations basically represent the relationship between fineness as d_{80} and different experimental design parameters. For three different size fractions of coal using Box-Behnken design were obtained as given below:

Table 4. Actual and coded values of three parameters for three responses

	Parameter			Experimental values of responses		
Run no.	A (min)	B	C (g)	$y1_{d_{80}}$ (μm)	$y2_{d_{80}}$ (μm)	$y3_{d_{80}}$ (μm)
1	8(+1)	15(0)	100(-1)	1700	770	250
2	6(0)	20(+1)	200(+1)	1900	880	350
3	8(+1)	10(-1)	150(0)	1900	875	325
4	4(−1)	15(0)	100(−1)	1800	800	300
5	4(−1)	20(+1)	150(0)	1875	860	340
6	6(0)	15(0)	150(0)	2000	900	320
7	8(+1)	20(+1)	150(0)	1750	725	275
8	6(0)	10(−1)	100(−1)	1850	800	275
9	4(−1)	15(0)	200(+1)	2250	950	400
10	6(0)	15(0)	150(0)	2000	900	320
11	4(−1)	10(−1)	150(0)	2150	920	375
12	8(+1)	15(0)	200(+1)	1975	880	350
13	6(0)	10(−1)	200(+1)	2100	910	375
14	6(0)	15(0)	150(0)	2000	900	320
15	6(0)	20(+1)	100(−1)	1650	700	250

For −1.65 + 4 mm size fraction:

$$y1_{d_{80}} = 781.25 + 9.375A + 43.12B + 12.43C + 3.12AB - 0.4375AC - 4.34 \times 10^{-16}BC - 3.12A^2 - 2.75B^2 - 0.0225C^2 \quad (7)$$

For −0.7 + 1 mm size fraction:

$$y2_{d_{80}} = -30 + 72.5A + 44B + 5.27C - 2.25AB - 0.1AC + 7 \times 10^{-2}BC - 3.43A^2 - 1.65B^2 - 0.0145C^2 \quad (8)$$

For −0.25 + 0.65 mm size fraction:

$$y3_{d_{80}} = 295.62 - 39.68A + 1.12B + 1.67C - 0.37AB - 0AC + 0BC + 2.65A^2 - 0.075B^2 - 0.002C^2 \quad (9)$$

Using the parameters defined in Table 2, the experimental values were obtained by performing experiments and the predicted values were calculated using Eqs. (7–9) in Table 3. From the experimental results listed in Table 4 and Eq. (5), the second order response functions representing d_{80} can be expressed in model equations. Table 5 shows the validation of model developed by using Box-Behnken design with experimental values of three different size fractions. The responses of three different size fraction of fineness can be calculated by Eqs. (7–9) for different variables defined.

Table 5. Comparison of experimental and predicted d80d80 values

$y1_{d_{80}}$ (μm)		$y2_{d_{80}}$ (μm)		$y3_{d_{80}}$ (μm)	
Experimental	Predicted	Experimental	Predicted	Experimental	Predicted
1700	1728	770	756	250	250
1900	1925	880	866	350	355
1900	1896	875	875	325	328
1800	1828	800	806	300	304
1875	1878	860	860	340	348
2000	2000	900	900	320	325
1750	1753	725	745	275	286
1850	1825	800	813	275	281
2250	2221	950	963	400	411
2000	2000	900	900	320	325
2150	2146	920	900	375	374
1975	1946	880	873	350	358
2100	2131	910	916	375	389
2000	2000	900	900	320	325
1650	1618	700	693	250	248

The actual and predicted values of three different size fractions are depicted in Figs. 2, 3 and 4. Predicted values are quite close to the experimental values and the values of R^2 are 0.98, 0.97 and 0.99 for −1.65 + 4 mm size fraction, −0.7 + 1 mm size fraction and −0.25 + 0.65 mm size fraction, respectively.

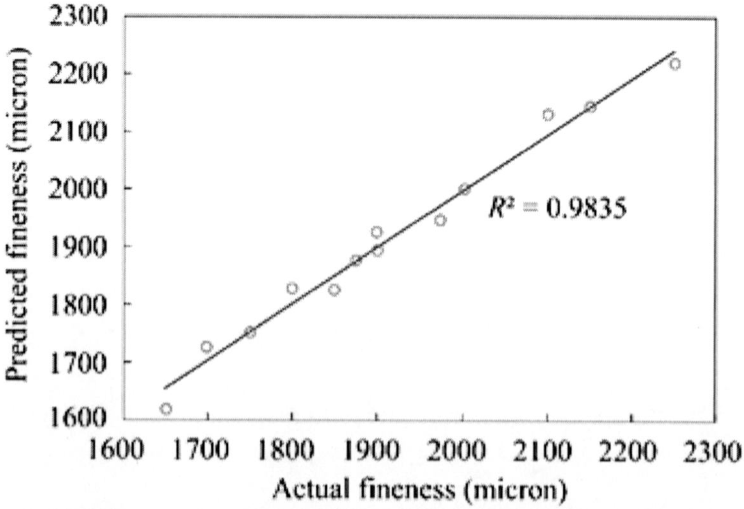

Figure 2. Experimental values versus predicted values of fineness for size fraction for −1.65 + 4 mm size fraction

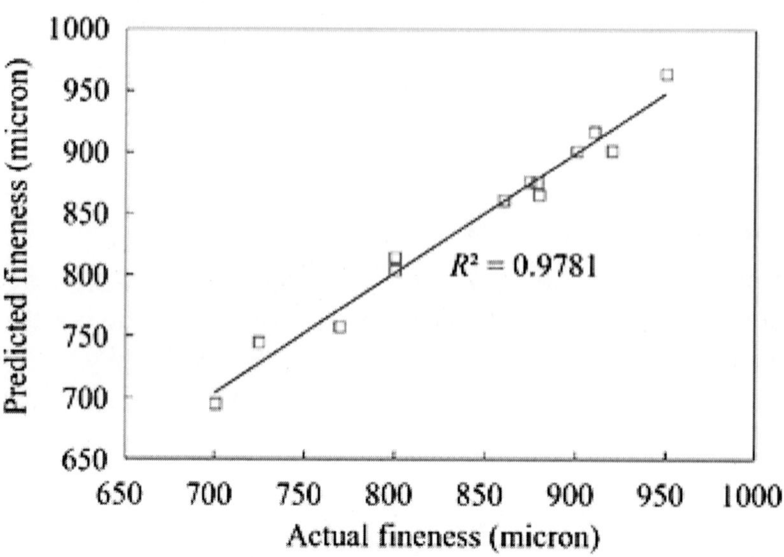

Figure 3. Experimental values versus predicted values of fineness for size fraction for −0.7 + 1 mm size fraction

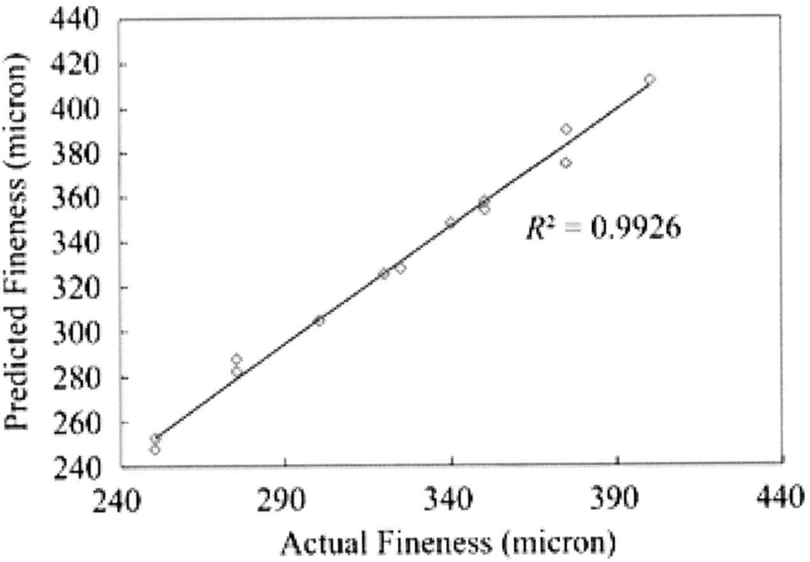

Figure 4. Experimental values versus predicted values of fineness for size fraction for −0.25 + 0.65 mm size fraction

In order to understand the effect of operating parameters on the d_{80} the results are shown in Figs. 5, 6, 7, 8, 9, 10, 11, 12 and 13. Figures 5, 6 and 7 show the 3D response surface plot for coarse size fraction (−1.65 + 4 mm). Figure 5 shows the effect of number of balls and grinding time on fineness. The grinding time and the number of balls on the coarse particle size have positive effect as the increased grinding time and increased number of balls enhance the fineness. However, individual effect of balls can also be observed as for less time increasing number of balls results in increased fineness. Figure 6 shows the effect of number of balls and amount of feed on response. The decreased amount of feed and increased number of balls are in favor of fineness. Figure 7 shows the effect of grinding time and amount of feed. It is revealed that increasing grinding time and decreasing amount of feed produces fine material.

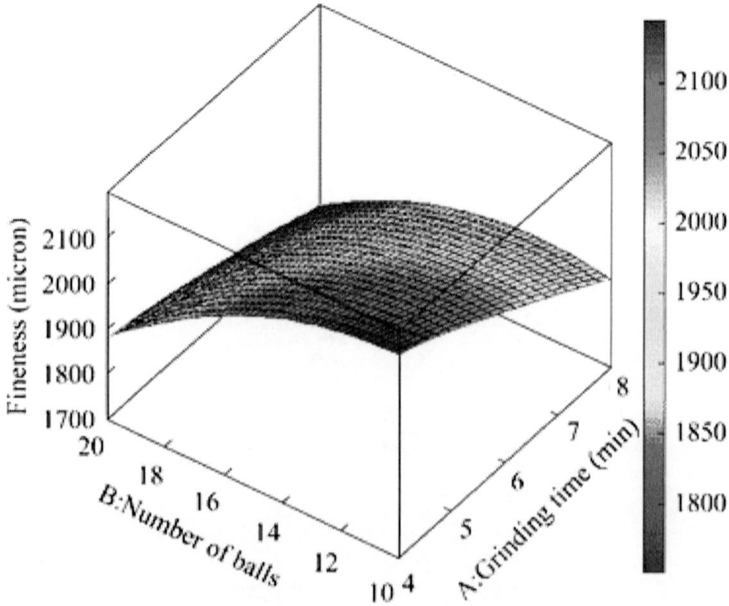

Figure 5. Response surface of fineness of coal to the variation of number of balls and grinding time at average amount of feed of 150 gm for −1.65 + 4 mm size fraction

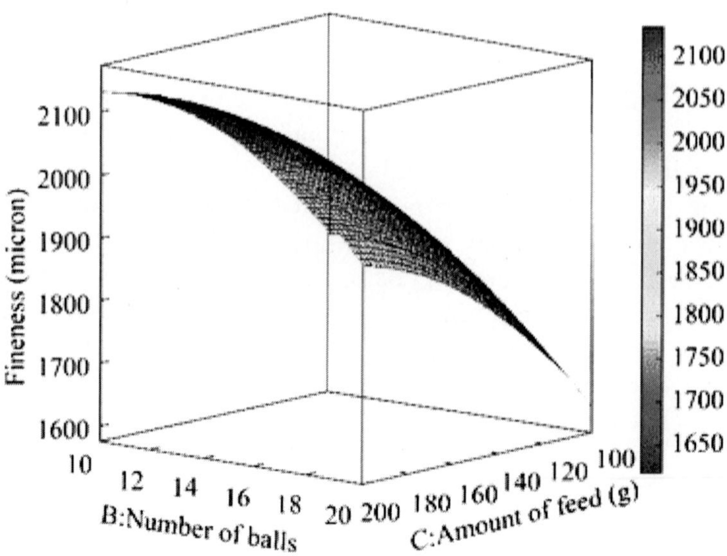

Figure 6. Response surface of fineness of coal to the variation of number of balls and amount of feed at average grinding time of 6 min for −1.65 + 4 mm size fraction

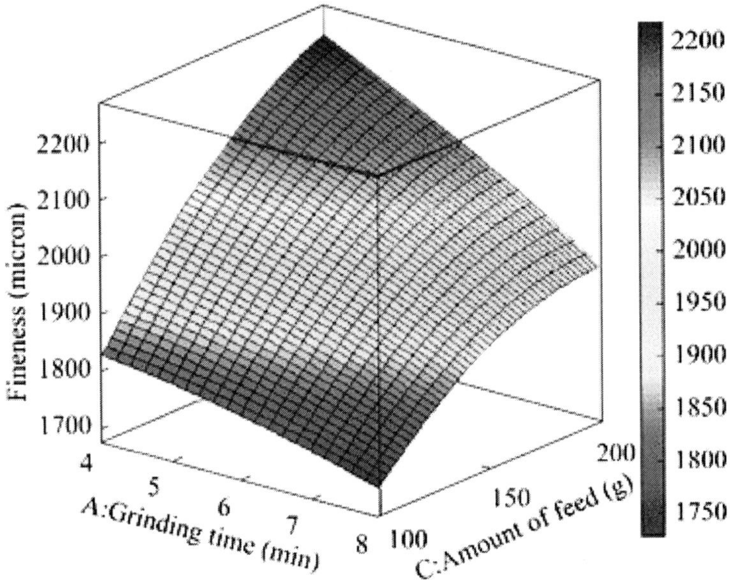

Figure 7. Response surface of fineness of coal to the variation of grinding time and amount of feed at average number of balls of 15 for −1.65 + 4 mm size fraction

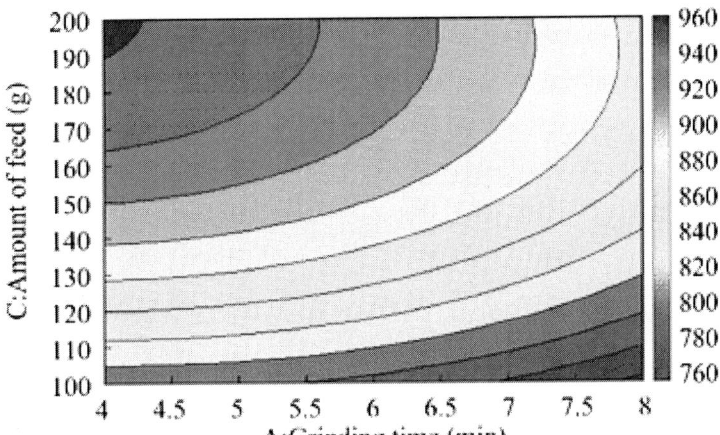

Figure 8. Contour plot response surface of fineness of coal to the variation of grinding time and amount of feed at average number of balls of 15 for −0.7 + 1 mm size fraction

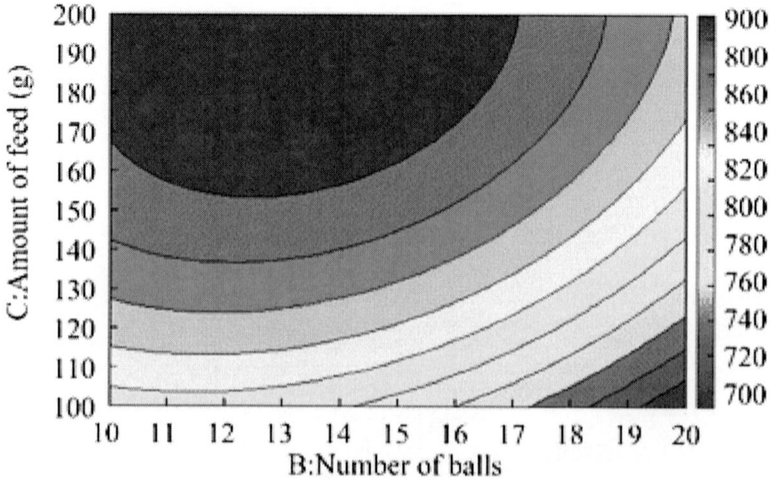

Figure 9. Contour plot response surface of fineness of coal to the variation of number of balls and amount of feed at average grinding time of 6 min for −0.7 + 1 mm size fraction

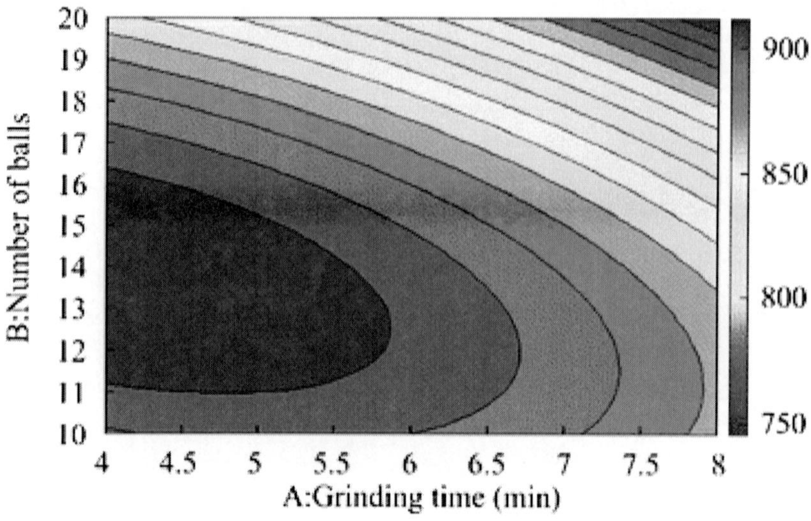

Figure 10. Contour plot response surface of fineness of coal to the variation of number of balls and grinding time at average amount of feed of 150 gm for −0.7 + 1 mm size fraction

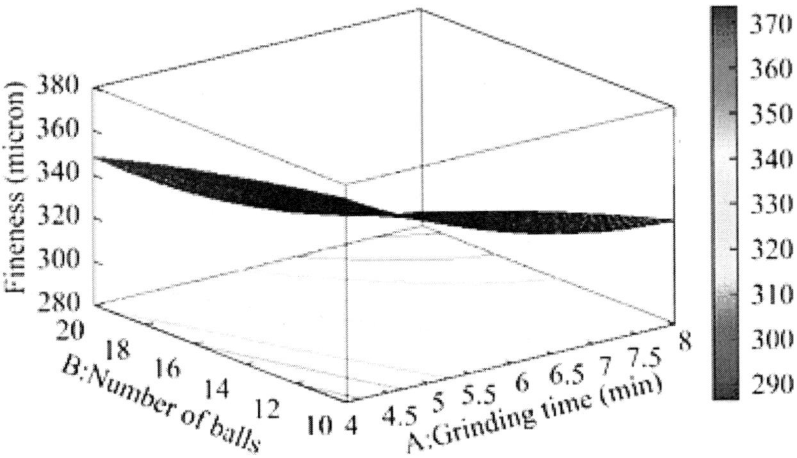

Figure 11. Contour and surface plot of response surface of fineness of coal to the variation of number of balls and grinding time at average amount of feed of 150 gm for −0.25 + 0.65 mm size fraction

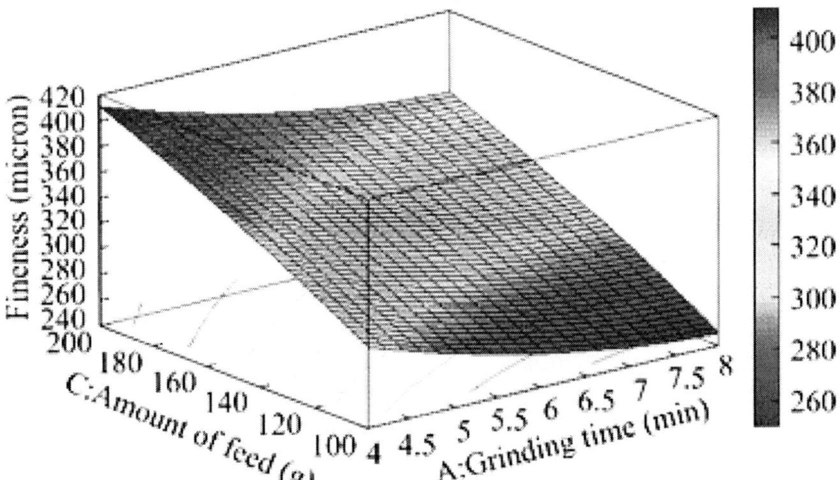

Figure 12. Contour and surface plot of response surface of fineness of coal to the variation of number of balls and amount of feed at average grinding time of 6 min for −0.25 + 0.65 mm size fraction

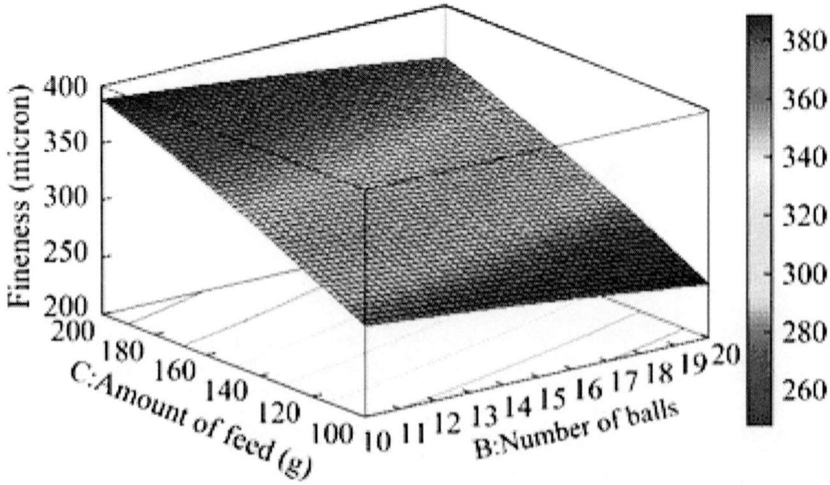

Figure 13. Contour and surface plot of response surface of fineness of coal to the variation of grinding time and amount of feed at average number of balls of 15 for −0.25 + 0.65 mm size fraction

Figures 8, 9 and 10 show the effect of different parameters on response in contour plot for the −0.7 + 1 mm size fraction. Figure 8 shows that decrease in amount of feed and increase in grinding time more fine material can be produced. This may be due to more impact of balls will be observed by the feed which results in more fine material. Figure 9 shows that number of balls used in grinding plays an important role. Increasing number of balls can produce finer particles. On the other hand if the fineness is fixed one can reduce the grinding time and energy can be saved. Figure 10 shows the effect of number of balls and grinding time on d_{80}. It is clear from figure that number of balls on fineness has dominant effect than grinding time. With more number of balls and less grinding time one can produce more fine powder. The figure reflects that process will be inefficient if ball mill is operated between 11 and 16 number of balls and 4 to 5.9 min grinding time.

Figures 11, 12 and 13 show the contour and 3D surface plot response of −0.25 + 0.65 mm size fraction. These figures show the similar effect of number of balls, amount of feed and grinding time on fineness as discussed for the other two size fractions. These contour plots can help finding fineness of coal particles at any zone of the experimental region. The results concludes that the small amount of feed stayed longer will result in fine product.

The optimal levels of different parameters studied like number of balls, grinding time, and amount of feed could provide a beneficial information to calculate the power requirement to grind a particular amount of coal. The response surface methodology provides a relationship among the parameters which can be utilized in finding optimal values.

5 CONCLUSIONS

In the present work, the application of response surface methodology (Box–Behnken design) for grinding of Indian coal of three different size fractions had been discussed. The three level factorial Box–Behnken experimental design model was investigated for three variables namely: grinding time, ball charge and amount of feed used for grinding of Indian coal. The three different model equations were obtained for three different size fractions of coal. MATLAB 7.1 software was used to model the equations. Predicted values of d_{80} for three different sizes were in good agreement with the experimental results in the range of R^2 value of 0.97–0.99. The model graphs were shown in 3D surface plots, contour plots as well as combination of contour and 3D surface plots were also drawn for better understanding. This study confirms that Box-Behnken design with response surface methodology can be applied for the grinding test of different size fraction of Indian coal. This technique is fast and efficient to show the final results of experiments and fewer number of experiments are required to extract the complex information of the system. Using surface response technique one can find optimum conditions for grinding of coal efficiently. This paper shows that this technique is economical as well as less time consuming.

Compliance with ethical standards

CONFLICT OF INTEREST

We confirm that none of the authors has any competing interest related to this manuscript.

REFERENCES

1. Aslan N, Cebeci Y (2007) Application of Box-Behnken design and response surface methodology for modeling of some Turkish coals. Fuels 86:90–97

2. Barranco R, Colechin M, Cloke M, Gibb W, Lester E (2006) The effects of grind quality on coal burnout in a 1 MW combustion test facility. Fuel 85:1111–1116

3. Edadnejad A, Karimi GR, Dehghani H (2013) Application of Response Surface methodology for modeling of ball mills in copper sulphide ore grinding. Powder Technol 245:292–296

4. Gunaraj V, Murugan N (1999) Application of response surface methodologies for predicting weld base quality in submerged arc welding of pipes. J Mater Process Technol 88:266–275

5. Harrison PC, Rowson N (1997) In: Proceedings of the 1997 Jubliee Research Event, Institution of Chemical Engineers, Nottingham, UK

6. Jankovic A (2001) Media stress intensity analysis for vertical stirred mills. Miner Eng 14(10):1177–1186

7. Kincl M, Turk S, Vrecer F (2005) Application of experimental design methodology in development and optimization of drug release method. Int J Pharam 291:39–49

8. Kwak JS (2005) Application of Taguchi and surface response surface methodologies for geometric error in surface grinding process. Int J Mach Tools Manuf 45:327–334

9. Lester E, Kingman S, Dodds C (2005) Increased coal grindability as a result of microwave pretreatment at economic energy inputs. Fuel 84:423–427

10. Prasher CL (1987) Crushing and Grinding Process Handbook. Wiley, Chichester

11. Tuzun MA (1994) A study of comminution in a vertical stirred ball mill, PhD Desertation. Chemical Engineering Department, University of Natal

Chapter 5

Progress in Developments of Dry Coal Beneficiation

Yuemin Zhao • Xuliang Yang • Zhenfu Luo • Chenlong Duan • Shulei Song

School of Chemical Engineering & Technology, China University of Mining & Technology, Xuzhou 221116, China

ABSTRACT

China's energy supply heavily relies on coal and China's coal resource and water resource has a reverse distribution. The problem of water shortages restricts the applications of wet coal beneficiation technologies in drought regions. The present situation highlights the significance and urgency of developing dry beneficiation technologies of coal. Besides, other countries that produce large amounts of coal also encounter serious problem of lack of water for coal beneficiation, such as American, Australia, Canada, South Africa, Turkey and India. Thus, dry coal beneficiation becomes the research hot-points in the field of coal cleaning worldwide in recent years. This paper systematically reviewed the promising research efforts on dry coal beneficiation reported in literature in last 5 years and discussed the progress in developments of dry coal beneficiation worldwide. Finally, we also elaborated the prospects and the challenges of the development of dry coal beneficiation.

Keywords: Dry coal beneficiation Air dense medium fluidized bed Density segregation Vibrated fluidized bed

1 INTRODUCTION

China's energy supply is heavily reliant on coal and this situation will last for decades to come. Now, China is the largest coal producer and consumer in the world. In 2012, China's coal production is up to 3.66 billion tons and coal consumption accounts for 68.4 % of its energy consumption. Thus, in order to improve energy efficiency and ensure the security of energy supply, China must increase the beneficiation ratio of run-of-mine coal. In 2013, the Chinese government released its toughest plan so far to combat air pollution, *Air Pollution Prevention Plan of Action*, in an effort to ease mounting public concern over air quality. In this plan, it is clearly pointed out that the beneficiation ratio of run-of-mine coal will be improved to more than 70 % till 2017 and apparently, a lot of continued efforts must be made to achieve this goal.

Meanwhile, more than 70 % China's coal reserves in regions of Northwest China and North China and unfortunately, these regions are almost dry and have prolonged coal weather each year, which puts many obstacles to the deployment of wet beneficiation technologies for coal cleaning. Thus, it is imperative to develop efficient dry coal beneficiation technologies. Besides, other countries producing large amounts of coal also encounter serious problem of lack of water for coal beneficiation, such as American, Australia, Canada, South Africa, Turkey and India. Thus, in recent years, dry coal beneficiation technologies have progressively become the research hot-points in the field of coal cleaning worldwide and a lot of promising research efforts have been reported in literature.

Overall, in the section of dry coal beneficiation, the most significant properties of coal that facilitate the separation process are density, shape, friction, electricity, and magnetite. According the different beneficiating mechanisms, dry coal beneficiation reported in recent literature are mainly classified into the following five categories: (I) air dense medium fluidized bed separation with/without external force field, (II) fluidized bed separation based on the difference of settling velocity, (III) compound dry separation, (IV) triboelectrostatic separation and (V) magnetic separation. These five categories of technologies have their own advantages respectively in processing certain coal or yielding clean coal product with certain quality. Generally speaking, technologies belong to the categories (I), (IV) and (V) usually have high separation efficiency but high operational cost, which are suitable for producing high-quality clean coal. In contrast, technologies in categories (II) and (III) don't need additional dense medium or power and usually have larger capacities and low processing cost, but low separation efficiency due to the original limitation of separation mechanism. These technologies are suitable for producing steam coal. Besides, the electrostatic separation technologies and the magnetic separation technologies have high requirements for coal properties and certain degree of dissociation, thus, the handling capacities are limited. Considering these limitation, these two types of beneficiation technologies are only used in very special purpose. In the following sections, the representative research works in terms of the aforementioned technologies in the last five years are reviewed in detail and finally, we discuss the prospects and the challenges of the development of dry coal beneficiation.

2 DRY BENEFICIATION WITH FLUIDIZED BED

Dry beneficiation of coal with an air dense-medium fluidized bed (ADMFB) uses a pseudofluid system of the mix of dense medium (such as magnetite powder) and fine coal powder as the separating medium and, within it, the feed coal conducts the sink-float processes based on Archimedes' principle. The light and heavy particles separate from each other by the separating density with light particles floating and heavy particles sinking (Chen and Yang 2003; Luo and Chen 2001; Luo et al. 2002a, 2003, 2007). The ADMFB separation differs in principle from pneumatic separating devices (using air as separating medium) such as air-tables and has comparatively high *E* values (Luo and Chen 2001).

2.1 Separation with air dense medium fluidized bed

China University of Mining and Technology (CUMT) has carried out numerous research works on ADMFB separation in the last about 30 years and developed a series of dry beneficiation techniques for coal cleaning including the bubbling fluidized bed (Luo and Chen 2001; Luo et al.2002a; Zhao et al. 2011a, 2011b), the vibrated fluidized bed (Luo et al. 2008; Yang et al. 2013a), and the magnetically fluidized bed (Fan et al. 2001, 2002, 2003; Luo et al. 2002b). Zhao et al. (2010) carried out systematic studies experimentally and numerically on fluidization characteristics of air dense medium fluidized bed with fine magnetite powder and pointed out that the increase of superficial air velocity will significantly improve the bed activity, but meanwhile, deteriorate the uniformity and stability of fluidization. Therefore, the superficial air velocity no more than 2.0 u_{mf} (minimum fluidization velocity) is suggested for coal beneficiation. In addition, a method of numerical simulation of the fluidization of ADMFB was proposed to study the bubbling characteristics and the particles motions and the simulation results were consistent with the experimental results, as shown in Fig. 1.

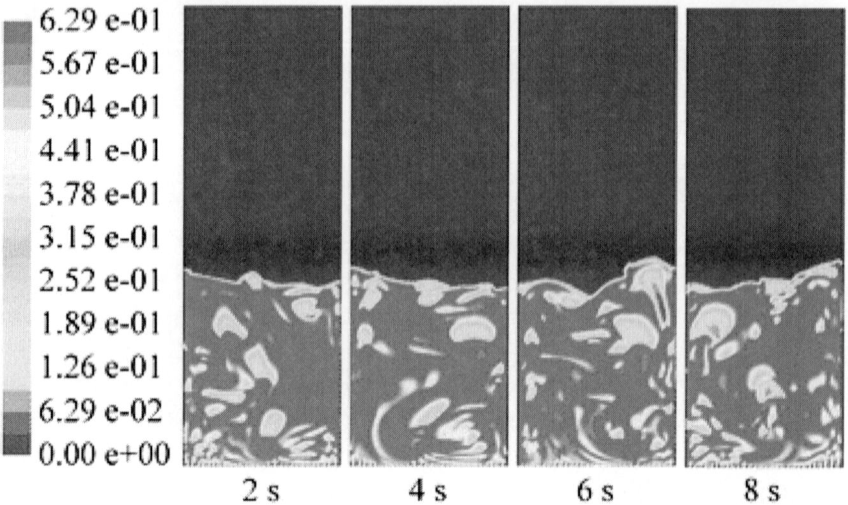

Figure 1. Bubbling and particles motions observed by numerical simulation of ADMFB (Zhao et al. 2010)

Furthermore, Zhao et al. (2011) established a modularized industrial-scale ADMFB separation system with a capacity of 40–60 t/h, as show in Fig. 2. This system uses the steel structure instead of the recent reinforced concrete structure, leading to a reduction of the construction coast, the area and the number of workers by 60 %, 75 % and 80 %, respectively. The separation results show that this modularized system can effectively separate 50–6 mm coal with an *E*value in the range of 0.05–0.07.

Figure 2. Modularized industrial-scale ADMFB separator

Luo et al. (2013) studied the effect of a secondary gas-distribution layer on the fluidization characteristics of an ADMFB for dry coal beneficiation and the experimental results show that the secondary layer with an optimal depth of 30 mm can significantly reduce the variance in both the bed pressure drop and the bed density, leading to an improvement in the fluidization quality, as shown in Fig. 3. It is also pointed out that the secondary layer can divide large bubbles passing through the region into small bubbles and can increase the pressure drop of the gas distributor. This work will provide a foundation for the design of the discharge structure of the bed and for calculating the capacity of the bed.

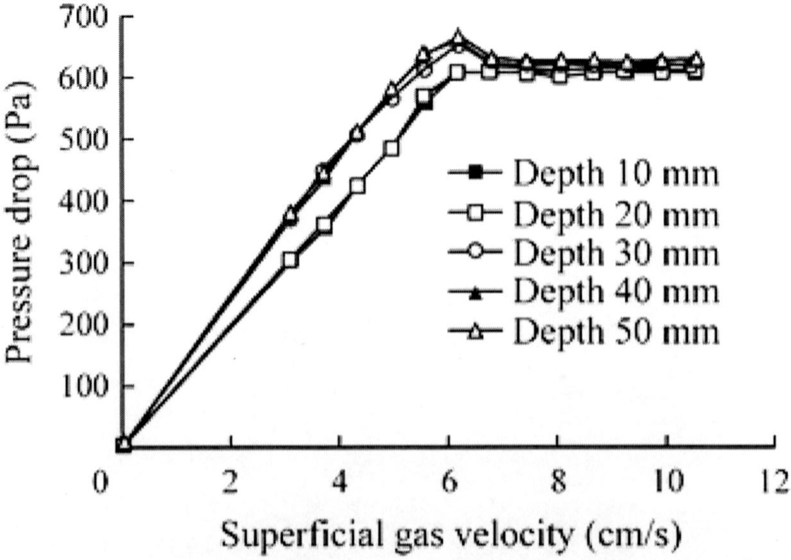

Figure 3. Fluidization curves of an ADMFB with different secondary layer depth (Luo et al. 2013)

In similarity with China, India is heavily dependent on coal for meeting its energy requirements in economic development (Sahu et al. 2011). In these years, many research works on using the ADMFB separation for Indian coal cleaning have been done by Indian scholars. Sahu et al. (2009, 2011, 2013) used an ADMFB separator to beneficiate India high ash coal with a size range of −25 + 6 mm and studied the stability of fluidization to make an expansion bed with a nanobubbling condition. It was pointed out that the stability of the fluidized bed had a significant effect on sharp separation between clean coal and gangue. A continuous ADMFB separator with a capacity of 600 kg/h was established, as shown in Fig. 4. The separation results show that the ash content is reduced from 40 % in feed coal to around 32 %–35.5 % in clean coal product with a 60 %–72 % yield and the separation efficiency *E* value is 0.12.

Figure 4. Continous ADMFB separator for beneficiating Indian high ash coal (Sahu et al. 2011)

Mohanta et al. (2011, 2012, 2013a, b) had made many efforts to optimize the operational parameters and the feed coal properties in order to achieving an optimum beneficiation performance of an ADMFB separator. It was pointed out that when the feed coal size is below 15 mm, the size effect on sharpness in separation becomes significant and the *E* values decrease faster. The experimental results indicate that for processing finer coal of size less than 15 mm, the ADMFB separator does not offer any real performance advantages better than the traditional process like heavy-medium cyclone and water-only cyclone. They also carried out a detailed investigation to examine the interdependencies among these parameters and their interactional effect on the separation performance of an ADMFB separator. The optimization results show that the optimum conditions were: superficial air velocity from 18.94 to 20.42 cm/s, bed height from 25.94 to 29.36, and coal to magnetite weight ratio of 0.02. A correlation for the theoretical prediction of minimum fluidization velocity was proposed in the following:

$$u_{\mathrm{mf}} = \frac{\mu}{d_{\mathrm{p}}\rho_{\mathrm{f}}}\left(\sqrt{(41.96)^2 + 0.049\mathrm{Ar}} - 41.96\right)$$

where u_{mf} is the minimum fluidization velocity; μ, d_p and ρ_f are the air viscosity, the particle diameter and the air density, respectively; Ar is the Archimedes number. The predictive adequacy and reliability was evaluated by adopting a statistical analysis approach on the data available in the literature and the results indicate that this new correlation is useful and accurate for predicting the minimum fluidization velocity and designing an ADMFB.

Researchers in University of Alberta in Canada (Prashant et al. 2010; Azimi et al. 2013a, b) have also done many works on ADMFB separation. Prashant et al. (2010) studied the effect of operating parameters such as the type of media, fluidizing velocity and separating time on beneficiation performance investigated under both batch and continuous ADMFB separators which are demonstrated in Figs. 5 and 6 respectively. The separation results show that the ash content of −6 + 2 mm feed coal is reduced from 25 % to 10 % with a yield of 80 %.

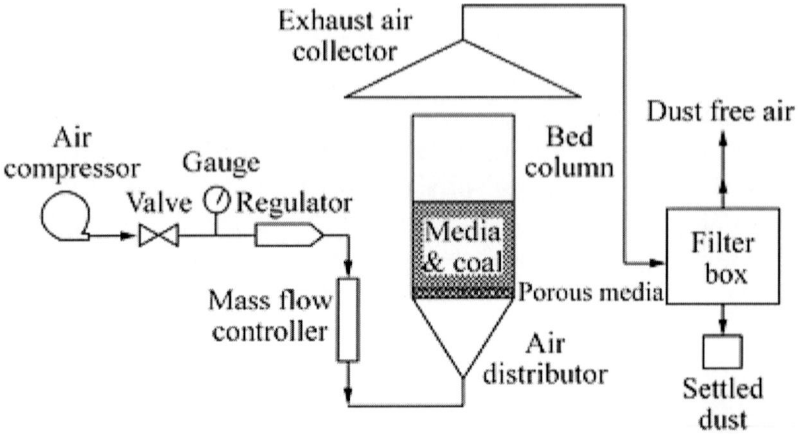

Figure 5. Schematics of a laboratory batch ADMFB separator (Prashant et al. 2010)

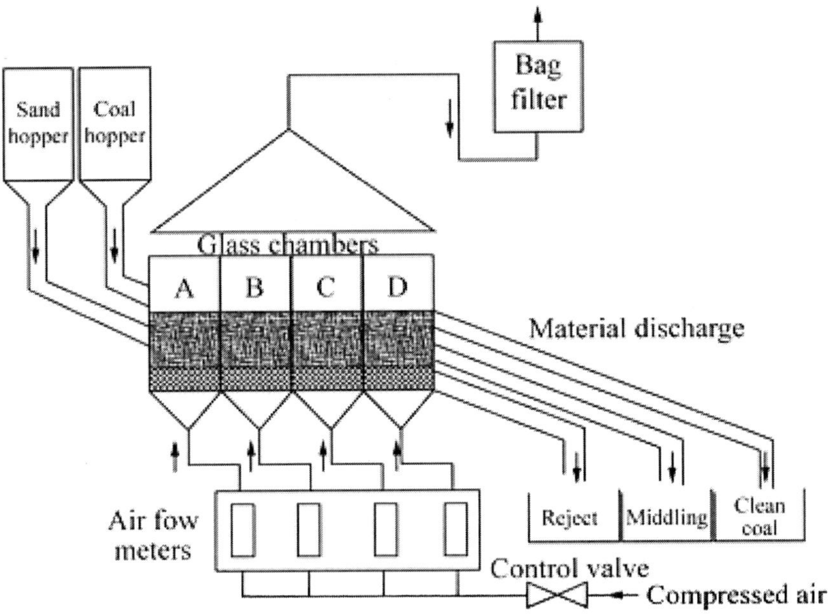

Figure 6. Schematics of a laboratory continuous ADMFB separator (Prashant et al. 2010)

Azimi et al. (2013a, b) used an ADMFB to separate low-rank coals in Canada and carried out a detail study of the factors including superficial air velocity, residence time, and bed height affecting the beneficiation performance of a low-ash lignite coal. The experimental results show that the optimum operating condition is superficial air velocity of 15 cm/s, residence time of 90 s and bed height of 15 cm and in this condition, the beneficiation results revealed that higher ash removal (23 %) and recoveries (86 %) are obtainable with a clean coal ash content, recovery, and separation efficiency of 10.6 %, 95.63 %, and 15.29 %, respectively. In addition, it was also pointed out that feeding more coal to ADMFB resulted in higher organic material recovery and lower ash content of the product and system separation efficiency.

2.2 Separation with vibrated fluidized bed

The conventional ADMFB separation is not available recently to separate fine coal smaller than 6 mm in size because of low separation efficiency and difficult in dense medium recovery and product purification. Luo et al. (2008) introduced the vibration energy into conventional ADMFB for separating dry fine coal particles and systemically studied the vibrational energy transmission, the interaction between vibration and air and the separation performance of vibrated air dense medium fluidized bed (VADMFB). The results show that the bubble formation in a VADMFB is efficiently restricted to small bubbles due to the cutting effect derived from the horizontal vibration component and the extruding effect from the vertical vibration component. An empirical correlation of the critical vibration frequency above which the bubble can be restrained and broken was proposed:

$$f > \left(\frac{6Q}{\pi}\right)^{-1/5} g^{3/5}$$

where *f*, *q* and *g* are vibration frequency, air flow rate and gravitational acceleration, respectively. The synergy between vibration and air make the fine coal bed can be fluidized with both micro-bubbles and a uniform bed density. The separation results show that the VADMFB has a good separation performance for −6 mm fine coal under optimal operating conditions with an *E* value of 0.07.

Macpherson and Galvin (2010) and Macpherson et al. 2010, 2011) in the University of Newcastle in Australia introduced the Reflux Classifier, whose water-based version had been successfully employed in industry for gravity separation of −2 + 0.25 mm coal, into the field of dry coal beneficiation. This new Reflux Classifier uses sand (−335 + 125 micron) as a dense-medium and uses vibration at two distinct levels to improve fluidization stability, as shown in Fig. 7. There are two parts: a vibrated fluidized bed in the lower section and a device consisting of parallel inclined channels in the upper section. The addition of an incline above the fluidized bed provides a more stable system allowing for grater separation efficiency and minimizing the effect of changing conditions. The density cut-point can be varied from 1,418 to

2,130 g/cm^3 by varying the underflow rate and the *E* value was within the range of 0.06–0.46 depending on particle size and gas rate.

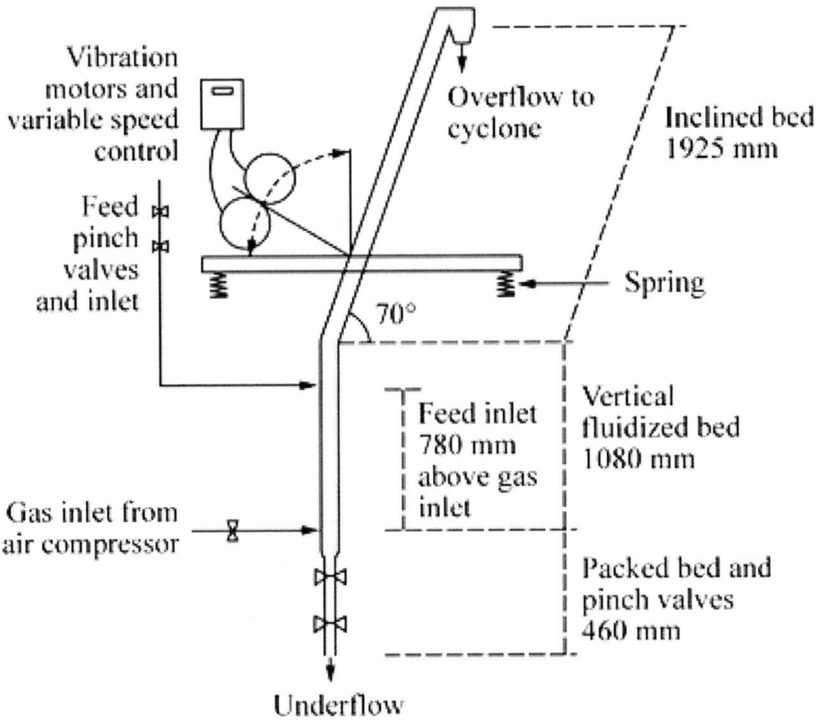

Figure 7. Schematic diagram of the reflux classifier using air-sand dense-medium and vibration (Macpherson et al. 2011)

2.3 Separation with magnetically stabilized air dense medium fluidized bed

Song et al. (2012a, b) in China University of Mining & Technology studied the continuous separation of −6 + 0.5 mm fine coal by using a crossflow magnetically stabilized air dense medium fluidized bed (MSADMFB) with the 0.074–0.045 mm magnetite beads as dense medium, as shown in Fig. 8. The fluidization results show that the range of steadily operational air velocity in MSADMFB is broader than that in a conventional ADMFB when they are at stable fluidization. The external magnetic field compels the magnetic grains to form parallel magnetic chains following the magnetic induction lines, which enlarge the bed voidage and form uniform air channels. Air bubbles cannot be produced when the air flows through the channels. Thus, a MSADMFB comparatively has a more stable bed density and smaller pressure fluctuation. The separation results show that the cross-flow MSADMFB can achieve successfully the separation of fine coal of −6 + 0.5 mm with a *E* value in the range of 0.068–0.095. It is also pointed out that the separation efficiency turns worse gradually with the decrease of the size of the fine raw coal.

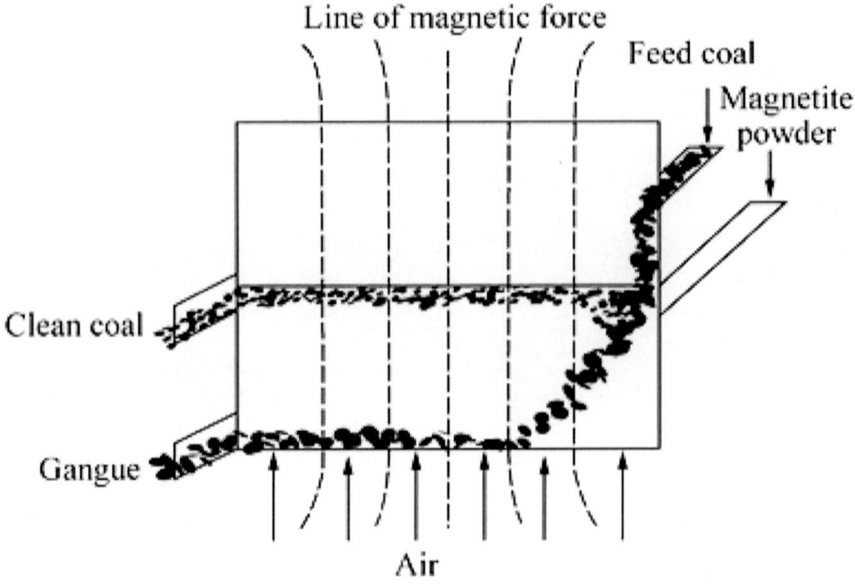

Figure 8. Schematic diagram of the crossflow magnetically stabilized air dense medium fluidized bed (Song et al.2012b)

3 FLUIDIZATION BENEFICIATION WITHOUT DENSE MEDIUM

For fine coal dry beneficiation, the fluidization separation using dense medium has several problem need to be solved, including the efficient recovery of fine dense medium and the purification of product. Meanwhile, the complex process and the relatively high operational costs due to the recovery and loss of dense medium also present great obstacle on the industrial applications of the fluidization separation with dense medium. Thus, some recent studies on fluidization beneficiation without artificial dense medium have been carried out in order to provide an efficient and economical solution for fine coal cleaning.

3.1 Vibro-fluidization beneficiation based on density segregation

Yang et al. (2013a, b) introduced the density segregation phenomenon in the field of gas–solid fluidization into the research on fine coal dry beneficiation and adopted a vibrated fluidized bed to foster the formation of fluidization environment that intensifies density segregation within the bed of fine coal, leading to an effective fine coal beneficiation performance. Figure 9demonstrates the schematic diagram of the experimental apparatus and Fig. 10 depicts the separation mechanism. Particles in the region of a bubble having considerable lower solid volume fraction than the surrounding bulk phase conduct sink-float processes and particles with higher density tend to sink preferentially over the lighter particles, which leads to local density segregation. Thus, after the function of periodic bubbling, the local density segregation evolves the density segregation of total bed, leading to stratification of the particulate bed with the layers of clean coal, middling and gangue distribute along the bed height from top to bottom. The experimental results show that the operational parameters including superficial air velocity, vibration intensity, bed height and fluidizing time have significant influences on the segregation performance. The separation results show that the probable error *E* values of −6 + 3 and −3 + 1 mm fine coal separated in a continuous separation system using a vibrated fluidized bed are 0.19–0.225 and 0.175–0.195, respectively, indicating that the fine coal separation in a vibrated gas-fluidized bed system is useful to and more effective for fine coal cleaning.

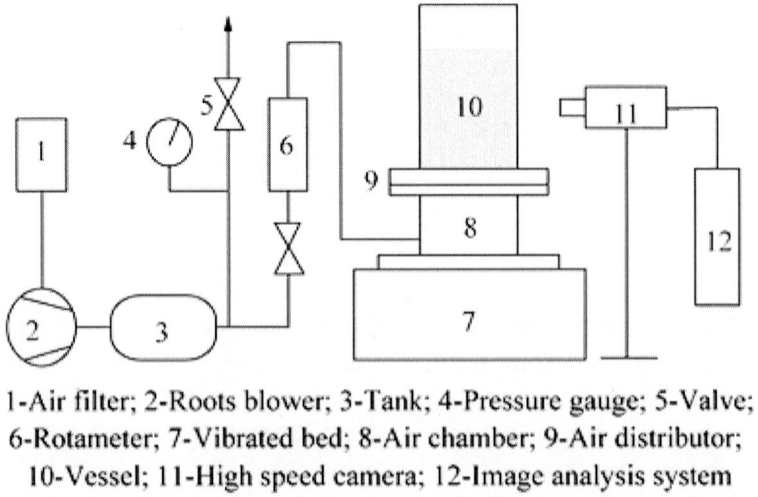

Figure 9. Schematic diagram of the experimental apparatus (Yang et al. 2013b)

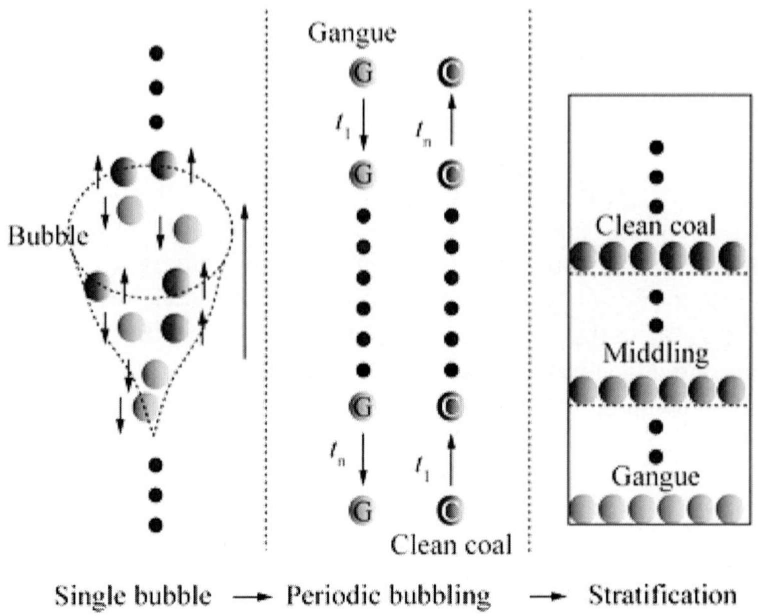

Figure 10. Separation mechanism of vibro-fluidization beneficiation based on density segregation (Yang et al.2013a)

3.2 Dry jigging

The air jig is a deep-bed separator that deploys the difference in hindered settling velocity of particles with different density, which is analogous to the proved wet jig. Air is supplied into the jig in two forms: a continuous flow and a superimposed pulsated airflow that provides the impetus for stratification and consolidation trickling. Sampaio et al. (2008) employed dry jigging to beneficiate Candiota coal in Brazil. This coal is low rank and can't be treated by any wet beneficiation technologies. The separation results show that the sulfur content reduces from 2.33 % in feed coal to 1.08 % in clean coal, providing a cheaper way than washing combustion gases. The detailed separation results are given in Fig. 11. Charan et al. (2011) used air jig manufactured by Allmineral Company in Germen to beneficiate Indian high-ash non-coking coals which contain high amounts of mineral matter. The first commercial air jig with a capacity of 50 t/h was installed and commissioned to separate −40 + 5 mm high-ash coals with an ash content of 40 %. The separation results show that a clean coal product with an ash content of 33 % was produced.

4 COMPOUND DRY BENEFICIATION

Recently, researchers have studied and developed some innovative dry coal beneficiation technologies depending on not only the density difference, but the difference of other properties including coefficient of friction and shape. This compound dry beneficiation enhances the stratification process of coal and gangue in many aspects and leads to a stable and efficient separation performance.

4.1 **FGX dry separator**

The FGX dry separator is firstly developed by China's researchers and already has several hundred commercial installations. It consists of a perforated separating bed, a vibrating mechanism, and a hanging support mechanism, shown as in Fig. 12. The deck vibration frequency, longitudinal deck angle, feeder frequency, and baffle plate height all have significant effects on the separation performance. Coal particles conduct the spiral flip motions under the synergy of vibration and air

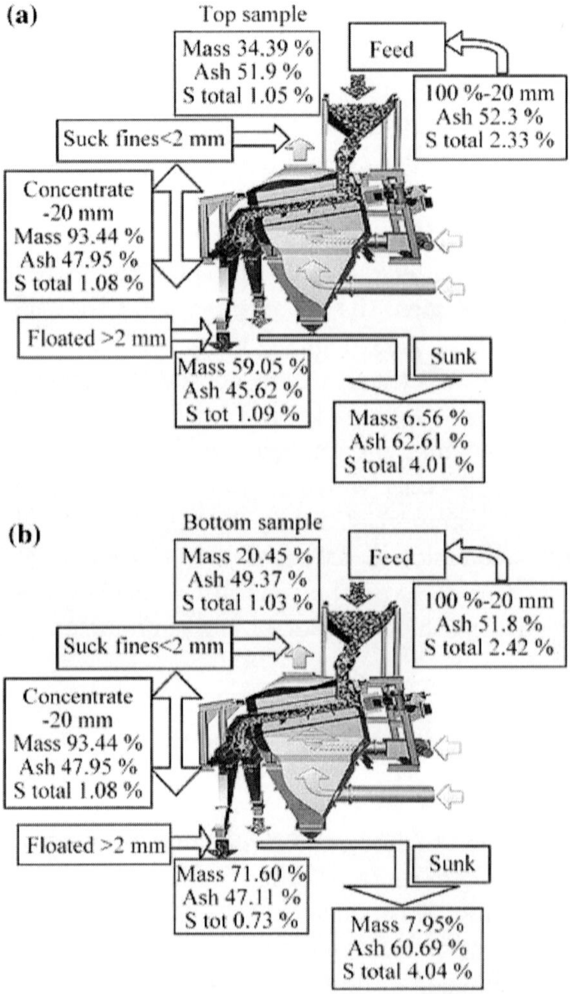

Figure 11. Detailed separation results of air jig studied by Sampaio et al. (2008)

flow. The lighter particles rise to the top of the particulate layer and are collected in the front section of the separating bed as the clean coal product. Meanwhile, the heavy particles stay in the bottom of the particulate layer and keep contact with the bed. Thus, with the function of vibration and riffles fitted on the bed, these heavy particles move forward to the narrow section of the bed and finally are discharged as gangue. Particles with intermediate density are discharged as middling. The FGX dry separator finally generates three products: clean coal product, middling, and gangue. Two dust collection systems are employed to clean the recycled air and to remove the dust from air before being emitted into the atmosphere.

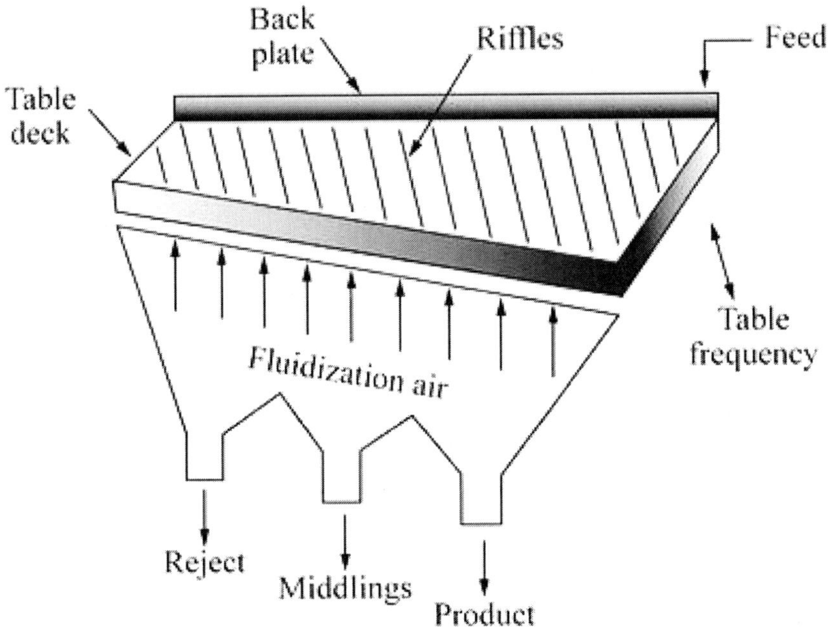

Figure 12. Schematic diagram of the FGX dry separator and its processing system (Honaker et al. 2008)

Zhang et al. (2011) used a FGX dry separator with a capacity of 10 t/h to clean raw coals extracted from Illinois mines in America. The separation performance of the FGX dry separator was evaluated by separating the −63.5 + 4.75 mm coal and the results show that the specific gravity of separation and the probable error *E* values are 1.98 g/cm3 and 0.17, respectively. A modified log-logistic partition model was established using experimental data reported in literature and in this study to predict the FGX dry separator coal-cleaning performance, given in the following:

$$PN = 100 * \frac{1}{1 + \exp^{a * \ln\frac{x}{b}}}$$

$$X = D_m / D_{50}$$

where a and b are fitting constants; D_m is the mean density and D_{50} is the separation density.

Honaker et al. (2008) deployed a 5 t/h pilot-scale FGX dry separator to produce clean coal having qualities that meet contract specifications and maximize the amount of high-density rock rejected prior to transportation and processing. The separation results clearly indicate that the FGX dry separator is able to reject more than 70 % of the high-density rock in a run-of-mine coal at high separation density values of around 1.8–2.2 g/cm^3 with an E valued 0.25.

4.2 Air table

The air table has a flat deck with the uniformly distributed apertures and riffles. Ambient air after filtering enters the deck and fluidizes the feed coal. The feed coal stratifies depending on the difference in the terminal velocity of the particles. Particles with smaller density move to the top of the particulate layer and then travel down the slope towards the discharge open. Particles with larger density settle to the bottom of the particulate layer travel in the direction of the deck's vibration along the riffles. The schematic diagram of this separation process is shown in Fig. 13.

Patil and Parekh (2011) used a dry air table for processing of −6.3 + 1.4 mm coal from a mine located in Western Kentucky. The separation results show that the ash content is reduced from 27 % to 10 % with a clean coal yield of 75 %–80 % and the combustible recovery is 95 % with the ash rejection of 77 %–80 %. In addition, the pyritic sulfur is reduced by about 33 % and the heat content increases from 23,997 to 29,595 kJ/kg, indicating excellent separation efficiency.

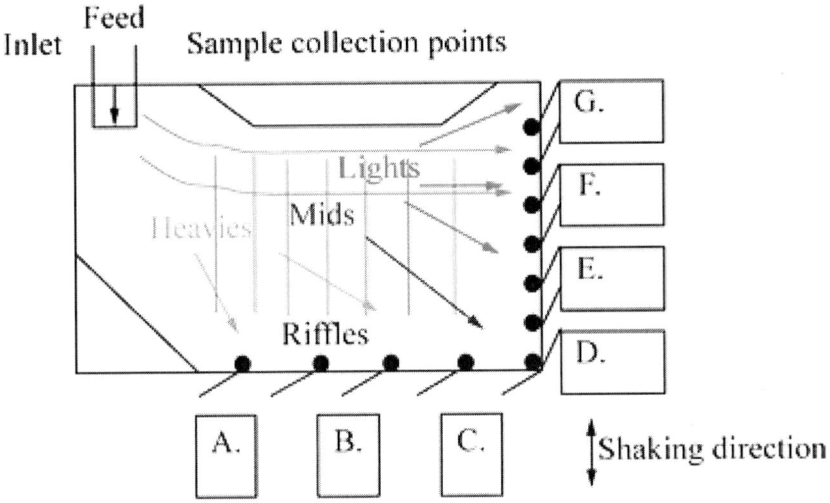

Figure 13. Detailed separation results of air jig studied by Sampaio et al. (2008)

Çicek (2008) developed a modified air table to beneficiate Turkish coal by a dry cleaning method and the separation results show that the coal samples having three size fraction of 8–5, 5–3 and 3–1 mm can be efficiently beneficiated by this modified air table with *E* values of 0.21, 0.14 and 0.09, respectively, indicating a satisfactory separation performance.

5 TRIBOELECTROSTATIC SEPARATION

The triboelectrostatic beneficiation is an advanced dry fine coal cleaning technology and can effectively process fine coal of less than 300 μm (Dwari and Hanumantha Rao 2008; Zhang et al. 2009). During the process of triboelectrostatic separation, coal particles are charged by contact or friction with other particles or with a third material including the walls of a container or pipe and then conduct free-fall process through an electric field that deflects the particles according to the magnitude and sign of their charge. The mechanism of triboelectrostatic separation determines that the feed coal must achieve a certain degree of dissociation and the coal particles and the gangue particles mainly exist in the form of monomer. Therefore, this technology has a comparative small handling capacity and is usually used for producing high value-added clean coal product.

Zhang et al. (2009) deployed the triboelectrostatic separation of mineral matters from clean coal with density of < 1.35 g/cm^3 and the samples of feed coal with a size smaller than 75 μm was separated and analyzed with XRF, sulphur and ash analyzer. Figure 14 depicted the principle of triboelectrostatic separator and the sampling rule. The separation results showed that a quantity of quartz of 69.31 %–71.27 %, kaolin of 75.66 %–81.93 % and pyrite of 86.74 %–90.52 % could be removed and also indicated that coal would be charged positively and quartz, kaolin, pyrite and calcite would be charged negatively, leading to efficient separation of coal and these mineral matters.

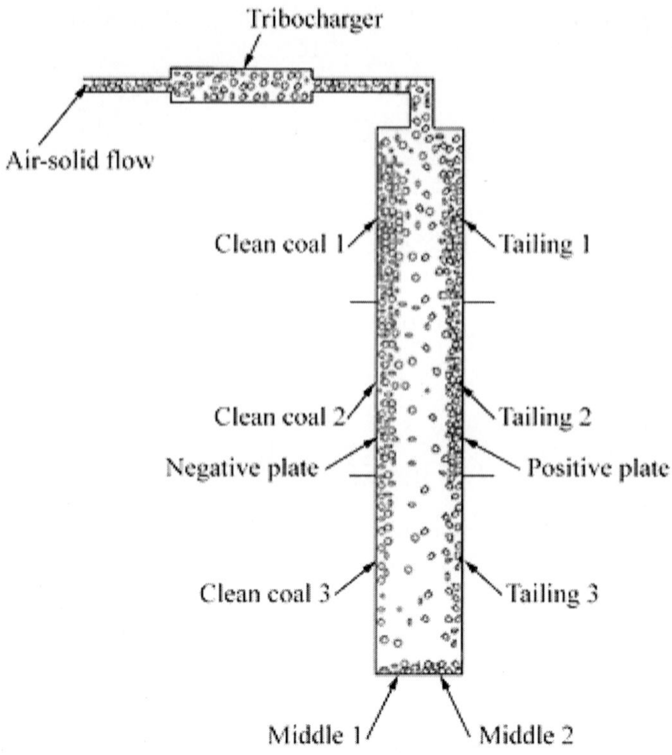

Figure 14. Principle of triboelectrostatic separator and sampling rule (Zhang et al. 2009)

Dwari and Hanumantha Rao (2008, 2009) used a new laboratory fluidized bed triboelectrostatic separator to beneficiate −300 μm Indian thermal non-coking coal from Ramagundam coal mines, shown as in Fig. 15. The tribocharger consisted of the internal baffles made up of copper metal and the coal and mineral particles charged with positive and negative polarities respectively. It was pointed out that the magnitude of particles charge is relevant with the efficiency of contact electrification in fluidized bed tribocharger. The separation results showed that the ash content can be reduced to 18 % and 33 % from 43 % in feed coal with an yield of about 30 % and 67 %, respectively.

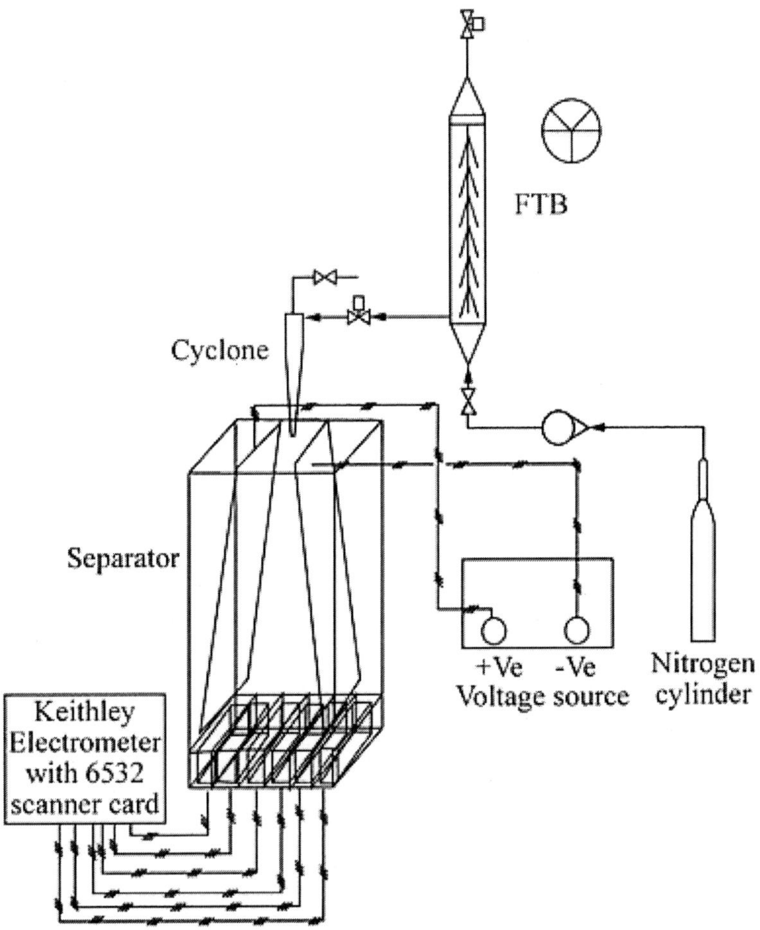

Figure 15. Schematic diagram of experimental setup for electrostatic separation (Dwari and Hanumantha Rao2009)

6 DRY MAGNETIC SEPARATION

Dry magnetic separation technology beneficiates coal powder depending on the different magnetic susceptibilities of coal and mineral materials under high gradient magnetic field. In similarity to triboelectrostatic separation technology, dry magnetic separation technology also requires that the feed coal has high degree of dissociation in order to facilitate the separating process of coal and gangue depending on magnetic susceptibilities difference. Meanwhile, dry magnetic separation technology also has comparative small handing capacity.

Jiao et al. (2009) used dry magnetic separation to improve desulphurization and deash efficiency of fine coal and investigated the relation between magnetic susceptibility and density, size as well as magnetic field intensity. The results showed that there is a linear correlation between magnetic susceptibility and density, as shown in Fig. 16. It was also pointed out that the magnetism of high density coal and low density coal was opposite at the cut density of 1.6 g/cm^3. Furthermore, the diamagnetism property of low density coal weakened and the paramagnetism of high density coal enhanced along with the size reduction of fine coal. The separation results showed that the desulphurization and deash rates achieved 55.52 % and 64.41 %, respectively.

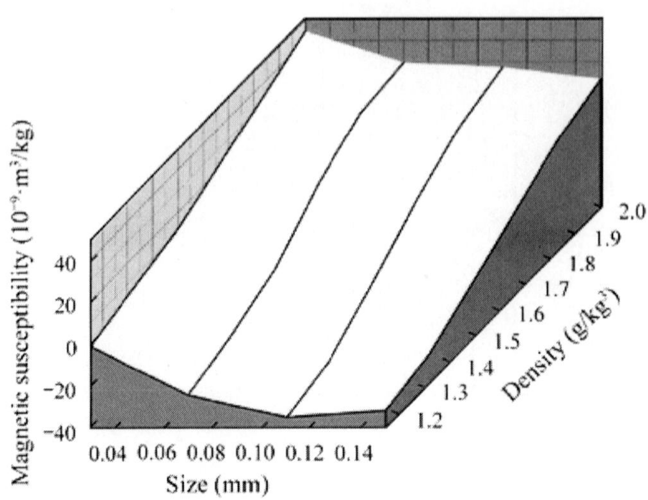

Figure 16. The correlation between magnetic susceptibility and densiy and size (Jiao et al. 2009)

7 PROSPECTS AND CHALLENGES

Dry coal beneficiation has an inherent advantage since it uses no process water and reduces the costs that are associated with process water, fines dewatering and slurry confinement. In particular, dry coal beneficiation is more suitable for coal cleaning in dry regions that lack available processing water. Meanwhile, the low-rank coal easily conducts the process of argillation when using wet coal beneficiation, yielding the difficulties in slurry treatment and product dewatering. Therefore, dry coal beneficiation is comparatively more suitable for low-rank coal cleaning and upgrading. Countries that heavily relay on coal for energy supply all progressively exploit and utilize low-rank coal to meet the increasing required energy consumption and protect the hard coal resource. Thus, it is imperative to develop high-efficiency dry coal beneficiation to support the utilization of low-rank coal.

However, there are a lot of works need to be done in order to form a reliable and stable solution that consists of several dry coal beneficiation technologies. Especially the separation stability and the separation efficiency of dry coal beneficiation technologies should be further improved in future. Furthermore, some demonstrational industrial dry coal beneficiation systems should be established for engineering verification and academic studies, providing an advanced research platform to carry out sufficient industry-university joint investigations. The significance of developing dry coal beneficiation should be paid sufficient attentions and the policy support and adequate financial support are required to facilitate the considerable progress in coal dry cleaning.

ACKNOWLEDGMENTS

The study is supported by the National Key Basic Research Program of China (2012CB214904), the National Natural Science Foundation of China (51221462; 51134022; 51174203), the Research Fund for the Doctoral Program of Higher Education (20120095130001) and the Natural Science Foundation of Jiangsu Province (BK20140209).

REFERENCES

1. Azimi E, Karimipour S, Rahman M, Szymanski J, Gupta R (2013a) Evaluation of the performance of air dense medium fluidized bed (ADMFB) for low-ash coal beneficiation. Part 1: effect of operating conditions. Energy Fuels 27(10):5595–5606

2. Azimi E, Karimipour S, Rahman M, Szymanski J, Gupta R (2013b) Evaluation of the performance of air dense medium fluidized bed (ADMFB) for low-ash coal beneficiation. Part 2: characteristics of the beneficiated coal. Energy Fuels 27(10):5607–5616

3. Charan GT, Chattopadhyay US, Singh KMP, Kabiraj SK, Haldar DD (2011) Beneficiaron of high-ash, Indian non-coking coal by dry jigging. Miner Metall Process 28(1):21–23

4. Chen QR, Yang YF (2003) Development of dry beneficiation of coal in China. Coal Prep 23(1–2):3–12

5. Çicek T (2008) Dry Cleaning of Turkish Coal. Energy Sources Part A 30(7):593–605

6. Dwari RK, Hanumantha Rao K (2008) Non-coking coal preparation by novel tribo-electrostatic method. Fuel 87(17–18):3562–3571

7. Dwari RK, Hanumantha Rao K (2009) Fine coal preparation using novel tribo-electrostatic separator. Miner Eng 22(2):119–127

8. Fan MM, Chen QR, Zhao YM, Luo ZF (2001) Fine coal (6-1 mm) separation in magnetically stabilized fluidized beds. Int J Miner Process 63(4):225–232

9. Fan MM, Chen QR, Zhao YM, Guan YP, Li B (2002) Magnetically stabilized fluidized beds for fine coal separation. Powder Technol 123(2–3):208–211

10. Fan MM, Chen QR, Zhao YM, Luo ZF, Guan YP (2003) Fundamentals of a magnetically stabilized fluidized bed for coal separation. Int J Coal Prep Utili 23(1):47–55

11. Honaker RQ, Saracoglu M, Thompson E, Bratton R, Luttrell GH, Richardson V (2008) Upgrading coal using a pneumatic density-based separator. Int J Coal Prep Utili 28(1):51–67

12. Jiao HG, Cui JY, Liu P, Chen QR, Tie ZX (2009) Experimental study of magnetic property and dry magnetic separation of fine coal. J China Univ Min Technol 38(1): 131–134, 148

13. Luo ZF, Chen QR (2001) Dry beneficiation technology of coal with an air dense-medium fluidized bed. Int J Miner Process 63(3):167–175

14. Luo ZF, Chen QR, Zhao YM (2002a) Dry beneficiation of coarse coal using an air dense medium fluidized bed (ADMFB). Coal Prep 22(1):57–64

15. Luo ZF, Zhao YM, Chen QR, Fan MM, Tao XX (2002b) Separation characteristics for fine coal of the magnetically fluidized bed. Fuel Process Technol 79(1):63–69

16. Luo ZF, Zhao YM, Tao XX, Fan MM, Chen QR, Wei LB (2003) Progress in dry coal cleaning using air-dense medium fluidized beds. Coal Prep 23(1–2):13–20

17. Luo ZF, Zhu JF, Fan MM, Zhao YM, Tao XX (2007) Low density dry coal beneficiation using an air dense medium fluidized bed. J China Univ Min Technol 17(3):306–309

18. Luo ZF, Fan MM, Zhao YM, Tao XX, Chen QR, Chen Z (2008) Density-dependent separation of dry fine coal in a vibrated fluidized bed. Powder Technol 187(2):119–123

19. Luo ZF, Tang LG, Dai NN, Zhao YM (2013) The effect of a secondary gas-distribution layer on the fluidization characteristics of a fluidized bed used for dry coal beneficiation. Int J Miner Process 118:28–33

20. Macpherson SA, Galvin KP (2010) The effect of vibration on dry coal beneficiation in the reflux classifier. Int J Coal Prep Utili 30(6):283–294

21. Macpherson SA, Iveson SM, Galvin KP (2010) Density based separations in the reflux classifier with an air–sand dense–medium and vibration. Miner Eng 23(2):74–82

22. Macpherson SA, Iveson SM, Galvin KP (2011) Density-based separation in a vibrated Reflux Classifier with an air-sand dense-medium: tracer studies with simultaneous underflow and overflow removal. Miner Eng 24(10):1046–1052

23. Mohanta S, Chakraborty S, Meikap BC (2011) Influence of coal feed size on the performance of air dense medium fluidized bed separator used for coal beneficiation. Ind Eng Chem Res 50(18):10865–10871

24. Mohanta S, Daram AB, Chakraborty S, Meikap BC (2012) Characteristics of minimum fluidization velocity for magnetite powder used in an air dense medium fluidized bed for coal beneficiation. Part Part Syst Charact 29(4):228–237

25. Mohanta S, Chakraborty S, Meikap BC (2013a) Optimization process of an air dense medium fluidized bed separator for treating high-ash non-coking indian coal. Miner Process Extr Metall Rev 34(4):240–248

26. Mohanta S, Rao CS, Daram AB, Chakraborty S, Meikap BC (2013b) Air dense medium fluidized bed for dry beneficiation of coal: technological challenges for future. Part Sci Technol 31(1):16–27

27. Patil DP, Parekh BK (2011) Beneficiation of fine coal using the air table. Int J Coal Prep Utili 31(3–4):203–222

28. Prashant D, Xu ZH, Szymanski J, Gupta R, Boddez J (2010) Dry cleaning of coal by a laboratory continuous air dense medium fluidized bed separator. In: Honaker RQ (ed) Processding of the international coal preparation congress 2010. SME, Lexington, pp 608–616

29. Sahu AK, Biswal SK, Parida A (2009) Development of air dense medium fluidized bed technology for dry beneficiation of coal_A review. Int J Coal Prep Utili 29(4):216–241

30. Sahu AK, Tripathy A, Biswal SK, Parida A (2011) Stability Study of an air dense medium fluidized bed separator for beneficiation of high-ash Indian coal. Int J Coal Prep Utili 31(3–4):127–148

31. Sahu AK, Tripathy A, Biswal SK (2013) Study on particle dynamics in different cross sectional shapes of air dense medium fluidized bed separator. Fuel 111:472–477

32. Sampaio CH, Aliaga W, Pacheco ET, Petter E, Wotruba H (2008) Coal beneficiation of Candiota mine by dry jigging. Fuel Process Technol 89(2):198–202

33. Song SL, Zhao YM, Luo ZF, Tang LG (2012a) Motion behavior of particles in air-solid magnetically stabilized fluidized beds for separation. Int J Min Sci Technol 22(5):725–729

34. Song SL, Zhao YM, Luo ZF, Tang LG, Yang XL (2012b) Fine coal particles separation by air-solid magnetic fluidization. Meitan Xuebao/J China Coal Soc 37(9):1586–1590

35. Yang XL, Zhao YM, Luo ZF, Chen ZQ, Song SL, Duan CL (2013a) Energy transfer mechanism of vibro-fludization and its separation for fine coal. Zhongguo Kuangye Daxue Xuebao/J China Univ Min Technol 42(2):266–270

36. Yang XL, Zhao YM, Luo ZF, Song SL, Duan CL, Dong L (2013b) Fine coal dry cleaning using a vibrated gas-fluidized bed. Fuel Process Technol 106:338–343

37. Zhang XX, Duan DY, Tian B, Wang JS, Deng F, Li HS, Ma RX (2009) Research on the triboelectrostatic separation of minerals from coal. Procedia Earth and Planet Sci 1(1):845–850

38. Zhang B, Akbari H, Yang F, Mohanty MK, Hirschi J (2011) Performance optimization of the fgx dry separator for cleaning high-sulfur coal. Int J Coal Prep Utili 31(3–4):161–186

39. Zhao YM, Luo ZF, Chen ZQ, Tang LG, Wang HF, Xing HB (2010a) The effect of feed-coal particle size on the separating characteristics of a gas-solid fluidized bed. J S Afr Inst Min Metall 110(5):219–224

40. Zhao YM, Tang LG, Luo ZF, Liang CC, Xing HB, Wu WC, Duan CL (2010b) Experimental and numerical simulation studies of the fluidization characteristics of a separating gas-solid fluidized bed. Fuel Process Technol 91(2):1819–1825

41. Zhao YM, Li GM, Luo ZF, Liang CC, Tang LG, Chen ZQ, Xing HB (2011a) Modularized dry coal beneficiation technique based on gas-solid fluidized bed. J Cent South Univ Technol 18(2):374–380

42. Zhao YM, Liu XJ, Liu KL, Luo ZF, Wu WC, Song SL, Tang LG (2011b) Fluidization characteristics of a gas-paigeite-powder bed to be utilized for dry coal beneficiation. Int J Coal Prep Utili 31(3–4):149–160

Chapter 6

Beneficiation of an Indian Non-Coking Coal by Column Flotation

N. Vasumathi[1] • T. V. Vijaya Kumar[1] • S. Ratchambigai[1] • S. Subba Rao[1] • S. Prabhakar[1] • G. Bhaskar Raju[1]

[1] CSIR-National Metallurgical Laboratory Madras Centre, CSIR Madras Complex, Taramani, Chennai 600 113, India

ABSTRACT

Beneficiation of non-coking coal is gaining ground in India. It not only reduces the volume of inert content to be transported to the power plant and also lowers the wear in the boiler houses. For special applications such as the fuel for integrated gasification combined cycle plant (IGCC), the ash content in the coal should preferably be below 15 %. Indian coals are characterized by high inter-grown ash content mainly due to 'drift origin' of Gondwana formation in Permian age. This warrants fine grinding of non-coking coal in order to liberate the ash forming minerals from coal macerals. A non-coking coal sample of vitrinite type from India was ground to 44 µm (d_{80}) and subjected to column flotation to improve its quality. The non-coking coal analyzing 34.6 % ash, 26.2 % volatile matter, 1.3 % moisture and 37.9 % fixed carbon could be upgraded to a concentrate/froth of 14.83 % ash at 72.18 % yield by optimizing collector and frother dosages and flotation column operating parameters, namely, froth depth, superficial feed velocity and superficial air velocity. The concentrate produced by this process is suitable as fuel for IGCC in coal-to-electricity route.

Keywords: Non-coking coal Washability Beneficiation Column flotation Coal gasification

INTRODUCTION

Coal, accounting for 63 % of the world's total energy content, is the world's most abundant fossil fuel resource. The world supply of crude oil and natural gas is limited and, at projected rate of exploitation is expected to be depleted within the next two or three decades. If worldwide economic development is to continue, alternate fuel resources (such as coal) need to be more fully exploited to meet the anticipated world energy requirement. India is in the threshold of techno-economic washing of coal. So far, large scale beneficiation was resorted to coking coals only for steel making purposes. Now, the stage is set to attract investors to take up beneficiation of non-coking coals also for power plants.

With a growing concern over energy security and sustainability, coupled with concerns about climate change and greenhouse gas emissions from coal combustion, the long term generation of coal based thermal power by India will require the use of cleaner coal and clean coal technologies (CCT). Coal beneficiation is the first and most cost effective step towards satisfying this requirement (Yenigalla 2013). Beneficiation of non-coking coal in India was not given due importance till last 10–12 years due to its low value or not able to meet the cost of the process (Misra et al. 2003). Studies conducted by the International Energy Agency have shown that 1 % increase in ash (generally after passing the 10 % ash level) results in a 1.2 %–1.5 % decrease in boiler availability and a decrease of 0.3 % in boiler efficiency (Konar et al. 1997). To cut down the consumption of oil in existing oil-fired furnaces, coal–oil and coal–water mixture can be used for which low-ash coals recovered from non-coking coals by deep coal cleaning may be of great help. Coal-to-electricity route spans many promising technologies which include integrated gasification combined cycle plant (IGCC). Analysis of data on gasification process has shown that low ash coals perform better for

IGCC. The percentage of reserves of inferior non-coking coal is 71 % of total Indian non-coking coal reserves. Therefore, the share of inferior coal production will further increase.

Most Indian coals have a considerable proportion of fixed ash due to 'drift origin' of Gondwana formation in Permian age (Khoury 1981). Most of the thick seams are banded in nature. The bands themselves are usually carbonaceous in nature and very rarely consist of pure sandstones (Mukherjee et al. 1982). These coals are difficult to clean by conventional methods. For applications such as IGCC, these coals need to be ground finer in order to liberate these finely disseminated or inter grown foreign matter. Most of the physical separation techniques have limitations in terms of efficiency of separation at this finer size of grinding. Flotation is the only versatile process in use for fine coal processing in India. Conventional mechanical flotation cells cannot remove all the liberated mineral matter in a single stage. If conventional cells are used, the entrained particles can be removed by refloating the concentrate several times but the cost of multi-stage flotation is often prohibitive. Of late, column flotation has been found to be superior to conventional cells because of a more extended collection zone and improved particle bubble contact efficiency. Keeping this in view, investigations were carried out on a non-coking sample to reduce the ash content from 34.60 % to 15 % in a flotation column.

2 MATERIALS AND METHODS

2.1 Material characterization

The non-coking coal used in this study was sourced from North Karanpura (Jharkhand, India). The following coal characterization studies were carried out preceding the flotation studies.

2.2 Proximate analysis

The proximate analysis of the coal sample is as shown in Table 1. It could be observed that the high ash content in this coal warrants processing for its utilization.

Table 1. Proximate analysis (air dry basis) of non coking-coal sample (%)

A_{ad}	M_{ad}	V_{ad}	FC_{ad}
34.6	1.3	26.2	37.9

2.3 **Petrographic analysis**

The petrographic analysis of the coal sample was carried out and the results are as shown in Table 2. Its mean maximum reflectance was found to be 0.71 %.

Table 2. Petrographic analysis of non-coking coal sample (%)

Vitrinite	Semi-vitrinite	Liptinite	Inertinite	Mineral matter
38.7 (46.5)	0.1 (0.2)	11.9 (14.4)	32.4 (38.9)	16.9

The values in parenthesis are on 'mineral-free' basis. Earlier researchers established the following floatability order of the petrographic components (Brady and Gauger 1940; Arnold and Aplan 1989; Gosiewska et al. 2002; Ding 2009).

Vitrinite > Exinite > Inertinite > Inorganic mineral matterVitrinite > Exinite > Inertinite > Inorganic mineral matter

This floatability order indicates the theoretical possibility of the separation of different petrographic components of the present coal sample.

2.4 Size analysis

The size analysis of coal was carried out and the results are as shown in Table 3.

Table 3. Size analysis of coal (d_{80}: 2108 µm)

Size (mm)	Weight retained (%)	A_{ad} (%)	Ash distribution (%)
+6	1.00	39.12	1.14
−6 + 2.8	27.52	24.93	20.00
−2.8 + 1.68	22.82	31.25	20.79
−1.68 + 0.85	25.17	38.16	28.00
−0.85 + 0.5	14.43	46.25	19.46
−0.5 + 0.212	6.71	39.18	7.66
−0.212	2.35	43.08	2.95

From the above table, it is clear that ash is distributed predominantly in the size range of −6 + 0.5 mm. The percent ash distribution value in the individual size fractions in this size range is around 20 %. The coal was further subjected to crushing in rolls for liberation of ash forming minerals.

2.5 Washability study

The coal was sieved at 0.5 mm and the −0.5 mm fraction of coal was segregated. The +0.5 mm coal was subjected to heavy media separation (HMS) in successive baths of increasing specific gravity. The medium of desired density is obtained by using organic liquid mixtures of bromoform, tetrachloroethylene and benzene. From the HMS study, the washability curves are plotted as shown in Fig. 1. From these, it is evident that 70 % of the coal size fraction (-6 + 0.5 mm) could be obtained as float with 15 % ash content by subjecting it for heavy medium separation at 1.70 specific gravity. The rejects amount to 30 wassaying 70 % ash (as shown in black dotted lines).

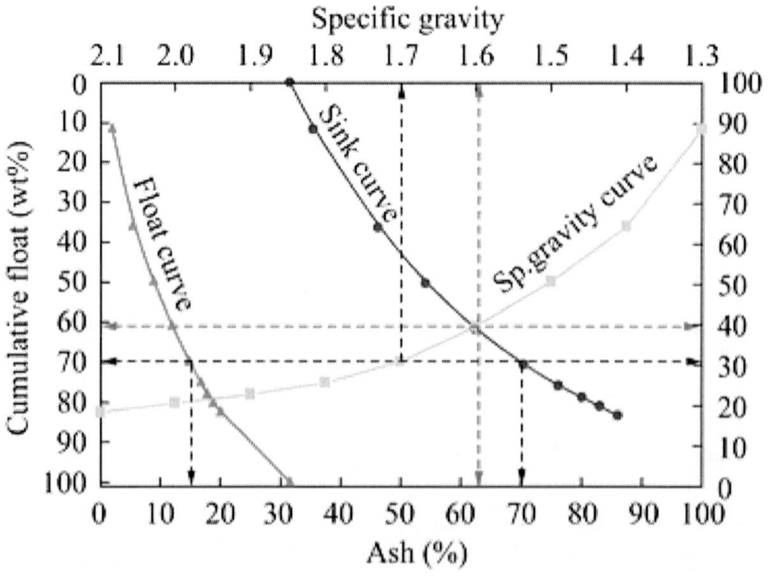

Figure 1. Washability curves for -6 + 0.5 mm fraction of coal

2.6 Flotation column

The automated laboratory flotation column with an internal diameter of 75 mm, designed and developed by CSIR-NML Madras Centre was used in the present study to know the effect of its operating parameters on the process. The schematic details of laboratory column were shown in Fig. 2. Electronically controlled metering pumps were used to feed and discharge the slurry. Slurry/froth interface was maintained using differential pressure transmitter (DPT). The output signal of the DP cell was looped with the stroke controller of the tailings pump so that the pumping rate could be automatically varied to maintain the interface level at a fixed froth depth. Sintered ceramic tube air sparger is used with 30 % porosity. Purge rotameter with a differential pressure regulator was used to control the flow of air to the column from a compressor. The column was filled with water and stabilized at required froth depth. The coal slurry at ~10 % solids by weight was conditioned initially with diesel (collector) in a conditioner for 3 min. In the second stage, MIBC (frother) was added and conditioned for another 3 min. This slurry was pumped to the column. The column was allowed to run for a minimum period of 3–4 residence times. Samples were drawn under near-steady state conditions. Both the process parameters and column operating conditions were recorded before collecting the samples. Samples were analyzed for ash content by standard method.

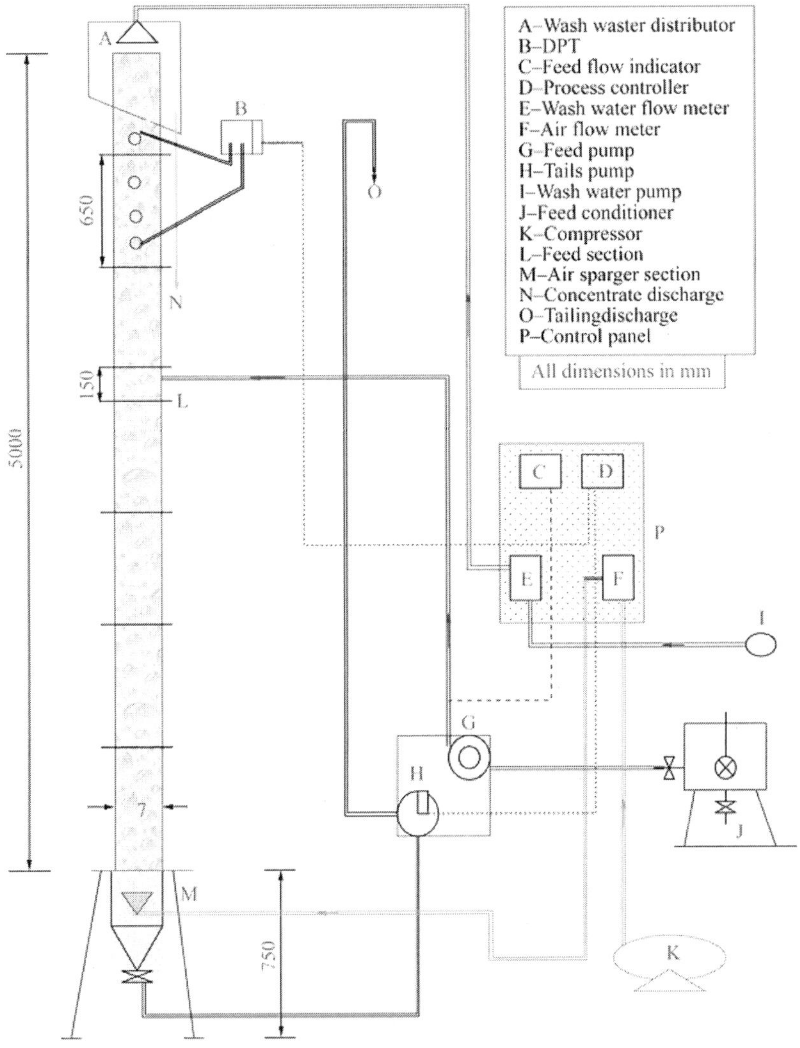

Figure 2. Schematic diagram of CSIR-NML laboratory flotation column

3 RESULTS AND DISCUSSION

Mechanized coal mining techniques generate large amount of fine particles that have to be recovered in coal washing plants. Fine coal processing has always been a problematic and costlier than the cleaning of coarse coal (Andrew Mohns 1997). Froth flotation technique is an effective and best available technique for cleaning fine coal with particle size less than approximately 0.5 mm (Bensley and Nicol 1985; Yoon et al. 2002; Koca et al. 2010). Froth flotation, which exploits the difference in surface properties of minerals, is recognized as an efficient method for coal cleaning and processing of minerals. The efficiency in separating the impurities from coal by flotation is basically determined by the relative hydrophobicity of coal and the associated gangue (Laskowski 2001; S Dey et al. 2013). In froth flotation, coal particles are often subjected to a suitable hydrocarbon oil treatment (Boylu and Laskowski 2007) to alter their hydrophobicity, enhance recovery, and/or improve selectivity (Peng 1996). Conventionally, the collector used in most of the Indian coal washeries is diesel oil in combination with different frothers. The dosage of collector and frother has significant effect on flotation performance (Dey and Pani 2012). Therefore, optimization of grind, collector and frother dosages was carried out followed by that of the operating parameters of flotation column.

3.1 Grind optimization studies

To study the maximum achievable separation of combustible material, the coal sample was subjected to stage crushing to obtain −1.68 mm product which was then fed to the ball mill. A representative sample was drawn for particle size and ash analysis and results are tabulated as Table 4.

Table 4. Size analysis of feed to ball mill (d_{80}: 1102 μm)

Size (μm)	Weight retained (%)	A_{ad} (%)	Ash distribution (%)
+ 850	44.15	33.49	43.67
−850 + 425	33.87	35.40	35.41
−425 + 212	10.24	31.12	9.41
−212 + 106	6.35	30.43	5.71
−106	5.39	36.42	5.80

The above sample was subjected for wet grinding in a laboratory ball mill and the grind variation studies were carried out for various grinding times (15, 20, 25, 30 and 35 min) at the pulp density of 60 % solids by weight. Batch flotation tests were carried out on ground coal samples to optimize the 'mesh of grind' to recover the maximum combustibles. Bench scale flotation tests were performed in a D12 Denver Flotation Machine. The ground samples were subjected to flotation using diesel as collector and MIBC as frother. The combustible recovery (CR), ash rejection (AR) and efficiency index (EI) were calculated by using the following standard formulae (Vasumathi et al. 2013).

%CR=[Mc(100−Ac)Mf(100−Af)]×100%CR=[Mc(100−Ac)Mf(100−Af)]×100

%AR=[1−(McAcMfAf)]×100%AR=[1−(McAcMfAf)]×100

EI =(CR+AR)−100EI =(CR+AR)−100

where A_c is the ash in concentrate, %; A_f is the ash in feed, %; M_c is the mass of concentrate andM_f is the mass of feed.

The results in Table 5 reveal that it is possible to obtain a froth/concentrate of 56 % yield with less than 12 % ash by flotation at d_{80} of 44 μm corresponding to 30 min. gGrind. Also, greater yield of 77 % of froth with 19.7 % ash content can be achieved without cleaning the rougher concentrate. Hence, it could be possible to obtain required 15 % ash with maximum yield at this grind by optimizing the reagent dosages.

Table 5. Flotation of coal at different grinding times (roughing followed by cleaning)

Grinding time (min)	Product	Yield (%)	A_{ad} (%)	CR (%)	AR (%)	EI
15	F C	8.31	5.91	11.65	98.51	10.16
	Cl.2-T	33.85	11.50			
	Cl.1-T	34.46	34.61			
	P T	23.38	71.05			
20	F C	51.51	13.88	68.95	79.96	48.91
	Cl.T	28.43	45.11			
	P T	20.07	78.22			
25	F C	42.67	12.56	57.19	84.59	41.78
	Cl T	31.67	37.27			
	P T	25.67	68.59			
30	F C	56.10	11.77	72.99	79.49	52.48
	Cl T	20.91	41.08			
	P T	23.00	73.90			
35	F C	51.86	12.31	68.40	80.95	49.35
	Cl T	26.10	42.09			
	P T	22.04	73.24			

Note FC final concentrate, *Cl T* cleaner tailings, *PT* primary tailings,diesel is 2.49 kg/t, MIBC is 0.66 kg/t, head ash of coal is 34.60 %

3.2 Column flotation studies

The collector and frother dosages and column operating parameters were optimized on laboratory scale flotation column. The grind of the coal was maintained the same (d_{80}: 44 µm) in all the following tests.

3.2.1 Optimization of collector (diesel) and frother (MIBC) dosages

The test conditions and the results on the effect of variation of diesel and MIBC dosages are given in the Tables 6 and 7, respectively.

Table 6. Effect of diesel on column flotation of coal

Diesel (kg/t)	Product	Yield (%)	Ash (%)	Ash distribution (%)	EI
2.55	Concentrate	73.94	16.78	37.31	54.9
	Tailings	26.06	80.00	62.69	
2.87	Concentrate	78.10	17.34	43.25	53.1
	Tailings	21.90	81.14	56.75	
3.19	Concentrate	77.66	21.28	50.97	39.52
	Tailings	22.34	71.17	49.03	
3.51	Concentrate	88.78	21.35	67.13	30.18
	Tailings	11.22	82.72	32.87	

Note MIBC dosage is 0.65 kg/t, froth depth is 600 mm, superficial air velocity is 0.78 cm/s, superficial feed velocity is 0.57 cm/s, head ash of coal is 34.60 %

Table 7. Effect of MIBC on column flotation of coal

MIBC (kg/t)	Product	Yield (%)	A_{ad} (%)	Ash distribution (%)	EI
0.33	Concentrate	53.46	13.26	23.24	43.48
	Tailings	46.54	50.30	76.76	
0.49	Concentrate	78.26	21.31	47.91	46.56
	Tailings	21.74	83.40	52.09	
0.57	Concentrate	76.11	21.82	46.60	45.86
	Tailings	23.89	79.65	53.40	
0.65	Concentrate	88.78	21.35	67.13	30.18
	Tailings	11.22	82.72	32.87	

Note Diesel dosage is 2.87 kg/t, froth depth is 600 mm, superficial air velocity is 0.78 cm/s, superficial feed velocity is 0.57 cm/s , head ash of coal is 34.60 %

It could be observed that as the diesel dosage is increased from 2.55 kg/t to 3.51 kg/t, the yield of the froth increased from 73.94 % to 88.78 % with ash in it increasing from 16.78 % to 21.35 %. At the same time, efficiency index of separation decreases from 54.90 to 30.18. This could be attributed to the non-selectivity of diesel at higher dosages and physical entrainment/entrapment of ash forming minerals in the froth resulting in higher yield and ash content in the froth.

The dosage of frother plays vital role as it affects the bubble size distribution and mean bubble size. In flotation, the bubble is the driving force and it determines the separation efficiency. The flotation results at various MIBC dosages and at a constant diesel dosage are presented in Table 7. The results clearly indicate that MIBC plays very important role in coal flotation. At constant diesel dosage, flotation concentrate yield increased as the MIBC dosage increased.

3.3 **Optimization of column operating parameters**

The following column parameters namely air flow rate, feed flow rate, froth depth, wash water and % solids of feed were optimized for non-coking coal flotation.

3.3.1 Effect of air flow rate

Generally, superficial air velocity should be as large as possible to ensure a high throughput. At the same time, if the superficial air velocity is too high, the flow pattern will be disturbed and there is every possibility to lose bubbly swarm. It is to be noted that column should operate in a bubbly flow regime, where gas hold-up varies linearly with air rate. A proper combination of gas rate and bubble size will generally provide a gas holdup in the flotation pulp in the range of 15 %–18 %.

The effect of superficial air rate on the flotation of coal is presented in Table 8. It is observed that as the superficial air rate increases there is no increase in the clean coal yield. But at higher superficial air rates the ash content in clean coal decreased marginally. This may be due to the reduction in gangue entrainment. The effect of superficial air rate on entrainment in froth may be explained as follows: the ash content in the clean coal is reduced at higher superficial air velocity because of increase in the bubble size in the froth zone. In the present investigation it was observed that the superficial air velocity of 0.78 cm/s was found to be optimum to obtain better grade and recovery.

Table 8. Effect of air rate on column flotation of coal

Superficial air velocity (cm/s)	Product	Yield (%)	Ash (%)	Ash distribution (%)	EI
0.39	Concentrate	78.63	20.14	46.88	47.95
	Tailings	21.37	83.96	53.12	
0.59	Concentrate	77.95	21.72	46.66	49.12
	Tailings	22.05	87.79	53.34	
0.78	Concentrate	78.10	17.34	43.25	53.10
	Tailings	21.90	81.14	56.75	
0.98	Concentrate	73.14	18.74	39.31	51.92
	Tailings	26.86	78.77	60.69	

Note Diesel dosage is 2.87 kg/t, MIBC dosage is 0.65 kg/t, froth depth is 600 mm, superficial feed velocity is 0.57 cm/s, head ash of coal is 34.60 %

3.3.2 Effect of feed flow rate

Change in feed flow rate usually affects grade and recovery. In the case of coal flotation, the increase in volumetric feed flow rate results in decrease of residence time of slurry which will affect the recovery. Enough residence time is to be maintained so that entire coal will be collected and separated. Feed flow velocities were varied from 0.33 to 0.87 cm/s by manipulating the speed of the variable frequency driven feed pump and the results were presented in Table 9.

Table 9. Effect of Feed velocity on column flotation of coal

Superficial feed velocity (cm/s)	Product	Yield (%)	A_{ad} (%)	Ash distribution (%)	EI
0.33	Concentrate	80.39	17.78	48.57	45.10
	Tailings	19.61	77.18	51.43	
0.41	Concentrate	79.58	16.99	45.56	48.39
	Tailings	20.42	79.11	54.44	
0.49	Concentrate	76.67	18.41	43.23	49.65
	Tailings	23.33	79.46	56.77	
0.57	Concentrate	78.10	17.34	43.25	53.10
	Tailings	21.90	81.14	56.75	
0.65	Concentrate	79.26	21.19	49.98	44.09
	Tailings	20.74	81.04	50.02	
0.87	Concentrate	78.12	20.28	45.49	50.04
	Tailings	21.88	86.75	54.51	

Note Diesel dosage is 2.87 kg/t, MIBC dosage is 0.65 kg/t, froth depth is 600 mm, superficial air velocity is 0.78 cm/s, head ash of coal is 34.60 %

As the superficial feed velocity increases, generally the weight recovery would decrease owing to the decrease in pulp residence time. On the other hand the quality of the concentrate generally increases due to less residence time of froth in the froth zone. From the results it is observed that the yield of clean coal is not much affected by increasing the superficial feed velocity which may be due to the natural floatability of coal. But at higher feed velocities, the ash content is increased in the clean coal because of less residence time of froth in the cleaning zone. Generally, good coal doesn't require larger residence times, but in the present case, the quality of clean coal is better at higher residence time i.e., at feed velocity of 0.41 cm/s. The calculated residence time at this feed velocity is about 18–20 min.

3.3.3 Effect of froth depth

Floatability characteristics are also another factor which decides the height of the flotation zone. If the material is having higher floatability like graphite and coal, the froth height increases and does not affect much on yield but simultaneously the concentrate grade increases as demonstrated by the present studies. Further, it has been seen that by increasing the froth depth, the entrained gangue minerals in the concentrate reduce significantly due to coalescence of bubbles, positive bias water and reduction of feed water in the concentrate. Experiments were conducted at different froth depths ranging from 500 mm to 900 mm and the results obtained were presented in

Table 10. The rejection of entrained particles, in this case, inorganic ash forming minerals, depends on froth depth. Due to high superficial air and feed velocities, fine and ultra fine sized ash forming particles might be entrained into the froth phase. If there is no enough froth depth, these particles will be carried along with the clean coal and thus the quality may be affected and the froth depth of 600 mm is optimized for further experiments.

Table 10. Effect of froth depth on column flotation of coal

Froth depth (mm)	Product	Yield (%)	A_{ad} (%) (%)	Ash distribution (%)	EI
500	Concentrate	78.53	21.13	48.70	45.24
	Tailings	21.47	81.41	51.30	
600	Concentrate	78.10	17.34	43.25	53.10
	Tailings	21.90	81.14	56.75	
700	Concentrate	76.02	21.10	47.26	43.53
	Tailings	23.98	74.65	52.74	
800	Concentrate	79.35	16.41	44.54	49.21
	Tailings	20.65	78.52	55.46	
900	Concentrate	70.78	14.50	31.48	58.30
	Tailings	29.22	76.45	68.52	

Note Diesel dosage is 2.87 kg/t, MIBC dosage is 0.65 kg/t, superficial feed velocity is 0.57 cm/s, superficial air velocity is 0.78 cm/s, head ash of coal is 34.60 %

It could be seen from the above results that the clean coal yield reduced at higher froth depths. This could be due to froth drop back into slurry phase across the froth/slurry interface at relatively higher froth depth. At the same time the ash content in the clean coal is reduced gradually as the froth depth increases.

3.3.4 Effect of feed % solids (pulp density)

In order to examine the effect of pulp density on the flotation of coal, tests were conducted at various pulp densities and the results of the same are given in Table 11.

Table 11. Effect of pulp density on column flotation of coal

% Solids by weight	Product	Yield (%)	A_{ad} (%)	Ash distribution (%)	EI
4.76	Concentrate	72.18	14.83	35.44	52.63
	Tailings	27.82	70.09	64.56	
9.09	Concentrate	78.10	17.34	43.25	53.10
	Tailings	21.90	81.14	56.75	
13.04	Concentrate	87.94	20.14	66.53	29.18
	Tailings	12.06	73.88	33.47	
16.67	Concentrate	63.17	19.50	33.36	47.25
	Tailings	36.83	66.80	66.64	

Note Diesel dosage is 2.87 kg/t, MIBC dosage is 0.65 kg/t, froth depth is 600 mm, superficial feed velocity is 0.57 cm/s, superficial air velocity 0.78 cm/s, head ash of coal is 34.60 %

The decrease in selectivity in flotation of clean coal was observed when the pulp density is increased from 4.76 % to 16.67 %. But the efficiency of separation is better between 5 % and 9 % solids.

4 CONCLUSIONS

A non-coking coal of vitrinite variety and analyzing 34.6 % ash, 26.2 % volatile matter, 1.3 % moisture and 37.9 % fixed carbon was subjected for flotation and column flotation studies. Concentrate/froth of 14.83 % ash at 72.18 % yield could be achieved at optimum process parameters such as 'mesh of grind' (d_{80}: 44 µm), collector dosage (diesel: 2.87 kg/t) and frother dosage (MIBC: 0.65 kg/t) and flotation column operating parameters like froth depth (600 mm), superficial feed velocity (0.57 cm/s), superficial air velocity (0.78 cm/s) and feed pulp consistency of 9 % solids by weight. Out of these parameters, froth depth and slurry feed percent solids have significant effect on the process. The concentrate produced by this process is suitable as fuel for IGCC in coal-to-electricity route.

ACKNOWLEDGMENTS

The authors are thankful to The Director, CSIR-NML for his encouragement and support for publishing this work.

REFERENCES

1. Andrew Mohns C (1997) Effect of particle size on coal flotation kinetics (PhD thesis). Queen's University, Kingston:pp1–4

2. Arnold BJ, Aplan FF (1989) The hydrophobicity of coal macerals. Fuel 68(5):651–658

3. Bensley CN, Nicol SK (1985) The effect of mechanical variables on the flotation of coarse coal. Coal preparation 1:189

4. Boylu F, Laskowski JS (2007) Rate of water transfer to flotation froth in the flotation of low-rank coal that also requires the use of oily collector. Int J Miner Process 83:125–131

5. Brady GA, Gauger AW (1940) Properties of coal surfaces. J ind eng chem 32:1599–1604

6. Dey S, Pani S (2012) Effective processing of low-volatile medium coking coal fines of Indian origin using different process variables of flotation. Int J Coal Prep Util 32:253

7. Ding LP (2009) Investigation of bituminous coal hydrophobicity and its influence on flotation. Energy fuels 23:5536–5543

8. Gosiewska A, Drelich J, Laskowski JS, Pawlik M (2002) Mineral matter distribution on coal surface and its effect on coal wettability. J Colloid Interface Sci 247(1):107–116

9. Khoury DL (1981) Coal Cleaning Technology. NDC, Park Ridge

10. Koca S, Bektas Y, Koca H (2010) Contact angle measurement on lignite surface. Proceedings of the XXV International Mineral Processing Congress (IMPC) Brisbane 2049

11. Konar BB, Banerjee SB, Chaudhuri SG, Choudhury A, Das NS, Sen Kalyan (1997) In: Coal Preparation. Allied Publishers Ltd New Delhi, CFRI Golden jubilee monograph

12. Laskowski JS (2001) Coal flotation and fine coal utilization. Elsevier Science, Amsterdam, p 225

13. Mukherjee AK, Chatterjee CN, Ghose S (1982) Coal Resources of India:it's formation, distribution and utilisation. Fuel Sci Technol 1:19–34

14. Peng FF (1996) Surface energy and induction time of fine coals treated with various levels of dispersed collector and their correlation to flotation responses. Energy fuels 10:1202–1207

15. Shobhana Dey, Gyana Manjari Paul, Santosh Pani (2013) Effective flotation of weathered coal using frother blend. In: Coal preparation technology CSIR - CIMFR Jharkhand

16. Vasumathi N, Vijaya Kumar TV, Subba Rao S, Prabhakar S, Bhaskar Raju G, Shiva Kumar S, Raman Uma (2013) Eco-friendly and cost effective reagent Sokem 590C for coal flotation. Int J Eng Res 2(7):418–433

17. Yenigalla GR (2013) Coal Beneficiation: A road to clean coal Technology. In CSIR - CIMFR Jharkhand India: Coal Preparation Technology Proceedings

18. Yoon RH, Luttrell GH, Asmatulu R (2002) Extending the upper particle size limit for coal flotation. J South African Inst Mining Metallurgy 102(7):411–415

Chapter 7

Exposure of Fluoride in Coal Basin

Khageshwar Singh Patel[1*], Ankit Yadav[1], Keshaw Prakash Rajhans[1], Shobhana Ramteke[1], Reetu Sharma[1], Irena Wysocka[2], Irena Jaron[2]

[1]*School of Studies in Chemistry/Environmental Science, Pt. Ravishankar Shukla University, Raipur, India*

[2]*Polish Geological Institute, Rakowiecka, Warsaw, Poland*

ABSTRACT

Coal is a dirty fuel contaminated with F^- and other elements. Several million tons of coal are burnt in Korba basin, central India to generate electricity with pouring fluorine and other elements into the environment. The water is contaminated with F^- and other chemicals beyond the permissible limits. The contaminated water is consumed by the human and animals by excreting the balanced toxicants through the stool and urine. Several folds higher concentration of F^- in urine (44 mg/L) and stool (266 mg/kg) samples of the cattle are observed. The prevalence of fluorosis diseases in cattle of the Korba basin is reported.

Keywords: Coal, Fluoride, Contamination, Toxicity

1. INTRODUCTION

Coal is a dirty fuel contaminated with toxic elements at the traces [1] [2] . The fluoride content in the coals is ranged from 20 - 500 mg/kg [3] . The environments in coal burning areas were severely contaminated with F^- and other toxicants [4] -[9] . The human and other animals living in the coal burning areas were affected by chronic endemic fluorosis diseases due to the excessive intake of $F^-$from air, water and food [10] - [12] . Several states of the country are suffered with fluorosis diseases due to intake of contaminated groundwater [13] - [20] . In the present work, the contamination of ground and surface water of the largest coal basin of India with F^- and other elements is described. The exposure assessment of the F^- toxicity in the domestic animals is described.

2. MATERIALS AND METHODS

2.1. Study Area

Korba basin, CG, India (22˚21'0"N and 82˚40'48"E) has rich deposit of coal over ≈ 1.0×10^4 km^2area. It is a power capital of Chhattisgarh state, India. The basin is composed of four blocks: Kartala, Katghora, Korba and Pali with population of » 1 million distributed over 710 cities, towns and villages. A huge amount of coal > 10000 MT/Yr has been consumed by various units of thermal power plants for generation of ≈ 40000 KW electricity. At least 12 open and underground coal mines are in operation with annual production of ≈3 BT coal by discharging > 100 BT/Yr liquid effluents in the environment. In addition, the Asia biggest Aluminum Plant (BALCO) is running in the Korba basin.

2.2. Sample Collection

The main river flowing through Korba basin is the Hasdeo River. The Bango dam has been constructed across this river through which water flows into canals. Several pit lakes are found to exist in the Korba basin due to mining out of the coal. The ground and surface water samples were collected from 30 and 6 locations of the basin in month of January, 2013 [21] (Figure 1). A 1-L cleaned narrow mouth polyethylene bottle was used for the collection. The container was rinsed three times with the sample water and filled with the water up to the mouth.

The physical parameters i.e. temperature (T), pH, dissolved oxygen (DO), reduction potential (RP) and electrical conductivity (EC) of the water were measured at the spot by using HANNA sensors.

The first morning stool (1 kg) of the domestic animals (n = 20) was collected manually. The stool sample was stored in the polyethylene bottle and dried in oven at 60°C till the dryness. They were crushed and sieved out the particles of mesh sizes ≤0.1 mm. A 5.0 g stool sample was extracted out with 25 mL hot deionized water (50°C) for 6 hr to measure the F^- and pH values.

The first morning urine sample (100 mL) was collected in a plastic bottle containing 0.2 g EDTA. Total 20 (4 × 5) urine samples of domestic animals (i.e. cattle, buffalo, sheep and goat) were collected in January, 2013. The samples was shipped to the laboratory in insulated container at about 4°C and stored at −20°C until use.

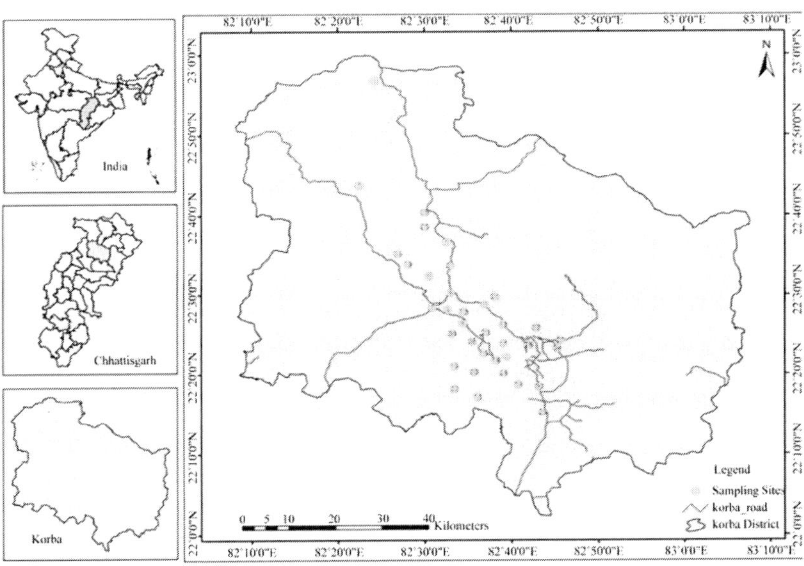

Figure 1. Representation of sampling locations in Korba basin.

2.3. Analysis

The Metrohm ion meter-781 was used for the measurement of F^- by using TISAB-III buffer in 1:1 ratio. The buffer was prepared by dissolving 300 g sodium citrate, 22 g 1,2-cyclohexanediamine-N,N,N,N-tetraaceticacid and 60 g NaCl in a volume of 1-L with the de-ionized water with subsequent adjustment of pH value to 5.0 - 5.5). The Dionex ion chromatography-1100 was used for monitoring of ions i.e. Cl^-, NO_3^-, SO_4^{2-}, Na^+, K^+, Mg^{2+} and Ca^{2+}. The Thermos ICP-OES and ICP-MS (Polish Geological Institute, NRI, Central Chemical Laboratory, Warsaw) were employed for analysis of the elements.

3. RESULTS AND DISCUSSION

3.1. Geology and Hydrology

The rock of the studied area was formed in the Archaean to Cenozoic ages. They consist of granite gneiss and granitoids covering with medium to coarse grained arkosic sandstone, a few pebble beds, conglomerate and shale of coal. The ground water occurs under phreatic, semi-confined and con-fined conditions control of the inter granular pore spaces in the shallow weathered zones and joints, fractures and caverns in deeper horizons. The water level depth is varied from 3.4 - 11.6 mbgl (meter below ground level) during the post-monsoon period in the aquifers.

3.2. Physico-Chemical Characteristics of Groundwater

The physico-chemical parameters of the groundwater are presented in Tables 1-3. The age and depth of 30 tube wells was ranged from 4.0 - 130 Yr and 4 - 18 m with mean value of 7 ± 1 Yr and 11 ± 1 m, respectively. A slight variation in temperature of the groundwater was recorded, ranging from 29.5 - 31.2°C with mean value of 30.6 ± 0.1°C. The pH value of the water was varied from 5.9 - 8.2 with mean value of 7.1 ± 0.2. The groundwater was found to be acidic in few locations (i.e. Surabahar, Sutara, Turali, Lalmatiya, Jenara, Dhelawdih, Banki Mongara, Dipka, Kusmunda, Bhilai, Hardibazar and Rumgara) due to the existence of higher content of the acids. The EC,TDS, DO and RP value of the groundwater was varied from 296 - 964 µS/cm, 414 - 990 mg/L, 5.6 - 6.7 mg/L and 199 - 241 mV with mean value of 497 ± 56 µS/cm, 628 ± 61 mg/L, 6.0 ± 0.1 mg/L and 218 ± 5 mV, respectively. The value of DO, EC and TDS was found moderately higher to the recommended value of 4.0 mg/L, 300 µS/cm and 500 mg/L, respectively [22] [23] . However, the RP value was observed to be at least 3-times less than the recommended value of 600 mV.

The concentration of F^-, Cl^-, NO_3^-, SO_4^{2-}, Na^+, K^+, Mg^{2+}, Ca^{2+}, Al, SiO_2, Fe, Mn and Zn was ranged from 2.1 - 12, 11 - 92, 10 - 90, 10 - 72, 10 - 70, 0.6 - 8.7, 3.0 - 44, 8.0 - 78, 7.2 - 45, 9.0 - 70, 0.4 - 8.2, 0.2 - 1.5 and 1.0 - 2.4 mg/L with mean value of 5.5 ± 0.9, 34 ± 7, 28 ± 7, 21 ± 5, 24 ± 6, 3.7 ± 0.8, 11 ± 3, 31 ± 6, 16 ± 3, 31 ± 6, 2.2 ± 0.6, 0.5 ± 0.1 and 1.5 ± 0.1 mg/L, respectively. Similarly, concentration of other metals: Li, Be, Rb, Co, Ni, Cu, Sn, Sb, Mo, Cd, La, Ce, Pb and U was observed in microgram levels, ranging from 4.7 - 36, 1.1 - 6.7, 1.4 - 38, 1.1 - 7.7, 2.4 - 27, 2.3 - 22, 1.2 - 11, 0.9 - 9.8, 0.1 - 5.0, 0.11 - 0.33, 0.1 - 2.1, 0.1 - 0.8, 1.1 - 29 and 1.0 -12 µg/L with mean value of 16 ± 3, 2.1 ± 0.5, 16 ± 5, 3.3 ± 0.8, 9.2 ± 2.4, 8.0 ± 2.1, 3.0 ± 1.0, 3.2 ± 0.8, 1.0 ± 0.4, 0.19 ± 0.02, 0.4 ± 0.2, 0.4 ± 0.1, 7.0 ± 2.4 and 2.5 ± 1.0 µg/L, respectively. Among them, the highest content of Cl^- was observed, may be due to leaching from the coal. The occurrence trend of 27 elements in the water was found in following decreasing order: Cl^- > Ca » SiO_2 < NO_3^- < Na^+ > SO_4^{2-} > Al > Mg > K^+ > F^- > Fe > Zn > Mn >> Li » Rb > Ni > Cu > Pb > Sb ≈ Co > Sn > U > Be < Mo > La » Ce > Cd. The concentration of F^- in the groundwater of the studied area was found to be comparable to F^- contents reported in the water of other locations of the country [5] - [18] .

Factor analysis was used for 17 variables of the water. Four factors were extracted which explain 83.27% of the total variance. A 58.49% of the total variance was explained by Factor-1, showing strong positive loadings on EC, TDS, F^-, Cl^-, SO_4^{2-}, Na^+, Mg^{2+}, Fe, Mn, Zn, Be, Rb, Co, Ni, Cu, Cd, Pb, La, Ce and Ca^{2+}. They were related to the mineralization of the groundwater which involved also weathering of gypsum and fluoride bearing minerals such as CaF_2. They were correlated well in the water as shown in Table 4, Table 5. A 10.93% of the total variance was accounted by factor-2, pyrite with strong positive loadings on NO_3^- and K^+, related to coal burning and mining activities. A 7.78% of the total variance was explained by factor-3, describing a strong loading on SiO_2.

A 6.06% of total variance was explained by factor-4 related to the age factor of tube wells.

Table 1. Physical characteristics of groundwater in Korba basin, January 2013.

S. No.	Location	pH	EC, μS/cm	TDS, mg/L	DO, mg/L	RP, mV
1	Pauli	7.4	296	745	5.9	199
2	Surabahar	6.6	527	627	5.8	228
3	Sutara	5.9	398	563	5.8	241
4	Katghora	7.8	388	509	5.9	202
5	Konkona	7.3	514	589	5.7	221
6	Turali	6.6	964	813	6.2	230
7	Lalmatiya	6.4	312	588	6.1	237
8	Jenara	6.8	464	574	5.6	199
9	Lakhanpur	7.1	633	611	5.8	215
10	Dhelawdih	6.7	515	628	6.1	212
11	Churri	7.2	629	738	6.7	201
12	Gopalpur	7.7	519	922	5.9	241
13	Darri	7.2	523	650	5.8	202
14	Banki Mongara	6.8	634	722	5.8	221
15	Dipka	6.9	511	441	5.9	230
16	Ghordeva	7.2	534	487	5.7	199
17	Balgi	7.0	713	933	6.2	228
18	Kuchana	7.3	391	633	6.1	241
19	Kusmunda	6.7	704	926	5.6	202
20	Bhilai	6.8	362	455	5.8	221
21	Dadar	7.0	322	464	6.1	230
22	Hardibazar	6.9	367	434	6.7	237
23	NTPC	7.4	380	471	5.7	199
24	Barpali	7.6	342	414	6.1	215
25	Manikpur	7.2	650	821	6.2	212
26	Korba	7.3	756	990	6.1	201
27	Balco	7.1	410	497	5.6	221
28	Rumgara	6.9	390	449	5.8	230
29	Urga	8.2	418	680	6.1	237
30	Bilaikhund	7.8	347	462	6.3	199

Table 2. Concentration of major elements in groundwater during January 2013, mg/L.

S. No.	F^-	Cl^-	NO_3^-	SO_4^{2-}	Na^+	K^+	Mg^{2+}	Ca^{2+}	Al	SiO_2	Fe	Mn	Zn
1	2.8	32	28	14	11	6	16	52	17	64	3.7	0.51	1.4
2	4.5	18	11	12	17	2	12	36	15	11	1.2	0.31	2.4
3	3.7	39	17	12	12	5	6	18	14	23	2.0	0.55	2.2
4	5.8	53	14	15	14	2	7	22	15	19	1.1	0.33	1.3
5	10	39	35	14	22	2	9	30	9.1	23	1.0	0.17	2.0
6	4.6	92	23	72	17	5	6	17	10	18	4.1	0.69	1.6
7	3.5	32	21	24	23	6	16	51	7.2	9	1.1	0.38	1.7
8	3.8	39	21	44	17	2	8	22	18	12	2.8	0.68	1.2
9	3.8	39	30	22	28	2	9	24	16	26	1.1	0.93	2.0
10	4.5	50	31	10	16	3	11	36	11	35	0.4	0.22	1.3
11	6.0	67	31	26	18	3	11	36	11	35	0.4	0.22	1.3
12	4.8	39	25	30	24	1.6	22	66	19	70	4.2	0.48	1.2
13	3.6	11	10	13	18	1.3	5	22	17	12	1.4	0.29	2.0
14	8.4	39	15	16	60	2.7	18	36	16	25	2.3	0.52	1.8
15	11	18	13	10	16	1.5	5	32	17	21	1.2	0.31	1.1
16	4.6	28	31	15	10	5.9	8	22	10	25	1.1	0.16	1.7
16	4.6	28	31	15	10	5.9	8	22	10	25	1.1	0.16	1.7
17	5.8	67	90	40	38	3.3	14	48	45	20	4.7	0.65	1.3
18	3.7	11	19	13	15	4.8	6	18	7.5	10	1.2	0.36	1.4
19	7.9	43	19	20	53	1.3	26	78	21	68	3.2	0.64	1.0
20	6.2	14	27	14	18	8.5	3	10	18	29	1.3	0.88	1.7
21	2.9	14	28	14	17	3.9	4	18	12	39	0.5	0.21	1.1
22	2.9	14	28	14	17	3.9	4	8	12	39	0.5	0.21	1.1
23	2.1	21	13	16	10	1.7	8	10	30	42	1.4	0.35	1.3
24	3.5	14	15	12	14	3.3	4	14	26	29	1.7	0.35	1.1
25	12	38	83	50	65	3.9	44	60	18	55	8.2	1.53	1.8
26	7.3	70	78	40	58	3.9	16	50	11	58	5.9	0.86	1.2
27	7.7	28	31	15	10	5.9	8	22	10	25	1.3	0.16	1.7
28	3.5	14	18	16	11	6.7	6	24	7.6	40	1.7	0.27	1.0
29	9.2	18	17	10	70	0.6	12	36	27	23	1.8	0.32	1.8
30	3.9	18	19	15	10	8.7	5	16	24	31	2.1	0.35	1.0

Table 3. Concentration of trace elements in groundwater during January 2013, µg/mL.

S. No.	Li	Be	Rb	Co	Ni	Cu	Sn	Sb	Mo	Cd	La	Ce	Pb	U
1	18	2.4	3.8	1.9	7.4	6.5	1.6	2.3	0.66	0.25	0.24	0.51	4.7	2.8
2	20	2.4	37	2.9	13.3	4.8	3.4	4.7	1.76	0.13	0.22	0.23	2.7	1.6
3	18	2.5	3.2	2.2	10.6	3.0	2.7	3.4	2.08	0.22	0.15	0.20	1.2	1.4
4	17	1.5	4.9	1.7	5.1	8.6	1.7	2.8	2.05	0.14	0.52	0.71	5.6	1.4
5	12	1.5	25.5	5.0	3.9	8.8	2.0	6.0	0.28	0.17	0.27	0.08	9.0	1.3
6	11	1.5	8.3	4.2	10.0	4.1	2.6	4.6	2.10	0.13	0.24	0.31	1.9	1.6
7	9	1.5	21.1	3.0	9.8	4.7	2.2	3.3	1.11	0.20	0.56	0.67	7.3	1.6
8	19	1.8	2.2	1.4	2.9	2.9	1.5	1.0	0.65	0.19	0.20	0.13	4.3	3.8
9	36	2.2	33	7.6	11.4	6.0	2.9	9.8	1.20	0.20	0.41	0.70	3.6	1.4
10	11	5.2	38	7.7	26.6	18	11	1.7	0.06	0.33	2.11	0.80	9.6	1.6
11	11	5.2	28	7.7	23.8	18	10	1.9	0.08	0.33	1.11	0.60	9.1	1.6
12	16	1.7	3.4	1.6	6.1	5.1	1.3	2.1	0.56	0.21	0.19	0.44	4.3	2.1
13	18	1.7	34	2.4	11	3.8	2.8	4.2	1.49	0.11	0.17	0.2	2.5	1.2
14	16	1.8	2.9	1.8	8.8	2.4	2.2	3.1	1.76	0.19	0.12	0.17	1.1	1.1
15	15	1.1	4.4	1.4	4.2	6.8	1.4	2.5	1.74	0.12	0.41	0.61	5.1	1.1
16	11	1.1	23	4.2	3.2	6.9	1.6	5.4	0.24	0.14	0.21	0.07	8.3	1.0
17	10	1.1	7.5	3.5	8.3	3.2	2.1	4.1	1.78	0.11	0.19	0.27	1.7	1.2
18	8.6	1.1	19	2.5	8.1	3.7	1.8	3.0	0.94	0.17	0.44	0.58	6.7	1.2
19	17	1.3	2	1.2	2.4	2.3	1.2	0.9	0.55	0.16	0.16	0.11	3.9	2.9
20	33	1.6	30	6.3	9.4	4.7	2.4	8.8	1.02	0.17	0.32	0.61	3.3	1.1
21	10	3.7	35	6.4	22	14	9.1	1.5	0.05	0.28	1.66	0.69	27	1.2
22	10	3.7	35	6.4	23	14	9.1	1.5	0.05	0.28	1.66	0.69	29	1.2
23	29	1.9	1.4	1.2	2.4	2.3	1.6	1.1	0.87	0.19	0.14	0.22	2.6	10
24	13	1.5	28	1.9	7.7	18	1.7	1.2	0.24	0.25	0.19	0.31	20	1.1
25	21	1.3	1.5	1.6	5.5	9.0	1.8	1.1	0.55	0.17	0.11	0.17	4.9	4.1
26	25	6.7	13	1.3	2.7	5.7	1.7	1.2	4.95	0.11	0.12	0.23	5.2	9.0
27	6.1	1.1	23	4.2	3.2	6.9	1.6	5.4	0.24	0.19	0.21	0.27	8.3	1.0
28	4.7	1.1	1.5	2.9	10	22	2.4	3.2	1.23	0.16	0.14	0.25	6.8	12
29	20	1.2	7.3	1.1	7.3	21	1.2	1.2	0.22	0.31	0.31	0.58	8.4	1.2
30	9.2	1.1	1.4	1.1	5.4	3.8	1.5	1.9	0.55	0.16	0.13	0.19	1.5	1.2

Table 4. Correlation matrix of elements in groundwater.

	F^-	Cl^-	NO_3^-	SO_4^{2-}	Na^+	K^+	Mg^{2+}	Ca^{2+}	Al	SiO_2	Fe	Mn	Zn
F^-	1												
Cl^-	0.18	1											
NO_3^-	0.36	0.47	1										
SO_4^{2-}	0.13	0.72	0.52	1									
Na^+	0.64	0.24	0.46	0.28	1								
K^+	−0.26	−0.12	0.09	0.01	−0.36	1							
Mg^{2+}	0.51	0.27	0.52	0.39	0.68	−0.20	1						
Ca^{2+}	0.42	0.34	0.40	0.25	0.59	−0.28	0.83	1					
Al	0.04	0.01	0.26	0.07	0.26	−0.25	0.13	0.13	1				
SiO_2	0.10	0.12	0.30	0.10	0.29	−0.01	0.54	0.56	0.09	1			
Fe	0.37	0.44	0.69	0.70	0.57	0.01	0.74	0.58	0.29	0.48	1		
Mn	0.32	0.32	0.56	0.61	0.54	0.05	0.63	0.35	0.21	0.29	0.78	1	
Zn	0.16	−0.02	−0.06	−0.06	0.12	−0.10	0.09	−0.06	−0.16	−0.44	−0.05	0.14	1

Table 5. Correlation matrix of trace elements in groundwater.

	Li	Be	Rb	Co	Ni	Cu	Sn	Sb	Mo	Cd	La	Ce	Pb	U
Li	1													
Be	0.11	1												
Rb	0.00	0.38	1											
Co	−0.04	0.42	0.75	1										
Ni	−0.22	0.57	0.63	0.74	1									
Cu	−0.31	0.28	0.29	0.35	0.51	1								
Sn	−0.24	0.68	0.61	0.76	0.95	0.51	1							
Sb	0.32	−0.25	0.39	0.48	−0.05	−0.25	−0.15	1						
Mo	0.30	0.27	−0.25	−0.33	−0.29	−0.39	−0.32	0.07	1					
Cd	−0.12	0.41	0.30	0.44	0.65	0.64	0.66	−0.31	−0.60	1				
La	−0.26	0.58	0.58	0.69	0.86	0.50	0.93	−0.19	−0.37	0.67	1			
Ce	0.04	0.30	0.39	0.49	0.59	0.39	0.53	0.05	−0.17	0.52	0.69	1		
Pb	−0.34	0.31	0.51	0.41	0.52	0.59	0.59	−0.27	−0.44	0.55	0.67	0.40	1	
U	0.16	0.18	−0.38	−0.29	−0.24	0.12	−0.16	−0.28	0.34	−0.18	−0.23	−0.28	−0.15	1

Table 6. Physical parameter of surface water during January 2013.

S. No.	Location	pH	EC, µS/cm	TDS, mg/L	DO, mg/L	RP, mV
1	Churri	6.7	220	430	7.8	182
2	Rumgara	7.2	331	409	6.7	202
3	Kusmunda	7.5	493	413	5.8	210
4	Balco	7.3	172	391	5.9	208
5	Manikpur	7.4	210	494	8.1	176
6	Krishi Nagar	6.8	304	437	7.8	182

Table 7. Chemical parameter of surface water during January 2013.

S. No.	F^-	Cl^-	NO_3^-	SO_4^{2-}	Na^+	K^+	Mg^{2+}	Ca^{2+}	Al	SiO_2	Fe	Mn	Zn
1	2.3	21	38	20	11	8.3	8.1	20	1.0	7.7	0.9	0.4	0.28
2	1.8	11	17	10	46	15	5.2	16	1.2	7.7	0.4	0.4	0.39
3	2.4	14	13	16	47	13	6.0	16	1.3	3.1	0.6	0.3	0.45
4	2.2	18	17	30	16	7.1	5.1	18	1.4	9.2	1.3	0.7	0.29
5	3.1	27	25	40	20	6.0	6.2	20	1.6	7.7	1.4	1.1	0.31
6	2.6	14	30	20	18	9.2	9.1	28	1.3	6.1	1.5	1.2	0.42

Table 8. Concentration of trace elements in surface water during January 2013.

S. No.	Li	Rb	Cr	Co	Ni	Cu	Sb	Pb	U
1	16	13	5.1	2.7	5.2	6.6	3.8	7.4	0.92
2	22	25	4.2	3.2	5.8	9.6	3.4	6.3	0.74
3	23	18	4.6	3.9	4.8	7.5	3.4	5.7	0.89
4	15	19	4.4	4.2	6.4	7.9	4.4	6.2	0.97
5	14	14	4.8	2.5	4.0	6.4	2.4	4.5	0.68
6	18	16	8.1	5.9	7.9	14	1.4	2.8	0.32

3.3. Physico-Chemical Characteristics of Surface Water

The physico-chemical characteristic of the surface water is shown in Tables 6-8. The value of T, pH, DO, RP, EC and TDS of water (n = 6) in the post monsoon period was ranged from 20.8˚C - 22.2˚C, 6.7 - 7.5, 5.8 - 8.1 mg/L, 176 - 210 mV, 172 - 493 µS/cm and 391 - 494 mg/L with mean value of 21.6˚C ± 0.4˚C, 7.3 ± 0.3, 7.0 ± 0.8 mg/L, 193 ± 12 mV, 429 ± 29 µS/cm and 429 ± 29 mg/L, respectively. The concentration of F^-, Cl^-, NO_3^-, SO_4^{2-}, Na^+, K^+, Mg^{2+}, Ca^{2+}, Al, SiO_2, Fe, Mn and Zn was ranged from 1.8 - 3.1, 11 - 27, 13 - 38, 10 - 40, 11 - 47, 6.0 - 15, 5.0 - 9.0, 16 - 28, 3.1 - 9.2, 0.06 - 0.26, 0.4 - 1.5, 0.3 - 1.2 and 0.3 - 0.5 mg/L with mean value of 2.4 ± 0.3, 18 ± 4.7, 23 ± 8, 23 ± 9, 26 ± 13, 10 ± 3, 6.5 ± 1.3, 20 ± 4, 1.3 ± 0.2, 6.9 ± 1.7, 1.0 ± 0.4, 0.7 ± 0.3 and 0.4 ± 0.1 mg/L, respectively. Similarly, concentration of other metals: Li, Rb, Cr, Co, Ni, Cu, Sb, Pb and U was observed at microgram levels, ranging from 14 - 23, 13 - 25, 4.2 - 8.1, 2.5 - 5.9, 4.0 - 7.9, 6.4 - 14, 1.4 - 4.4, 2.8 - 7.4 and 0.3 - 1.0 µg/L with mean value of 18 ± 3, 18 ± 4, 5.2 ± 1.2, 3.7 ± 0.9, 5.7 ± 1.1, 8.7 ± 2.3, 3.1

± 0.9, 5.5 ± 1.3 and 0.8 ± 0.2 μg/L, respectively. They were found to occur in following decreasing order: NO_3^- > Ca > Na^+ > Cl^- > SO_4^{2-} > K^+ > Mg > F^- > Al > Fe > Mn > Zn > Li > Rb > Cu > Pb > Ni > Cr > Co > Sb > U. The concentration of F^- in the surface water of the studied basin was found to be higher than other locations of the country [19] [20] .

The factor analysis was applied and four factors were extracted. Factor-1 was accounted for 48.70% of the total variance, related to anions, Ca^{2+} and Mg^{2+}. It could be correlated to complex processes such as weathering of fluoride bearing minerals (AlF_3, CaF_2, MgF_2), gypsum, carbonate minerals and anthropogenic activities i.e. coal burning and leaching, aluminum plant effluents, etc. Factor-2 was accounted for 22.15% of the total variance, related to EC, TDS and Na^+ which determined the mineralization of pond water. Factor-3 was accounted for 16.23% of the total variance having strong loadings of Fe and Mn, related to the speciation of Fe and Mn in the water. Factor-4 was accounted for 9.23% of the total variance, negatively loaded with the pH values. This factor was in the inverse relationship with the other factors.

3.4. Water Quality

The concentration of F^-, Al, Mn and Fe was found to be several folds higher than recommended value of 1.0, 0.03, 0.10 and 0.30 mg/L, respectively. The higher values of EC, TDS, RP, F^-, Cl^-, NO_3^-, Mg^{2+}, Ca^{2+}, SiO_2, Al, Fe, Ni, Zn, Sb, Pb and U in the groundwater than the surface water was observed, may be due to leaching from the coal (Figure 2, Figure 3). The main sources of the contaminants in the water of the studied area expected are coal mine leachates and the Aluminum and Thermal power plant effluents discharged into the environment.

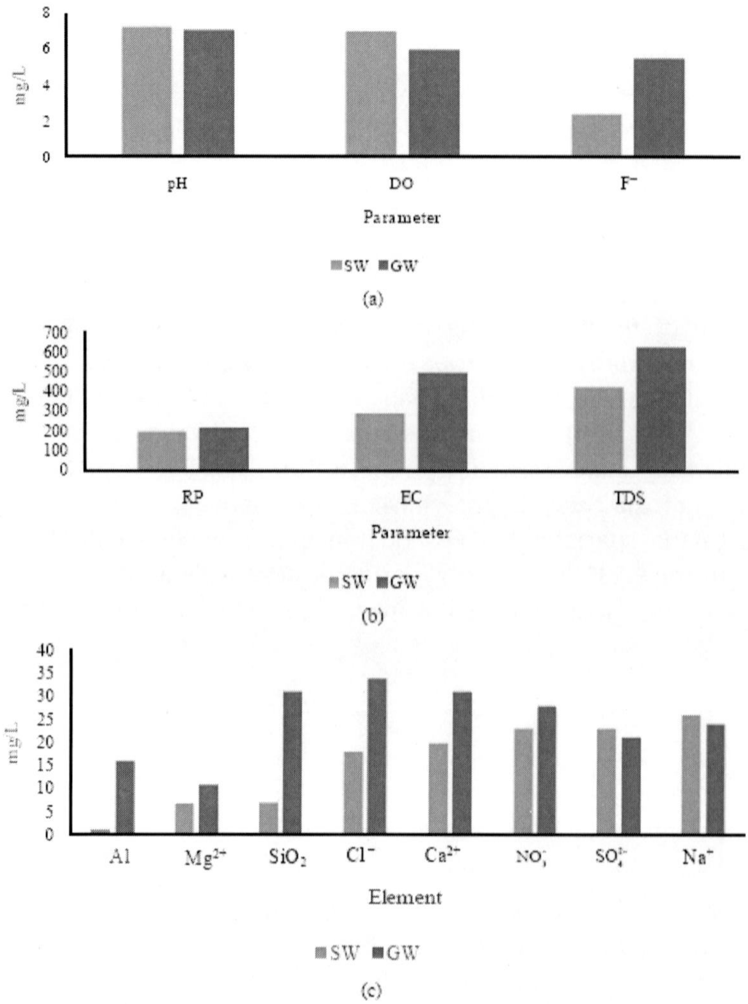

Figure 2. Distribution of pH and elements in ground and surface water in the post monsoon period, January, 2013.

3.5. Exposure Assessment

The toxic elements i.e. F^-, Al, Mn, Fe, Cu, Cd, Pb and U are exposed to human and animals through the contaminated water. Among them, the concentration of F^- is dominated in ground and surface water. The stool and urine samples were reported as good indicator for the exposure assessment. In this study, the F^- content was analyzed in stool and urine

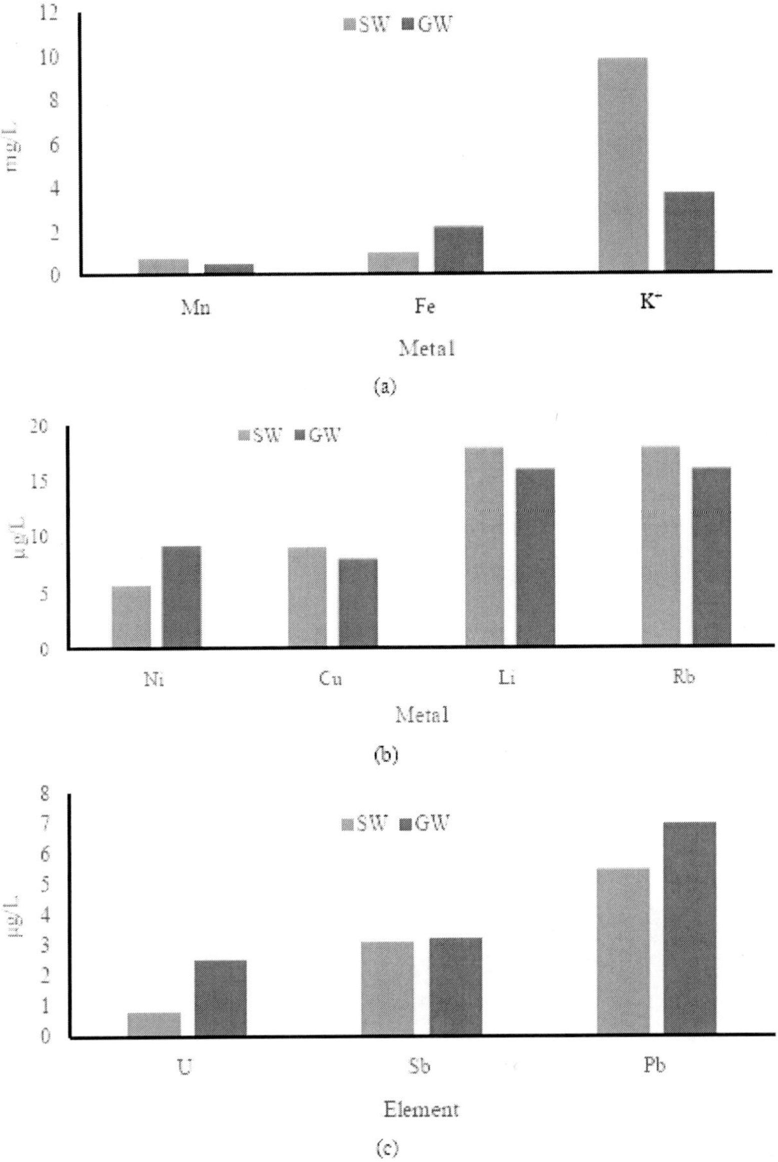

Figure 3. Distribution trace elementsin ground and surface water in the post monsoon period, January, 2013.

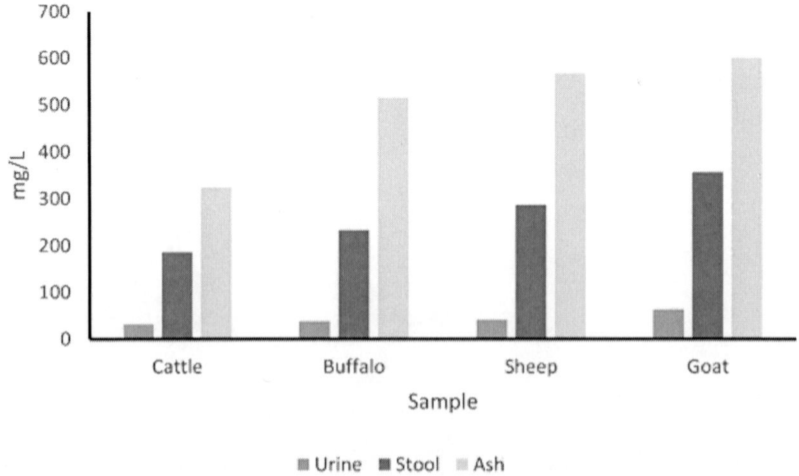

Figure 4. Comparison of F^- concentration in clinical samples.

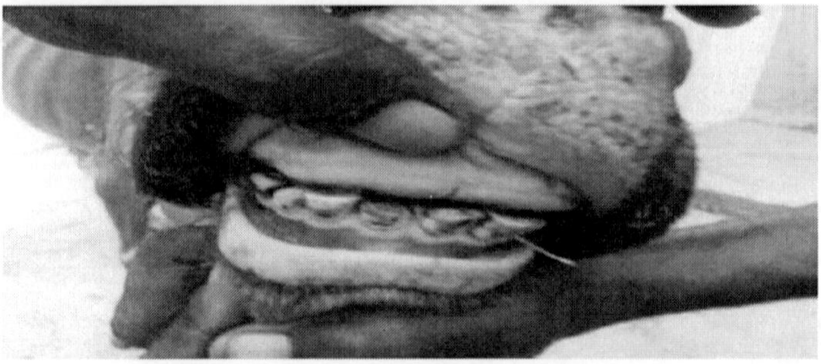

(a)

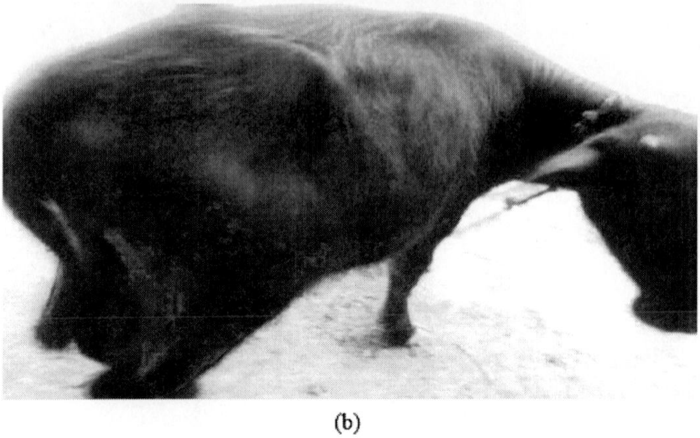

(b)

Figure 5. Dental (a) and hair (b) fluorosis in buffalo.

Table 9. Concentration of F^- in clinical samples during January 2013.

S. No.	Sample	Cattle	Buffalo	Sheep	Goat
1	Urine	32	38	42	63
2	Stool	186	233	287	356
3	Ash	324	516	567	602

samples of domestic animals i.e. cattle, buffalo, sheep and goat (Table 9). The F^- concentration in the urine and stool samples (5 × 4 × 2 = 40) was ranged from 32 - 63 mg/L and 186 - 356 mg/kg with mean value of 44 mg/L and 266 mg/kg (dried mass), respectively. The highest F^- concentration was observed in the goat clinical samples, which might be due to higher intake of the contaminated biomass and water (Figure 4). Several cases of fluorosis diseases in the domestic animals of the basin was observed and shown inFigure 5.

4. CONCLUSION

The water of the Korba basin is contaminated with elements (i.e. F^-, Al, Fe and Mn) by multiple sources (i.e. coal burning, Aluminum plant effluent, mine leachate, etc.). Fluoride is enriched and several folds higher in the animal urines than recommended limit of 4 mg/L with higher prevalence of fluorosis diseases. The domestic animals are severely affected with fluorosis diseases due to higher consumption of the contaminated food and water.

ACKNOWLEDGEMENTS

We are thankful to the UGC, New Delhi for award of the Rajiv Gandhi Research Fellowship to KPR and SR.

REFERENCES

1. Finkelman, R.B. (1999) Trace Elements in Coal Environmental and Health Significance. Biological Trace Element Research, 67, 197-204.

2. Ghosh, R., Majumder, T. and Ghosh, D.N. (1987) A Study of Trace Elements in Litho types of Some Selected Indian Coals. International Journal of Coal Geology, 8, 269-278.

3. Wu, D., Zheng, B., Tang, X., Li, S., Wang, B. and Wang, M. (2004) Fluorine in Chinese Coals. Fluoride, 37, 125-132.http://www.fluorideresearch.org/372/files/FJ2004_v37_n2_p125-132.pdf

4. Pandey, P.K., Pandey, M. and Chakraborty, M. (2013) Fluoride Mobilization due to Coal Mining in Parts of Chhattisgarh. Journal of Environmental Protection, 4, 385-388.

5. Gupta, S., Mondal, D. and Bardhan, A. (2012) Geochemical Provenance and Spatial Distribution of Fluoride in Groundwater in parts of Raniganj Coal Field, West Bengal, India. Archives of Applied Science Research, 4, 292-306.http://scholarsresearchlibrary.com/aasr-vol4-iss1/AASR-2012-4-1-292-306.pdf

6. Borah, J. (2011) Monitoring Fluoride Concentration and Some other Physico-Chemical Properties of Groundwater of Tinsukia District, Assam, India. International Journal of ChemTech Research, 3, 1339-1342.

7. Reza, R. and Singh, G. (2013) Groundwater Quality Status With Respect to Fluoride Contamination in Industrial Area of Angul District Orissa India. Indian Journal of Scientific Research and Technology, 1, 54-61.http://www.indjsrt.com/administrator/modules/category/upload/10-15.pdf

8. Kotoky, P., Barooah, P.K., Baruah, M.K., Goswami, A., Borah, G.C., Gogoi, H.M., Ahmed, F., Gogoi, A. and Paul, A.B. (2008) Fluoride and Endemic Fluorosis in the Karbianglong District, Assam, India. Fluoride, 41, 42-45.http://www.fluorideresearch.org/411/files/FJ2008_v41_n1_p072-075.pdf

9. Pathak, R.P., Pankaj, S., Sameer, V., Mahure, N.V., Rajeev, K. and Ratnam, M. (2012) Detection of Fluoride Contamination in the Surface and Sub-Surface Water near Thermal Power Station.

International Journal of Engineering and Science, 1, 44-47. http://www.researchinventy.com/papers/v1i1/G011044047.pdf

10. Guijian, L., Liugen, Z., Duzgoren-Aydin, N.S., Lianfen, G., Junhua, L. and Zicheng, P. (2007) Health Effects of Arsenic, Fluorine, and Selenium from Indoor Burning of Chinese Coal. Reviews of Environmental Contamination and Toxicology, 189, 89-106.

11. Ando, M., Tadano, M., Yamamoto, S., Tamura, K., Asanuma, S., Watanabe, T., Kondo, T., Sakurai, S., Ji, R., Liang, C., Chen, X., Hong, Z. and Cao, S. (2001) Health Effects of Fluoride Pollution caused by Coal Burning. Science of the Total Environment, 271, 107-116.

12. Fidanci, U.R. and Sel, T. (2001) The Industrial Fluorosis Caused by a Coal-Burning Power Station and Its Effects on Sheep. Turkish Journal of Veterinary and Animal Science, 25, 735-41. http://journals.tubitak.gov.tr/veterinary/issues/vet-01-25-5/vet-25-5-16-0007-6.pdf

13. Singaraja, C., Chidambaram, S., Anandhan, P., Prasanna, M.V., Thivya, C., Thilagavathi, R. and Sarathidasan, J. (2014) Geochemical Evaluation of Fluoride Contamination of Groundwater in the Thoothukudi District of Tamilnadu, India. Applied Water Science, 4, 241-250.

14. Singaraja, C., Chidambaram, S., Anandhan, P., Prasanna, M.V., Thivya, C. and Thilagavathi, R. (2013) A Study on the Status of Fluoride Ion in Groundwater of Coastal Hard Rock Aquifers of South India. Arabian Journal of Geosciences, 6, 4167-4177.h

15. Pandey, A.C., Shekhar, S. and Nathawat, M.S. (2012) Evaluation of Fluoride Contamination in Groundwater Sources in Palamu District, Jharkhand, India. Journal of Applied Sciences, 12, 882-887.

16. Kumar, A. and Kumar, V. (2015) Fluoride Contamination in Drinking Water and its Impact on Human Health of Kishanganj, Bihar, India. Research Journal of Chemical Sciences, 5, 76-84. http://www.isca.in/rjcs/Archives/v5/i2/13.ISCA-RJCS-2015-018.pdf

17. Ghosh, S., Chakraborty, S., Roy, B., Banerjee, P. and Bagchi, A. (2010) Assessment of Health Risks Associated with Fluoride-Contaminated Groundwater in Birbhum District of West Bengal, India. Journal of Environmental Protection Science, 4, 13 - 21.http://aes.asia.edu.tw/Issues/JEPS2010/GhoshS2010.pdf

18. Reddy, B.M., Sunitha, V. and Reddy, M.R. (2013) Fluoride and Nitrate Geochemistry of Groundwater from Kadiri, Mudigubba and Nallamada Mandals of Anantapur District, Andhra Pradesh, India. Journal of Agricultural Engineering and Biotechnology, 1, 37-42.http://www.academicpub.org/DownLoadPaper.aspx?PaperID=14051

19. Madhavan, N. and Subramanian, V. (2001) Fluoride Concentration in River Waters of South Asia. Current Science, 80, 1312-1319.http://www.iisc.ernet.in/currsci/may252001/1312.pdf

20. Mamatha, S.V. and Haware, D.J. (2013) Document on Fluoride Accumulation in Ground and Surface Water of Mysore, Karnataka, India. Current World Environment, 8, 259-265.

21. APHA, AWWA and WEF (2005) Standard Methods for the Examination of Water and Wastewater. 21st Edition, APHA, Washington DC, USA.

22. BIS (2012) Drinking Water—Specification (ICS 13.060.20) 2nd Edition, Bureau of Indian Standard, New Delhi. http://cgwb.gov.in/Documents/WQ-standards.pdf

23. WHO (2011) Guidelines for Drinking Water Quality. 4th Edition World Health Organization, Geneva, Switzerland.http://apps.who.int/iris/bitstream/10665/44584/1/9789241548151_eng.pdf

Chapter 8

Benefaction and Pyrolysis of Şırnak Asphaltite and Lignite

Yildirım I. Tosun

Mining Engineering Department, Şırnak University, Şırnak, Turkey

ABSTRACT

Depending on advanced technological developments in energy production the low quality coals needed the most economical technologies and even in order to make it possible to produce coalderived products. Compliance with environmental norms of coal pyrolysis or gasification of various type of coals, feasible combustion systems and energy production facilities are needed in today's modern technology, also enable the production of liquid and gaseous coal fuels. However, raw materials and chemical nature of them requires a variety of adaptation methods. This study examined the high sulfur and ash types of Kütahya, Denizli, Aydin, Soma lignite, Sirnak asphaltite and lignite. The representative samples were taken from local areas of the lignites. Fundamentally, the conditions regarding better desulfurization way, the high quality pyrolysis lignite oil production, high value light oil, coal tar and gas products were determined at the goal of high fuel producing yield.

Keywords: Coal Pyrolysis, Desulfurization, Oil Char Production, Semi-Pyrolysis

1. INTRODUCTION

Lignite consumption in energy production is increasing in parallel with growing energy needs today. In terms of the available reserves and production amount, lignite natural resources are limited. The highest amount of coal resources are consumed in energy production in Turkey, as the most economical means [1] [2] . In Turkey approximately 80% (36 million tons) of total lignite production of almost 45 million tons was utilized in primary energy production. The clean coal fuels such as coal liquids, gases will also provide high value-added products for the benefit of the country and even will make a very positive growth and employment. Turkish Energy Ministry promotes to the highest priority researches and developments in coal cleaning and processing for evaluation low qualified coal resources in the southeastern Anatolian local region, for high coal reserves of Sirnak City and production of high quality asphaltite and lignite in Sirnak City. The most effective and cost-effective technologies are needed in today's modern technologies [3] [4] . Municipal authorities also seek to validate the feasibility of production of clean coal fuels such as coal liquids and gases and even injecting, storing and monitoring power station stack gases, CO_2 in the subsurface (geologic storage) as an approach to mitigate atmospheric emissions of CO_2. Coal industries promote technological developments required basic feasibility and the standard qualification tests[5] [6] . There are lots of signs for the co-production electricity from biomasses and lignite in various industrial fields even using irregular low capacity of cellulosic biomass wastes. Processing technologies using biomass should contribute to production of cleaner coal fuels [5] [6] . On the nature and characteristics of the based biomass and lignite, production of high value clean coal fuels should be commonly managed by pyrolysis and gasification of lignite and other cellulosic waste resources in the local region. Even, for this purpose, the country needs the further fuel cleaning technologies providing the essential clean oils and gases. For this reason, Kütahya Gediz, Denizli Çivril, Aydin, Tunçbilek, Soma, Sirnak lignites and Sirnak asphaltite were subjected to coal washing, pyrolysis and gasification and the production of coal liquids and gaseous fuels regarding energy development in the southeastern Anatolian region and further improve the diversification of fuels. Tunçbilek and Aydın washed lignite ash contents of clean coal products presented lower levels. For Tunç- bilek, Aydın lignites, various pyrolysis devices were used. Pyrolysis tests were carried out regarding the oil production ability and the efficiencies in the gasification of carbonized portion of coals, gas and liquid. The coal gas fuel production efficiencies were determined at the optimum values, and optimized pyrolysis parameters of the furnace equipment.

However, lignite-based technologies are examined on the basis of raw materials and contribute to the side of the forest waste biomass as raw material for cellulosic biomass can be processed together with the lignites. Some extent mixing was carried out in order to improve the pyrolysis quality of Sirnak asphaltite and lignite.

2. LITERATURE REVIEW

Considerable research on coal pyrolysis and gasification has been conducted over the years, but the pyrolysis results are widely dispersed because of the complex chemistry of coal [7] [8] . Time related coal-pyrolysis modelling assumes basically first-order kinetic equations, or less sensitive for heating rate. The other distributed activation model is dependent on the heating rate. The last two more advanced models need three and four constants, respectively, which basically depend on the coal properties but also cover to some extent, the effect of heat-and-mass transfer phenomena. That is the reason for the different values of the activation energy and preexponential factor cited in the literature and the lack of generally valid data. The same situation exists in the case of coal-char gasification. The reaction rate of char is influenced mainly by chemical and physical factors, which include coal type, catalytic effect of the ash and the specific surface area of char, which changes during the reaction course with the development of internal pores, and finally, their destruction [9][10] . In the case of the scaling-up procedure, the uncertainty of a complex model of the reacting system may be very high and it is reasonable under some conditions to use a methodology based on quasi-equilibrium conditions, which can be reflected at a larger scale. This approach assumes basically that the process itself, with all specific features, is a decisive factor for the path of the reactions of coal decomposition. Therefore a static model of char gasification and coal pyrolysis was developed: it is based on the assumption that the final process temperature is a decisive factor for the required volatile-matter content in the char being in a quasi-equilibrium state with respect to the gas temperature [11] [12] . To achieve this, it is necessary to create conditions of internal circulation of the transported coal and char in a riser, where the average concentration of solids amounts to 0.05 - 0.15 $m^3 \cdot m^3$, i.e. the conditions for residence time are long enough for the thermal decomposition of coal and intensive mixing so enhancing mass and heat transfers. On the other hand, the gasification of circulated char is mainly chemically controlled due to the relatively short residence time of the gas. Additionally, the characteristic feature for a given circulating fluid-bed reactor is a very narrow range of operational residence time, both for the gas and coal, which allows us to use the quasi-equilibrium approach. The model developed reflects the kinetic processes (transport phenomena of coal species, reaction rate) by constants of the developed equations, which depend mainly on the coal properties.

3. COAL PYROLYSIS IN A CIRCULATING FLUIDISED BED

A circulating fluidised-bed reactor was used as a coal pyrolysis reactor. The process was tested at a scale of 1 - 2 kg/h: collecting operational and design data to build an industrial installation. A technological diagram of the coal-pyrolysis process development unit is presented in Figure 1. As illustrated in Figure 1, where 1st unit is coal storage, 2nd unit is pyrolysis reactor, 3rd unit is coarse char cyclone, 4th unit is fine char cyclone, 5th unit is coarse char bag, 6th unit is circulating coal recycle.

In this research, representative specimens the different types of Turkish lignites; Kütahya Gediz, Denizli Çivril, Aydın, Tunçbilek, Soma, Şırnak asphaltite and lignite coals, were crushed and comminuted to minus 1 mm size by controlled screening. Air dried samples 1 - 2 kg from each different coal types were prepared and sealed in nylon bags. Proximate analysis of Turkish lignites and coals used in the experiments are given in Table 1. The qualities of processed coal products are ascertained by chemical and standard coal analysis of ASTM 3173- 317

4. RESULTS AND DISCUSSION

Screen analysis of Turkish lignite samples were made by standard Tyler Screens and particle size distributions and normal distributions of lignite samples are respectively illustrated in Figure 2. Coal particles in lignite samples were mainly distributed between 100 and 300 μm size fractions. High specific surface areas of 1160 cm^2/gr for lignite samples were found and also highly sufficient in order to react with alkali.

In the pyrolysis experiments with addition hydrated lime, reactor temperature changed between 100°C and 400°C and lignite samples mixed only by lime and soda and ash addition at equal rates. Produced coal specimens were subjected to chemical analysis for sulfur hold-up determination. Test results of pyrolysis by lime and addition of Turkish lignite are seen in Figure 3. Some Turkish lignite could not easily be desulfurized so that ash and sulfur contents of these lignite samples were high and further washing would improve pyrolysis results. Higher temperatures over 350°C decreased desulfurization rate due to high volatile lost from coal samples. Lime might not be dissolved in soda mixture in the concern of nonpenetrating and lower reacting by coal surfaces so that resulted in lower desulfurization rates.

From the point of view of pyrolysis experimentation, Sirnak asphaltite calorific value was significant. That char quantity in the experimental conditions is calculated on the basis of the gas composition in the ambient

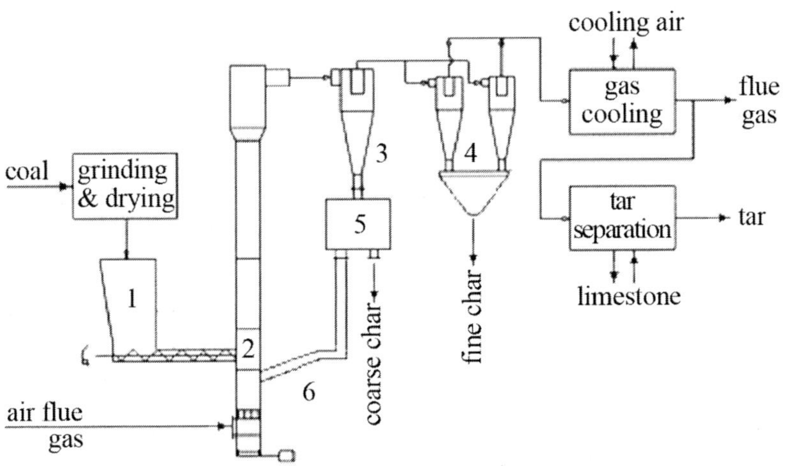

Figure 1. Fluidized coal-pyrolysis process.

Table 1. Proximate analysis of Turkish lignites (ADB: Air dried base; DB: Dried base; DAB: Dried ashless base).

Coal Type	Ash, % ADB	Moisture, % ADB	Total S, % DB	Volatile Matter, % DAB
Şırnak Asphaltite	42.3	0.1	7.1	62.6
Şırnak Lignite	24.3	18.1	3.1	52.6
Aydın Şahinali	19.3	8.1	1.1	49.6
Kütahya Gediz	22.0	1.7	3.6	42.7
Denizli Çivril	15.2	12.7	3.0	45.2
Tunçbilek	25.7	5.7	3.9	38.2
Soma Kisrakdere	13.8	14.0	2.2	40.4

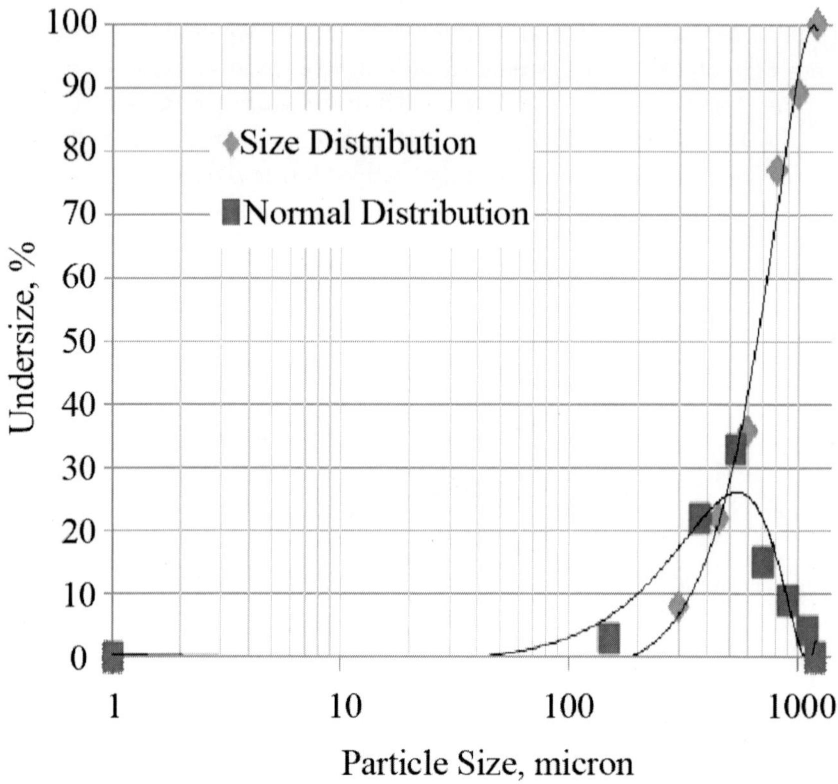

Figure 2. The particle size distributions and normal size distributions of Turkish lignites.

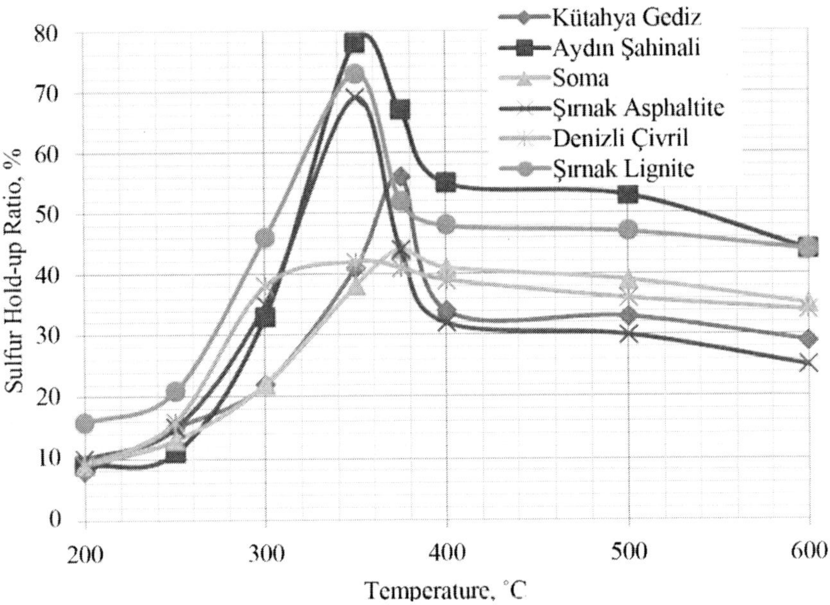

Figure 3. The sulfur hold-up rates in pyrolysis of Turkish lignites by hydrated lime and soda.

state. So neither the contained water vapor nor the condensing hydrocarbons are taken into account. Pyrolysis liquid and gaseous products of Sırnak asphaltite may equal to 5 - 20 g·m^3 tar and 5 - 10 g·m^3 of benzene, toluene, xylene in unit process gas. These components increase the gas calorific value by 320 - 1200 kJ and, in total, allow one to obtain a gas of calorific value of up to approximately 6000 kJ·m^3 for Sırnak asphaltite (Figure 4). Char yield of Şırnak asfaltite and the other lignites increased at lower pyrolysis temperatures.

Test results of pyrolysis by the marble waste on desulfurization are seen in Figure 5. The specimens of Sırnak asphaltite and lignite, Soma lignite were efficiently desulfurized by marble waste (6% MgO). Semi pyrolysis and pyrolysis medium with marble waste decreased sulfur contents of chars and liquid products.

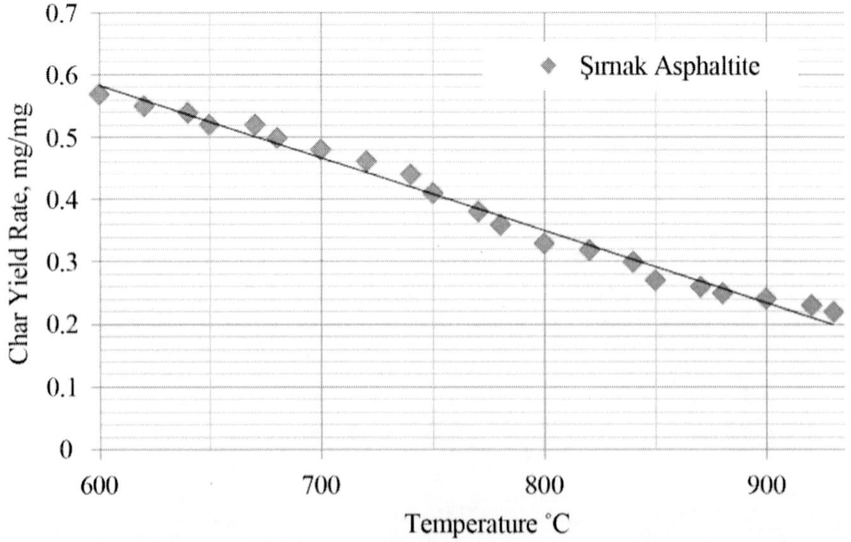

Figure 4. Comparison of experimental data and calculated results of char yield at pyrolysis temperatures.

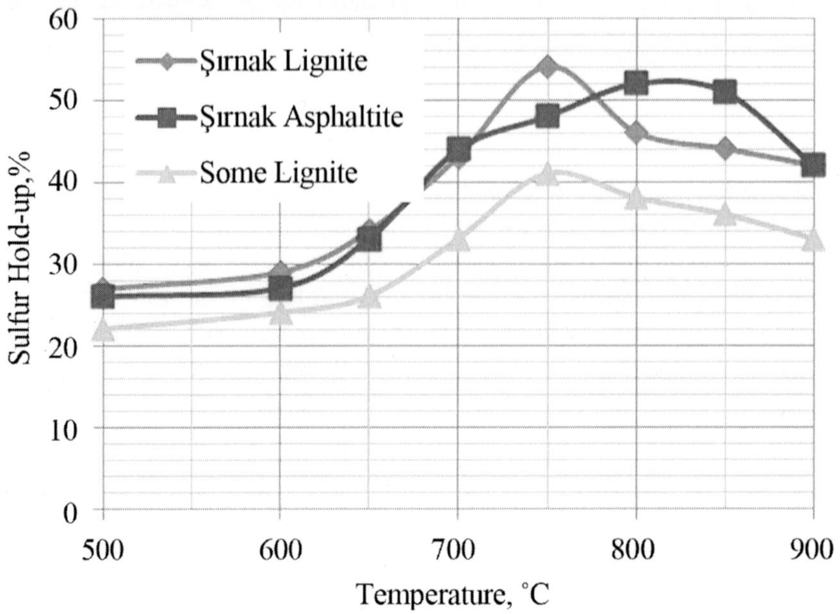

Figure 5. Sulfur hold-up rates in pyrolysis of Turkish lignites by marble waste.

5. CONCLUSIONS

Research institutions mainly should search characteristics of lignite-to-date processes and improve technological applications and even advance processes to carry out performance tests, depending on the nature of the raw materials necessary to develop efficiencies.

Pyrolysis of different types of Turkish lignite was successfully processed in terms of deashing and desulfurization.

At higher rates of desulfurization of different types of Turkish lignites could be obtained from the tests using alkali reagents and hydrated lime mixtures. It has been clearly determined that lime and hydrated lime are much beneficial in caustic leaching of different types of Turkish lignites. It was also distinctly ascertained that marble waste fine containing approximately 6% magnesia was much efficient as a desulfurizing agent in pyrolysis. Magnesia content was efficient as hydrous lime but much more advantageous in absorption stage of gaseous medium.

Şırnak asphaltite should be cleaned and high ash content reduced char and oil yield in pyrolysis. Benefaction from Turkish lignites in the various combustion systems, in order to receive clean energy with clean coal liquid and gaseous products must be generated in lower temperature pyrolysis. It is also adviced that the high amount of formation of flue gas will be managed at higher pyrolysis temperatures over 700°C and even extracts more environmental friendly gaseous products. Also in the research works production of clean energy with the design of the addition of high-quality coal and biomass waste mixtures are processed and biomass fuels could be an alternative clean fuel sources. Clean coal fuels produced by pyrolysis may be supplied in southeastern Anatolian region in Turkey. Hence, those clean alternative resources will further enhance the industrial development in the region. For this purpose, further washing of Kütayha Gediz, Denizli Çivril, Aydın, Tunçbilek, Soma lignites, Sirnak asphaltite and lignite may improve pyrolysis products. Thus, the higher performance can be achieved by a certain mixture of solid fuel additives. A preferable advanced design in pyrolysis could produce clean coal fuels in the local site of the country so significant that needs to obtain the highest quality coal fuels.

REFERENCES

1. Anonymous (2009) TKİ Lignite Report 2009, TC Energy Ministry Reports. Ankara, Türkiye.

2. Anonymous (2009) TTK Coal Report 2009, TC Energy Ministry Reports. Ankara, Türkiye.

3. Boynton, R.S. and Gutschick, K.A. (1983) Lime, Industrial Minerals and Rocks. AIME, New York

4. Culfaz, M., Ahmet, M. and Gürkan, S. (1996) Removal of Mineral Matter and Sulfur from Lignites by Alkali Treatment. Fuel Processing Technology, 47, 99-109.

5. Reimers, G.W. and Franke, D.W. (1991) Effect of Additives on Pyrite Oxidation, RI: 9353. Bureau of Mines, Washington DC. [Paper reference:2]

6. Tosun Y.I., Rowson, N.A. and Veasey, T.J. (1994) Bio-Column Flotation of Coal for Desulfurization and Comparison with Conventional and Column Flotation. 5th International Conference of Mineral Processing, Nevşehir, 1994. [Paper reference:2]

7. Wheelock T.D. (1979) Chemical Cleaning, Coal Preparation. 4th Edition, AIME, New York.

8. Yoon, R.H. (1991) Advanced Coal Cleaning, Part 2, Coal Preparation. 5th Edition, AIME, Colorado.

9. Donskoi, E. and McElwain, D.L.S. (1999) Approximate Modelling of Coal Pyrolysis. Fuel, 78, 825-835.

10. Wiktorsson, L.P. and Wanzl, W. (2000) Kinetic Parameters for Coal Pyrolysis at Low and High Heating Rates—A Comparison of Data from Different Laboratory Equipment. Fuel, 79, 701-716.

11. Fu W.-B. and Wang, Q.-H. (2001) A General Relationship Between the Kinetic Parameters for the Gasification of Coal Chars with CO_2 and Coal Type. Fuel Processing Technology, 72, 63-77.

12. Liu, G., Benyon, P., Benfell, K.E., Bryant, G.W., Tate, A.G. and Boyd R.K. (2002) The Porous Structure of Bituminous Coal Chars and Its Influence on Combustion and Gasification under Chemically-Controlled Conditions. Fuel, 79, 617-625.

Chapter 9

IR Spectroscopic Characterization of Lignite as a Tool to Predict the Product Range of Catalytic Decomposition

Valentin Cepus*, Markus Borth, Mathias Seitz

Department Engineering and Natural Sciences, Merseburg University of Applied Sciences, Merseburg, Germany

ABSTRACT

The catalytic pyrolysis of lignites is a technical process whose development is complex and time- consuming with the goal to maximize the yield of the desired low-volatile hydrocarbons of choice and to minimize the yield of solid residual products. Not every type of lignite is suitable for this process due to its particular chemical composition. In order to be able to predict which lignite specimen will be an especially promising raw material for the pyrolytic liberation of target products, the chemical classification by IR spectroscopic methods was investigated. MIR spectroscopy has been demonstrated to be a valuable tool to characterize the the molecular composition of lignites and to determine the concentrations of aliphatic and aromatic functional groups in lignite as well as alcoholic OH and other forms of bound oxygen. These data provide a comprehensive chemical characterization of the material and help to predict the composition of the chemical components liberated by catalytic decomposition. With a complementary NIR spec- troscopic approach, a chemometric method has been developed with which the elemental com- position of the lignites can be determined in a fast and pragmatic way leading to a prediction of the product range of a theoretical pyrolytic product range. Thus, this spectroscopic investiga- tion is a toolbox that can answer the question if the commercial exploitation of catalytic pyro- lysis of a lignite sample in question will make sense without preliminary conduction of long and time-consuming testing.

Keywords: Coal, Lignite, IR Spectroscopy, Characterization, MIR, NIR, Pyrolysis

1. INTRODUCTION

The raw materials coal and lignite contain an enormous potential for the supply of precious products in the chemical industry. The conversion of coal, which is a complex mixture of compounds, by a catalytic pyrolytical process yields the possibility to liberate liquid and gaseous hydrocarbons among other types of compounds that can be applied with high technical versatility. The conversion of coal and lignite by combination of chemical processes will thus be important for the future supply with educts and intermediates in the chemical industry. These special applications require optimized variants of combinations of the reaction system catalyst-coal for the efficient substantial exploitation. The following results and discussions have been collected by catalytic conversion of southeastern German lignites. From former experimental investigations, it is known that the catalyst zeolith ZSM-5 releases a maximum yield of unsaturated low-molecular hydrocarbons [1] [2] .

In order to predict the product spectrum of such a catalytic conversion, it is important to characterize the lignites as raw materials of these processes. This has been achieved by infrared spectroscopic (IR) investigations with local lignites. The evaluation comprises a well-defined choice of representative lignites from open-cast mining “Vereinigtes Schleenhain” and “Amsdorf”. For a first classification, the division into light and dark lignites is proved to be appropriate. Middle infrared (MIR) spectroscopy permits the definition of constitutional preconditions for suitable types of lignites. Besides the questions of raw material quality, the knowledge of the correlation between the IR spectrum of lignite tar and lignite coke with the chosen process parameters is of vital importance. Furthermore, the correlation between the quantitative determination of elemental composition with characterization in near infrared (NIR) has to be investigated. By identification of the characteristic spectroscopic information, the properties of the raw materials and anticipated product yield and quality after pyrolysis should thus be employed for an optimized exploitation of lignites.

2. MATERIALS AND METHODS

IR spectroscopic investigation of lignite in MIR (middle infrared) range required some sample preparation as pulverization in a mortar. 5 to 8 mg lignite was mixed with 1 g of KBr (for IR spectroscopy) and milled in an agate mortar. 5 spatula of this mixture were put into a 13 mm pelleting tool and pressed for 10 min at a load of 10 tons with a hydraulic press. For the preparation of tar 15 mg of this highly viscous liquid were dissolved in 2 ml of dichloromethane. 1 g of KBr were mixed with 300 µl of the dichloromethane solution, milled in an agate mortar, heated for 2 min in an oven at 50°C before pressing to a pellet as described above [3] .

IR measurements were conducted with a Thermo Nicolet Avatar 360 FT-IR spectrometer with DTGS detector. The IR spectra were recorded between 4000 and 400 cm^{-1} with a resolution of 8 $cm^{-1}$and coaddition of 32 scans in transmission mode. Acquisition of all measurements was conducted in triplicate. The recorded spectra were converted into absorbance spectra and baseline corrected. Data processing and calculations were performed with the spectroscopy software "OMNIC". The fitting processes were employed as described elsewhere [4] .

For NIR (near infrared) spectra the sample preparation was different. Pure samples and calibration mixtures were milled in a mortar and homogenized in a shaker for 20 min. The spectra were recorded on a Thermo Nicolet Nexus NIR spectrometer with a fibre-optic probe head in diffuse reflexion mode and an InGaAs detector between 4000 and 10,000 cm^{-1} at a spectral resolution of 8 cm^{-1} with accumulation of 128 scans and measured in triplicates. Calculation of the fit model with PLS regression was conducted using the software TQ Analyst.

To determine the hydrocarbon yield of different lignite samples with different compositions, a discontinuous rotary kiln was used, described elsewhere [1] . At standard reaction conditions (inert, 400°C, ZSM-5 powder as catalyst with modulus 240 and a catalyst to lignite (maf) ratio of 1:20), a mixture of lignite and catalyst was charged into the pre heated reactor. The volatile products pass a filter where tar is deposited. Condensable products were detained in two traps cooled at 0°C and −196°C. The product gases and liquids streams were measured and analyzed by IR spectroscopy and gas chromatography. The yield of hydrocarbons in the range of one to nine carbon atoms was used as a measure to quantify high quality products.

3. RESULTS AND DISCUSSION

3.1. Characterization of Raw Lignites with MIR Spectroscopy

The comparison of FTIR spectra of different lignite samples from open-cast mining "Schleenhain" (Figure 1) shows several spectroscopic features that can be found in all IR spectra. These comprise a broad O-H-stretching vibration at 3400 cm^{-1} of hydrogen bridges, C-H-stretching vibration of methylene groups between 3000 and 2800 cm^{-1}, C=O and C=C stretching vibrations around 1700 and 1600 cm^{-1}, and C-O-C/C-O vibrations at 1280/1090 cm^{-1}, respectively [5] . In contrast to this only in a part of the measured IR spectra the signal of the SiO-H and Si-OH stretching vibrations around 3700 to 3600 cm^{-1} and 1040 cm^{-1} can be detected [6] - [8] . Although the basic structure element of lignites is based on aromatic rings, no indication of an absorption band between 3100 to 3000 cm^{-1} of the aromatic C-H stretching vibration is found which gives a hint to a network structure in which most of the aromatic hydrogen atoms are substituted.

The comparison between the light and dark lignite samples shows that light lignites exhibit a more pronounced absorption between 3000 and 2800 cm^{-1}, revealing more aliphatic methylene groups than dark lignites. In contrast to this dark lignites show a stronger absorption at 1600 cm^{-1} which can be referred to aromatic C-C double bonds. In the lower wave number region light lignites show an absorption band at 720 cm^{-1} that is caused by out of plane deformation of at least 4 consecutive methylene groups hinting to longer-chain aliphatic units for light samples in contrast to dark samples where this absorption is not identified.

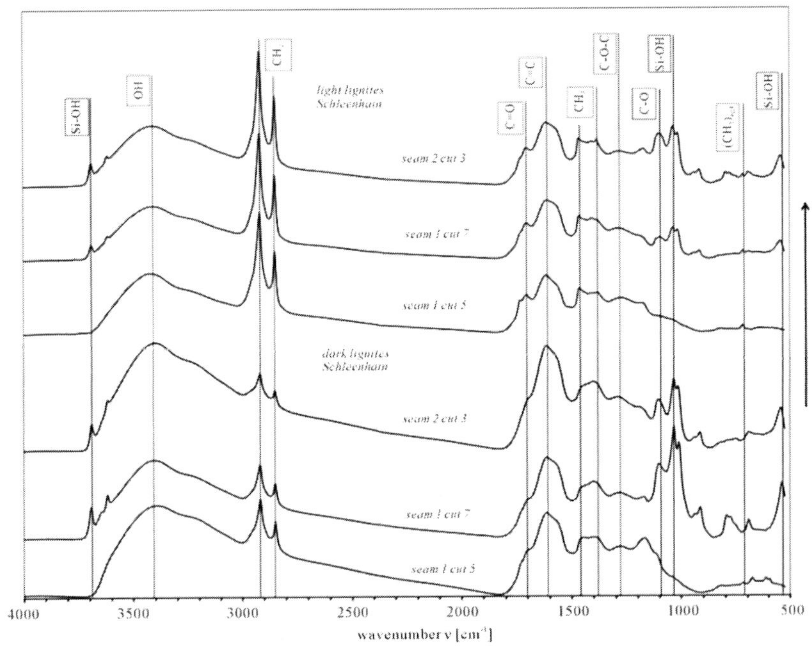

Figure 1. FTIR spectra of lignite from open cast mining "Schleenhain", comparison of light and dark Southeastern German lignites from different seams.

A closer investigation has been conducted with respect to the range of the aliphatic C-H stretching vibrations in the range between 3000 and 2800 cm^{-1} by deconvolution of the IR spectra as shown in Figure 2. The asymmetric and the symmetric CH_2 stretching vibrations at 2921 cm^{-1} and 2850 cm^{-1} lead to absorption bands with the highest observed absorbances. The resulting values for the absorbance of light lignites are twice as high as that of the dark lignites (Figure 3). The strong absorption bands of light lignites correlate with a higher observed yield of volatile hydrocarbons by pyrolysis of light lignites. Thus, a tendency towards higher alkane and alkene yield can be concluded from the presence of strong bands of CH_2 stretching vibrations. A more quantitative evaluation of the pyrolytic process can be achieved by integration of the areas of the deconvoluted bands at 2850

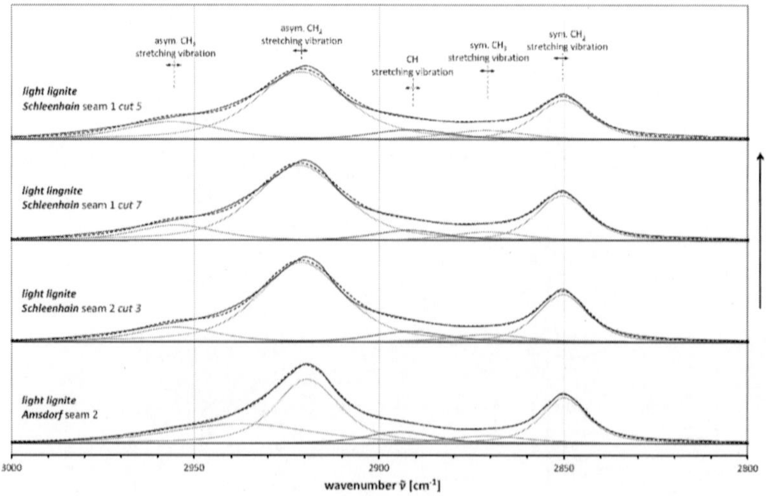

Figure 2. FTIR spectra of light lignite with Gauss-Lorentz deconvolution function bands contributing to the spectrum of the sample. Total spectrum ▬, deconvoluted spectrum ▬ ▬ ▬, methylene bands ▬, methyl bands ▬, methine bands ▬.

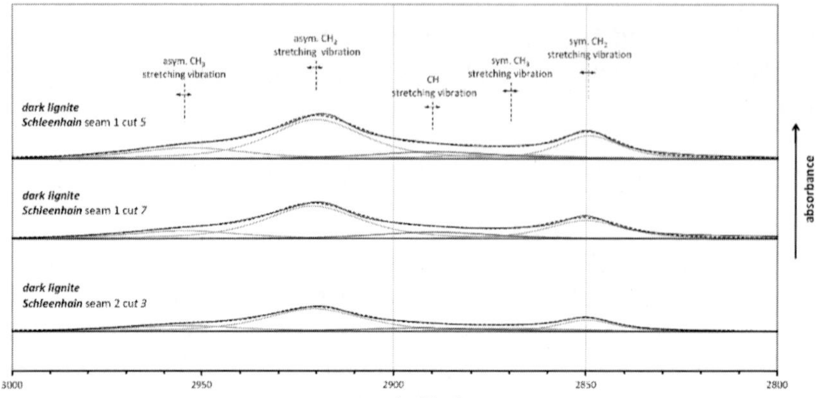

Figure 3. FTIR spectra of dark lignite with Gauss-Lorentz deconvolution function bands contributing to the spectrum of the sample. Total spectrum ▬, deconvoluted spectrum ▬ ▬ ▬, methylene bands ▬, methyl bands ▬, methine bands ▬.

cm^{-1} and 2921 cm^{-1}. For the light lignite "Schleenhain" of seam 2, the calculated areas of the spectrum hint to the highest yield among all investigated lignites and are in agreement with experimental results by pyrolysis. A yield of 11.8% waf (referring to water and ash free lignites) of saturated and unsaturated hydrocarbons in the range of one to nine carbon atoms could be achieved by pyrolysis of this type of lignite. In contrast to this by the conversion of dark lignite "Schleenhain" of seam 2 only a yield of 3.7% waf could be isolated.

In addition to the relatively broad absorption bands of methylene in this wavenumber interval, several relatively broad bands of asymmetric and symmetric CH_3 stretching vibrations at 2955 cm^{-1} and 2870 cm^{-1} can be observed. It can be derived that CH_3 absorption bands of light lignites are slightly stronger than of dark lignites. The absorbance of methine groups at 2890 cm^{-1} hints to the degree of branching of aliphatic molecular contributions in the lignite structure.

In the spectral region between 1800 and 1000 cm^{-1} a multitude of absorption can be detected which can be roughly assigned to categories of specific functional groups.

In Figure 4 this lower specific spectral region is explored for light lignites. For the characterization of structural changes of carbonyl compounds the fitted bands between 1645 - 1750 cm^{-1} are investigated [5] . The band pattern of all lignite spectra show at least three pronounced absorption bands of the carbonyl groups. Especially for light lignites a band can be detected at 1735 cm^{-1} that can be referred to aromatic esters. The band at 1705 cm^{-1} with the highest intensity in this region hints to the presences of carboxylic functions. Especially for the lignites from "Amsdorf" a rather complex structure of carbonyl compounds is expected due to the numerous bands that could be differentiated by deconvolution for this sample. The absorbance at 1650 cm^{-1} has been tentatively assigned to different aromatic ketones. The determination of aromatic structures has been supported by observations in the spectral region between 1500 - 1630 cm^{-1} standing for the C=C stretching vibration. Especially the absorption at 1615 cm^{-1} exhibits a stable value and in conjugation with C=O this band becomes strong and broad. By increasing carbonization there is the tendency of an increasing absorbance of this band which furthermore correlates with a higher coke yield. For the wavenumber region between 1365 and 1475 cm^{-1} spectra are recorded with characteristic signals of aliphatic C-H bending vibrations. The absorbance bands at 1460 cm^{-1} and 1375 cm^{-1} can be assigned to bending vibrations of the CH_3 group. In comparison to the bands of CH_2 bending relative ratios can be calculated to explain the degree of branching.

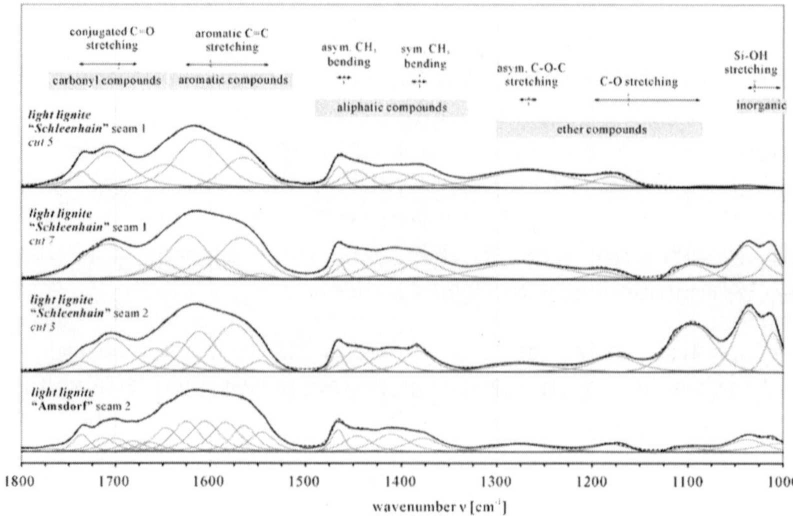

Figure 4. FTIR spectra of light lignite with Gauss-Lorentz deconvolution functions contributing to the spectrum of the sample. Total spectrum ▬., deconvoluted spectrum ▬ ▬ ▬, contributing bands of functional groups ▬.

3.2. MIR Characterization of Pyrolysis Products

Furthermore, the IR spectra of the raw lignite samples and the products of the pyrolytic process, coke and tar, are shown in Figure 5. The comparison between lignite and coke shows a noticeable agreement with respect to the pattern of the aromatic absorption bands. The intensity of the band of the carboxylic functional group at 1705 cm^{-1} is remarkably weaker in the spectrum of tar than in the spectrum of lignite and is hardly observable in the spectrum of the coke. Thus, we can determine a lower amount of carbonyl and carboxylic groups in tar than in lignite. Similar band patterns that have been recognized in IR spectra of the utilized lignite samples are also observable in the IR band pattern of the products. By comparison of these spectra a successful assignment of the functional groups can be achieved. In the spectral region between 1500 and 1630 cm^{-1} the characteristic stretching vibrations of the C=C groups highlight the complex aromatic constitution of coke.

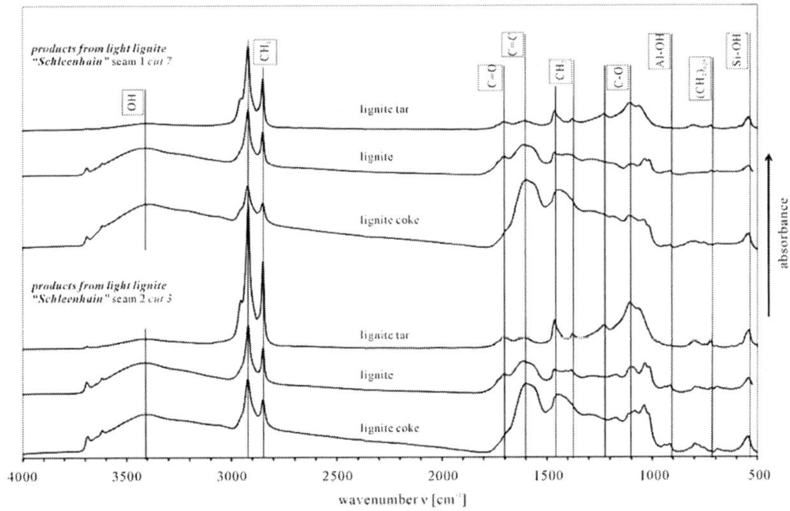

Figure 5. FTIR spectra of tar and coke from catalytic conversion of light lignites "Schleenhain" with zeolith ZSM-5, com- parison with products and utilized raw lignites.

This individual band pattern represents the highly aromatic state of the released coke, which is in agreement with the chemical structure documented in literature [9] . In contrast to this, the IR spectra of tar only show a broad band with low intensity at 1610 cm^{-1}, hinting to low amounts of aromatic compounds. The IR absorbance bands below 1500 cm^{-1} reveal the aliphatic CH bending vibrations. At 1465 cm^{-1} the asymmetric CH_3 bending vibration can be observed [8] . Due to the increase of this band it can be derived that tar from lignite of seam 2 contains a higher concentration of CH_3 groups. In general, the absorption of this CH_3 bending vibration appears as narrow intensive band in the spectrum of tar and lignite. As an additional spectral feature, the intensive and broad absorption in the region between 1050 and 1150 cm^{-1} is dominated by C-O stretching vibrations. Due to the simultaneous presence of the O-H stretching vibrations at 3400 cm^{-1} these bands can be assigned to the absorption bands of secondary alcohols. Thus, lignite tar contains a higher concentration of secondary alcohols than lignite coke. Furthermore it can be derived that the concentration of secondary alcohols in tar from lignite of seam 1 is higher. For the further characterization of phenol containing compounds the absorption at 1170 cm^{-1} has been investigated. This band is very pronounced for the lignite sample of seam 2. The conversion products also exhibit this absorption band, which is, however, a bit lower than for lignite. With rising concentration of phenol in the aqueous phase a decrease of this band can be observed in the IR spectrum of coke. For the band at 1035 cm^{-1} of the Si-OH stretching vibration which hints to the presence of inorganic silicate minerals, similar intensity of the signals are observable for the IR spectra of coke and the respective lignite sample out of which the coke was converted [10] . The lack of this band in the IR spectra of tar shows that there is no silicate mineral present in the produced tar.

3.3. NIR Chemometric Characterization of Raw Lignites

As NIR spectroscopy is a powerful tool for the chemometric characterization of complex mixtures this technique seems to be adequate for the aim of a fast method for evaluation of lignites being a promising raw material for catalytic conversion [11] . A time consuming approach for such a characterization is the elemental analysis of C, O, and H in raw lignites. For this purpose, a calibration has been conducted of 30 individual samples mixed with 5 different reference lignite samples that contain known concentrations of the named elements. For the chemometric calculations, all calibration samples were measured in triplicate in order to obtain an appropriate amount of data points. A set of 30 typical exemplary NIR spectra of investigated calibration samples are recorded of which one example is shown in Figure 6.

For the calculation of the concentration of carbon and hydrogen the whole set of untreated NIR spectra of the calibration references was utilized. Primarily the spectral region between 4310 and 5880 cm^{-1} contained the necessary information. This region is dominated by the absorption of the CH_3 groups as well as by overtones of the C=O group. In Figure 7 and Figure 8 the correlation diagrams between the

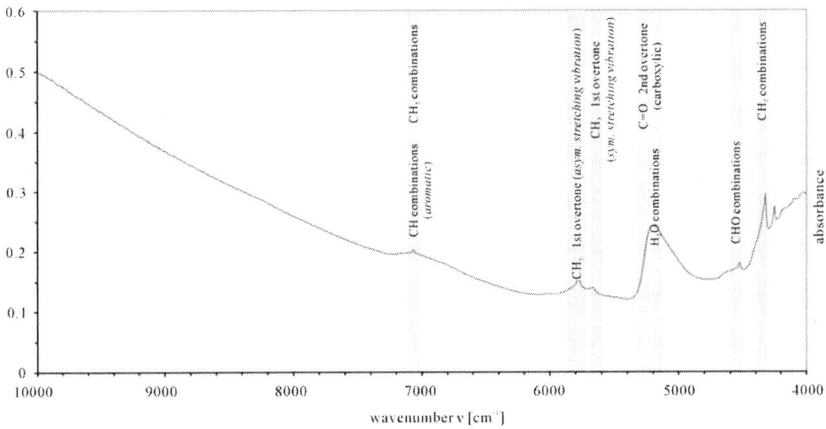

Figure 6. NIR spectrum of lignite calibration mixture based on dark lignite "Schleenhain" from seam 1 cut 7, example.

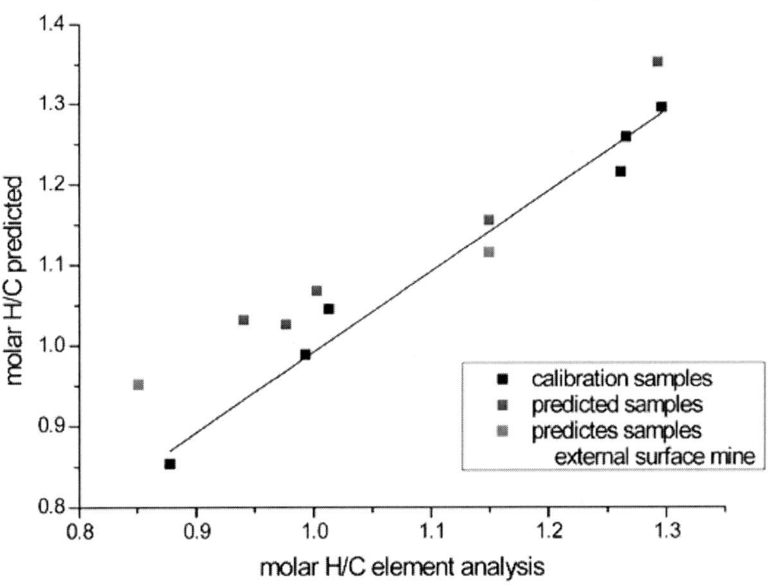

Figure 7. Correlation plot of the prediction of the molar H/C ratio by the PLS fit model based on NIR spectroscopic data with reference values obtained by elemental analysis for the calibration mixture based on dark lignite "Schleenhain" from seam 1 cut 7.

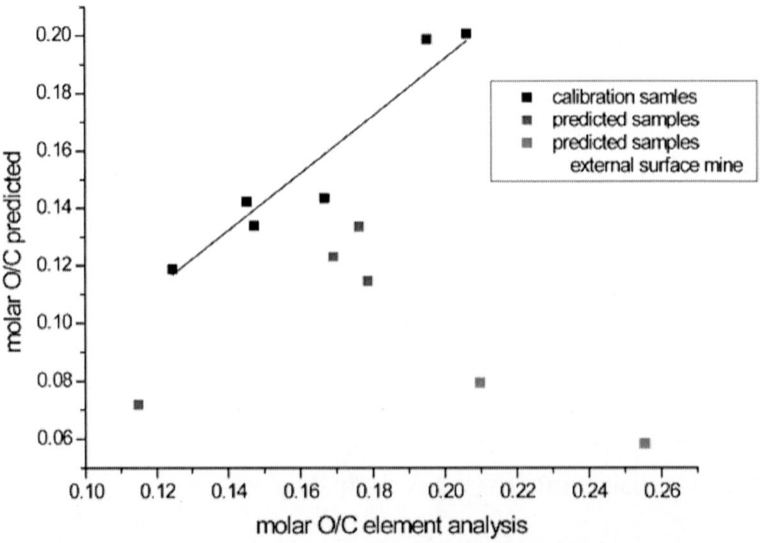

Figure 8. Correlation plot of the prediction of the molar O/C ratio by the PLS fit model based on NIR spectroscopic data with reference values obtained by elemental analysis for the calibration mixture based on dark lignite "Schleenhain" from seam 1 cut 7.

NIR spectroscopic determination of the elemental composition and the results of the elemental analysis are shown. It can be revealed that the elemental composition of lignites can be predicted by an NIR based calibration model. Especially for the elements C and H the prediction is accurate. However, it can be observed that the current fit model leads to an underestimation of the oxygen content, especially for the external samples. This difference can be explained by the fact that for the elemental analysis the determination of oxygen is a calculated result based on the balance calculation referring to the elemental analysis of C, N, H. Furthermore, the exact water content of lignites might affect the absorbance at 5200 cm^{-1}. This wavenumber is used for the determination of oxygen bound in carbonyl form.

4. CONCLUSION

Prediction of the Hydrocarbon Yield

For the production of hydrocarbons, the hydrocarbon yield of the catalytic pyrolysis of lignites depends strongly from the molar H/C and O/C ratio. Especially, lignites with a high H/C ratio and a low oxygen content show the highest amount of desired product, while special types of lignites that are oxygen rich and hydrogen poor release only few liquid and gaseous products. As only few extraordinary lignites occur in nature, the hydrocarbon yield can be described only as a function of the H/C ratio (Figure 9). Thus, the hydrocarbon yields of lignites with typical H/C ratios can be predicted well by NIR spectroscopy, even if the molar O/C ratio cannot be determined. As a result, NIR measurements can help to decide quickly, whether the lignite is suitable for hydrocarbon production with high yields of high-quality pyrolysis products.

5. SUMMARY

The catalytic pyrolysis of lignites requires a technical refinement process with the goal to maximize the yield of desired low-volatile hydrocarbons and to minimize the yield of solid products. In order to achieve this target, the chemical composition of the incoming raw materials was characterized with MIR spectroscopy. As a complementary approach, the NIR chemometric determination of the elemental composition of carbon, hydrogen, and oxygen was established. It could be demonstrated that it is directly correlated to the product distribution of the pyrolytic products. For this, the chemical and elemental composition of the raw materials and the reaction products MIR and NIR methods have been developed that are applicable in a simple and pragmatic way.

Characteristic absorptions in the MIR spectra of the investigated raw lignites reveal the molecular constitution, for instance the aromatic and aliphatic character of the hydrocarbon backbone and important functional groups

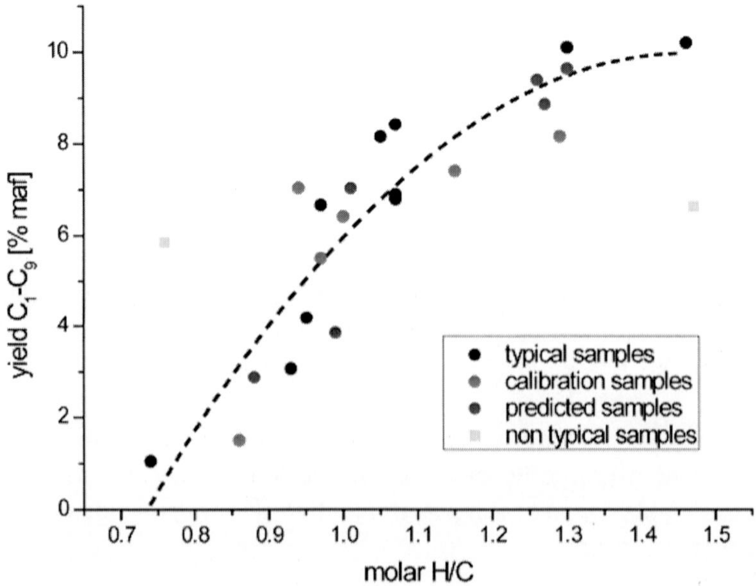

Figure 9. Low molecular hydrocarbon yield dependent of the molar H/C ratio for the NIR calibration samples, the predicted samples by the PLS fit model and all samples investigated in pyrolysis; extraordinary lignites occur as outliers.

such as ketones, carboxylic acids, and alcohols. The characterization of the raw lignites showed that light lignite types contained a higher amount of aliphatic groups in contrast to dark types that showed a higher concentration of aromatic groups. This fact helps explaining the observed product spectra by catalysed pyrolytic conversion and helps to select lignite types that are promising with respect to commercially viable product yields upon catalytic decomposition.

Using the MIR spectroscopic toolbox for the understanding of the molecular constitution, also the products of the pyrolytic conversion process have been investigated. The characterization of the obtained lignite tars showed that they consist largely out of aliphatic hydrocarbons and of secondary alcohols and phenols. In contrast to this, the obtained coke products contain the majority of aromatic compounds as released by lignite. It can be shown that from lignite to coke the complexity of the aromatic structures increases. MIR characterization of raw lignites and of the pyrolysis products tar and coke can be employed as a straightforward tool for the understanding of the molecular constitution suitable for industrial use and for screening and prediction of lignites of high commercial value for catalytic conversion with zeolith ZSM-5.

By a chemometric approach based on an NIR spectroscopic calibration of samples with a known elemental composition of carbon, hydrogen and oxygen, the ratio of the concentration of hydrogen to carbon can be successfully predicted. The use of a fibre-optic ATR probe head leads to a very simple acquisition of NIR spectra. The time requirement for the duration of the measurement including milling and preparation process as well as result calculation does not exceed 10 minutes for one sample and can thus be considered as relevant and suitable for industrial application and replaces the time-consuming effort of conventional elemental analysis yielding to a substantial prediction of expectable hydrocarbon products by catalysed pyrolysis.

ACKNOWLEDGEMENTS

The authors gratefully acknowledge the financial support from Federal Ministry for Education and Research, Germany, and the MIBRAG for allocating the lignites with different elemental compositions, used in this work

REFERENCES

1. Seitz, M., Heschel, W., Nagler, T., Nowak, S., Zimmermann, J., Stam-Creutz, T., et al. (2014) Influence of Catalysts on the Pyrolysis of Lignites. Fuel, 134, 669-676.

2. Chodak, M., Niklińska, M. and Beese, F. (2007) The Use of Near Infrared Spectroscopy to Quantify Lignite-Derived Carbon in Humus-Lignite Mixtures. Journal of Near-Infrared Spectroscopy, 15, 195-200.

3. Oikonomopoulos, I., Perraki, T. and Tougiannidis, N. (2010) FTIR Study of Two Different Lignite Lithotypes from Neocene Achlada Lignite Deposits in Greece. Bulletin of the Geological Society of Greece, 5, 2284-2293.

4. Schmalfeld, J. (2008) Die Veredlung und Umwandlung von Kohle: Technologien und Projekte 1970 bis 2000 in Deutschland. Deutsche Wissenschaftliche Gesellschaft für Erdol, Erdgas und Kohle e.V.

5. Ibarra, J., Munoz, E. and Moliner, R. (1996) FTIR Study of the Evolution of Coal Structure during the Coalification Process. Journal of Organic Geochemistry, 24, 725-735. [Paper reference:2]

6. Ibarra, J., Royo, C., Monzón, A. and Santamaría, J. (1995) Fourier Transform Infrared Spectroscopic Study of Coke Deposits on a Cr2O3-Al2O3 Catalyst. Elsevier Science, 9, 191-196.

7. Pretsch, E., Clerc, T., Seibl, J. and Simon, W. (1990) Tabellen zur Strukturaufklarung organischer Verbindungen mit spektroskopischen Methoden. 3. Auflage, Springer-Verlag Berlin Heidelberg, New York.

8. Ramaswamy, K. and Venkatachalapathy, R. (2006) Infrared Spectroscopic Study of Neyveli Lignite Samples from Mine Cut-I and II. Spectroscopy Letters: An International Journal for Rapid Communication, 24, 759-778.

9. de Aragao, B.J.G. and Messaddeq, Y. (2008) Peak Separation by Derivative Spectroscopy Applied to FTIR Analysis of Hydrolyzed Silica. Journal of the Brazilian Chemical Society, 8, 1582-1594.

10. Cerny, J. (2007) Aliphatic C-H Bond Responses in the 900 - 700 cm-1 Region of the FTIR Spectra of Coal Tars. Fuel Science and Technology International, 13, 807-818.

11. Seitz, M., Nagler, T., Welscher, J., Nowak, S., Zimmermann, J., Hahn, T. and Schwieger, W. (2014) Catalytic Cracking of Lignites. Erdol, Erdgas, Kohle, 130, 80-84.

Chapter 10

Coal Washing Scenario in India and Future Prospects

R. Venugopal1 • J. P. Patel2 • C. Bhar3

[1] *Dept. of Fuel and Mineral Engineering, Indian School of Mines, Dhanbad, Jharkhand 826004, India*

2 Washery Division, Bharat Coking Coal Limited, Saraidhela, Dhanbad, Jharkhand 828127, India

3 Dept. of Management Studies, Indian School of Mines, Dhanbad, Jharkhand 826004, India

ABSTRACT

Coal Washing Exploration in India dates back to 1900s; though, first coking coal washeries in India were installed after independence. At present, most of the coking coal washeries are owned by Public Sector Companies; whereas, most of the non-coking coal washeries are owned by Private Sector. Even after six decades of coal washing practices, there has not been significant development in the coal washing intelligentsia. Indian Coal Washing industry is still dependent on imported equipment, which has been designed to treat coal that is significantly different from Indian coal of drift origin. In this paper, authors have ventured into evolution of Indian Coal Washing Industry (with a focus on coking coal washing sector), its present condition and future prospect for growth. The paper emphasizes need for developing indigenous solutions to industrial challenges and highlights importance of increased coordination among academia-research institutions and coal industry.

Keywords: Coking coal washingIndian coal industry History of coal washing Future prospects

1 EVOLUTION OF COAL WASHING IN INDIA

After setting up Steel Plants in India (in 1907), Tata Iron and Steel Company (TISCO) started explorations to meet the demand of coking coal for their steel plants; and, first breakthrough was achieved by Mr. A. Farquhar during his washability studies of Jharia and Raniganj coal during 1918–1926. After ascertaining economic viability, in 1946, TISCO decided to set up Coal Washeries in West Bokaro (1951) and Jamadoba (1952) (Ghosh 1964; Sen 2015). Chance Cone separator (a dense media separator utilizing sand as media) was the first washing equipment installed in Jamadoba Coal Preparation Plant (JCPP) for treatment of Jamadoba and Bhowra Coal. Subsequently, Lodna Coal Company set up Feldspar Jig—based very simple Coal Washery in 1955. (Ghosh 1964; Sen 2015; Singh and Guha 1965; Sinha 1988).

Hindustan Steel Limited (HSL) was set up in 1954, for expansion of Steel Industry in India. Similarly, National Coal Development Corporation (NCDC) was founded for planned growth of coal mining industry in India. To meet demand of coking coal for HSL Steel Plants, Central Coal Washeries in the vicinity of existing Marshaling yards of Kargali, Durgapur, Dugda, Bhojudih and Patherdih were envisaged; and Kargali Washery was commissioned in 1958 by NCDC as first public sector washery in India. Subsequently, Durgapur, Dugda, Bhojudih and Patherdih Coking Coal Washeries were commissioned during the III and IV Five Year Plans. (CCL 2015; CIL 2015; Ghosh 1964; CCL 2015; AICC 1965; Singh and Guha 1965; Sarkar 1972; Das 1975; Agrawal 1976; CEMPDIL 2016; Konar et al. 1998).

Following the footprints, Indian Iron and Steel Company (IISCO), also set up Chasnalla Washery in 1968 (Sarkar 1972; Agrawal 1976; Konar et al. 1998; Rai and Tiwari 2009; Sen2015). Later on, NCDC installed Kathara and Swang Washeries in 1969 and 1970. After nationalization of coal industry in India, ownership of NCDC Washeries was transferred to Central Coalfields Limited (CCL), (CEMPDIL 2016).

After a decade, for administrative reasons, ownership of SAIL Washeries was transferred to Bharat Coking Coal Limited (BCCL). To cater for the growing demand of clean coal, Coal India commissioned seven more coking coal Washeries (Sudamdih, Moonidih, Nandan, Barora, Mohuda, Kedla and Madhuband) during 1980–2000 (Ministry of Coal 2014; Sen 2015; Sinha1988; Saxena 2015; NSO 2015). The most recent ones are Kedla and Madhuband Washeries of CCL and BCCL commissioned in 1997 and 1998 respectively.

Over the period, quality of coal has deteriorated due to exhaustion of superior grade coal in upper seams of BCCL and CCL Coal Belt (Sarkar et al. 2015). However, there has not been significant development in coal washing circuits, as the most recent washeries (Kedla and Madhuband) are Batac Jig-based washeries in contrast with the earliest coal washeries of HSL, which comprised of Heavy Media Separators (Rai and Tiwari 2009).

The demand of coal increased with the growth of Steel Industry. Keeping this in view, Coal India has planned to install six new coking coal washeries in BCCL with design capacity of 18.6 MTPY. These washeries will be installed and run on BOM scheme (Build Operate and Maintain). Three of these washeries are under construction. Five stage of Washing is being introduced in India with these washeries (Sarkar et al. 2015; Saxena 2015).

2 GROWING IMPORTANCE OF WASHING

India has proven coal reserves of total 125.91 billion tons, out of which only 18.4 billion tons is that of coking coal as shown in Table 1 (Source: Ministry of Coal).

Over the years, prime-coking coal reserves in the Jharia Coal Field have been exploited from the upper seams, leaving behind lower seams of inferior grade coal for further extraction.

Grade wise distribution of various seams of coal in BCCL (the only producer of prime-coking coal in India) is indicated in Table 2 (Sarkar et al. 2015).

Table 1. Geological reserves of coal in India as on 01.04.2014 (Quantity in MT)

Type of coal	Proved	Indicated	Inferred	Total
(A) Coking				
Prime coking	4614	699	0	5313
Medium coking	13,303	11,867	1879	27,049
Semi-coking	482	1004	222	1708
Sub-total coking	18,400	13,569	2101	34,070
(B) Non-coking	106,916	128,838	30,249	266,002
(C) Tertiary coal	594	99	799	1493
Total	125,909	142,507	33,149	301,565

Table 2. Grade wise distributions of coal seams/reserves in Bharat Coking Coal Limited

Demarcation	Coal seam	Thickness (m)	Grade of coal
Upper horizon	XVIII A/B/C/D	5.9	ST-II and W-I
	XVII	4.45	coking
	XVI A/B/C/D/E	10.58	
	XV	2.54	W-I and W-II
	XIV	0.56	coking
	XIII	3.94	
	XII	6.25	W-II coking
	XI/XII	6.83	
	IX/X, X	10.57	
Lower horizon	VIII A,B	7.84	NLW
	VIII	2.62	W-IV
	V/VI/VII	20.84	
	IV/(T)/IV (B)	8.4	Non-coking

Unlike metallurgical grade coal (ST-I & ST-II), inferior grade coal cannot be directly blended with imported coal for coke making. Increasing demand of growing steel industry and diminishing production of metallurgical grade coal has resulted in increasing import burden on Indian Steel Industry to meet its demand for quality coking coal (Ministry of Coal 2014; CIL2015). Trend of gross national demand for coking coal, domestic production and import during last six years has been shown in Fig. 1 (Ministry of Coal 2014; NSO 2015).

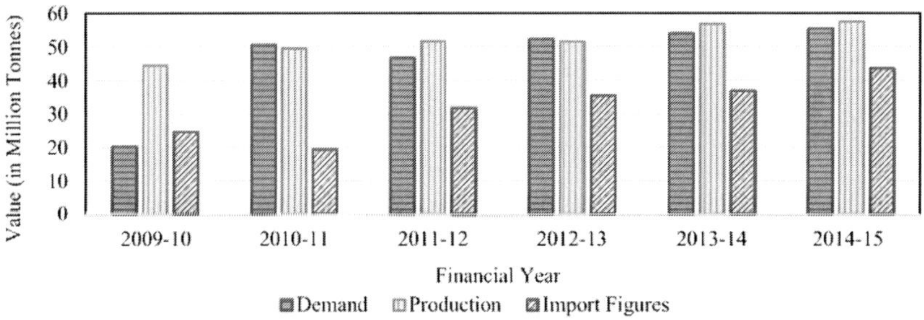

Figure 1. Trend of gross national demand, annual production and import of coking coal in India

Statistically, domestic production is enough to meet demand of coal for the Indian Steel Industry; still, import of coking coal has only been increasing. Inferior quality of Indian coking coal, and insufficient washing capacity are the principal reasons behind this huge demand–supply gap. Approx. 79 % of the total national demand is met by imported coal. Quality of Coking Coal produced during 2014–2015 is indicated in Table 3. (Ministry of Coal 2014; NSO2015).

Table 3. Grade of coking coal produced in India

Grade of coal	ST-I	ST-II	Semi coking	W-I	W-II	W-III	W-IV
Production in MT	0.50	0.456	0.13	0.115	2.228	12.499	41.973

It is evident that, about 95 % of total coal produced in India is of Washery Grade -III and IV. As per Coal Statistic Report, only 24 % of the total coking coal produced in India has been utilized in metallurgical sector; and remaining 76 % has been diverted to non-metallurgical sector. The figures are even dismal, if only public sector companies are considered, with a meager 14.4 % of the total coking coal utilization in metallurgical sector. The situation is alarmingly critical as we are not only losing valuable coking coal reserves to non-metallurgical sectors, it's costing Indian foreign currency reserves to the tune of US $5.2 billion in import of coking coal (Ministry of Coal 2014; CIL 2015).

3 EXISTING SCENARIO OF COKING COAL WASHING IN INDIA

So far, 25 coking coal Washeries have been installed and commissioned in India out of which only 19 are in operation at present with a total washing capacity of 33.17 MTY; out of these Washeries, 15 are operating in Public Sector. (Sen 2015; Saxena 2015; NSO 2015 ; Konar et al.1998; AICC 1965; Sinha 1988; Rai and Tiwari 2009). It is evident from Table 3 and Table 4 that the existing capacity of Washeries are not sufficient to handle total amount of coking coal produced in the country, and remaining quantity of coal is bound to get diverted to non-metallurgical sector. Moreover, it is also to be noted that only 13.79 MT of Coking Coal is being channelized to Metallurgical Sector, which amounts to only 41.6 % of the total washing capacity (Ministry of Coal 2014). This poor capacity utilization can be attributed to old plant machinery and obsolete washing circuit in most of the existing washeries, which have outlived their-designed life. Recurring losses in the coking coal washeries is another principal deterrent in capacity utilization (Sarkar et al. 2015). On further investigation, it was revealed that these washeries are operating at very poor yield; and in many of the public sector washeries fine coal circuit is not in operation (Sarkar et al. 2015; Alam and Patel 2012). Thus, actual utilization of coking coal in the metallurgical sector is even less, not more than 10.0 %. The authors are carrying out studies on Coking Coal Washeries of Bharat Coking Coal Limited to help improve their performance and ultimately maximize the optimum utilization of coking coal in the metallurgical sector.

Table 4. Existing Coking Coal Washeries in India and Washing Technologies

S. no.	Name of washery	Year of commissioning	Design capacity (MTPY)	Washing technology		
				Coarse coal	Small coal	Fine coal
1	Jamadoba	1952	0.9	–	−13 + 0.5 mm (HM Cyclone)	−0.5 mm (Flotation)
2	Kargali	1958	2.78	−300 + 30 mm (Coarse Coal Jig)	–	–
3	Durgapur[b]	1960	1.5	−80 + 20 mm (H M Bath)	−20 + 1.0 mm (Small Coal Jig)	−1 + 0.25 mm (Spiral) −0.25 (Flotation)
4	Bhojudih	1962	1.7	−75 + 25 mm (HM Bath)	−25 + 0.5 mm (Batac Jig)	−0.5 mm (Flotation[a])
5	Dugda-II	1968	1.8	–	−13 + 0.5 mm (HM Cyclone)	−0.5 mm (Flotation[a])
6	DCOP	1968	1.35	Not Available		
7	Chasnalla	1968	1.5	−80 + 20 mm (H M Bath)	−20 + 0.5 mm (Small Coal Jig)	−0.5 + 0.15 mm (Spiral) −0.15 (Clean Coal)
8	Kathara	1969	3.0	−80 + 13 mm (Drewboy Bath)	−13 + 0.5 mm (HM Cyclone)	−0.5 mm (Flotation[a])
9	Swang	1970	0.75	−80 + 20 mm (Deshaler Jig)	−20 + 0.5 mm (HM Cyclone)	−0.5 mm (WO Cyclone[a])
10	Sudamdih	1981	1.6	–	−37 + 0.5 mm (HM Cyclone)	−0.5 mm (Flotation[a])
11	West Bokaro-II	1982	1.8	–	−13 + 0.5 mm (HM Cyclone)	−0.5 mm (Flotation)
12	Moonidih	1983	1.6	–	−30 + 0.5 mm (HM Cyclone)	−0.5 mm (WO Cyclone[a])
13	Nandan	1984	1.2	−75 + 13 mm (Coarse Coal Jig)	−13 + 0.5 mm (Fine Coal Jig)	−0.5 mm (Flotation[a])
14	Rajrappa	1988	3.0	−80 + 16 mm (Batac Jig)	−16 + 0.5 mm (Fine Coal Jig)	−0.5 mm (Flotation[a])
15	Mohuda	1990	0.63	–	−25 + 0.5 mm (HM Cyclone)	−0.5 mm (Flotation[a])
16	Bhelatand	1994	0.86	–	−13 + 0.5 mm (HM Cyclone)	−0.5 mm (Flotation)
17	West Bokaro-III	1994	2.1	–	−13 + 0.5 mm (HM Cyclone)	−0.5 mm (Flotation)
18	Kedla	1997	2.6	−80 + 15 mm (Batac Jig)	−15 + 0.5 mm (Fine Coal Jig)	−0.5 mm (Flotation[a])
19	Madhuband	1998	2.5	–	−13 + 0.5 mm (Batac Jig)	−0.5 mm (Flotation[a])

[a] Fine coal circuit is not in operation

[b] Details pertaining to operation of fine coal circuit not available

Coal Washing Technology has evolved globally. Some countries have almost frozen their washing technology for a particular coal seam. This is not the case with Indian coal industry. The Washing Technology developed elsewhere, have been slightly customized to face the challenges of difficult washing characteristics of Indian Coal. As such, no washing equipment has been designed specifically to treat Indian Coal with difficult washing characteristics, which is not encountered in case of carboniferous coal. Details of washing schemes in existing coking coal washeries of India has been compiled in Table 4 (Sen 2015; Ministry of Coal 2014; Chakraborty et al. 2015; Sarkar et al. 2015; CCL 2012; Rai and Tiwari 2009). Heavy Media Cyclone is the most widely used separator in these washeries. Presence of high proportion of near-gravity-material (NGM) mandates cyclone washing (Rai and Tiwari 2009). Misplacement of coal in such washing schemes can be further improved with the aid of automation and control. However, most of the public sector washeries are yet to integrate automation and control in their existing circuits.

4 CHARACTERISTIC OF INDIAN COAL

Indian coals of drift origin (Gondwana Coal) can be characterized by high ash content with impurities disseminated in the coal bed. Very thin bands (5 mm–3 cm) of impurities are often encountered in between coal bands. Such coal has high near gravity material (NGM > 30 %) content. Liberation size of coal is much smaller; and significant improvement in yield is observed when coal is crushed down to as low as −6 mm to −3 mm and −1 mm. However, operating a coal washery with such small coal size is not feasible under present techno-economics. This makes Indian Coal difficult to wash, resulting in poor yield and low organic efficiency.

Indian industries find Indian Coal inferior to the imported coal mainly because of high ash content. However, Indian Coal can also be characterized by low sulfur (0.2 %–0.7 %), low iron, low chlorine, low toxic/rare earth elements content and macerals rich coal with high ash fusion temperature, and refractory nature of silica and alumina rich ash. It offers great potential for further investigation into coal washing technology dedicated for Indian Coal.

5 FUTURE OF COAL WASHING INDUSTRY

Government of India planned to expand coal-washing capacity in India by 13th Five Year Plan. Installation of 20 Coal Washeries in public sector with total installed capacity of 111.10 MT was envisaged. At present, CIL has taken up installation of 15 coal washeries on build, operate and maintain (BOM) Scheme.

Out of these, six are coking coal washeries, under the ownership of Bharat Coking Coal Limited with total design capacity of 18.6 MTY; and the remaining nine Washeries are Non-Coking Coal Washeries with total designed capacity of 74.5 MTY under the ownership of CCL (03 No—19.5 MTY), SECL (02 No—15 MTY) and MCL (04 No—40 MTY) subsidiaries of Coal India Limited (Saxena 2015; Sarkar et al. 2015).

The six coking coal washeries of Bharat Coking Coal Limited will be equipped with advanced automation and control circuits. Three of these washeries are under construction with the remaining under various stages of finalization. Typical cyclone—Cyclone, Jig—Cyclone combination has been used. deshaling cyclone and teeter bed separator (TBS) are being used for the first time in India.

A brief description of washing technology proposed in these washeries is presented in Table 5(Sarkar et al. 2015; Sapru 2015).

Table 5. Upcoming coking coal washeries in coal India limited

Name of washery	Design capacity	Washing technology		
		Coarse coal	Small coal	Fine coal
Madhuband	5.0 MTY	−75 + 13 mm (Deshaling cyclone)	−13 + 1 mm (HM Cyclone)	−1 + 0.25 mm (TBS) −0.25 mm (Flotation)
Patherdih	5.0 MTY	−75 + 13 mm (Jig)	−13 + 0.5 mm (HM Cyclone)	−0.5 mm (Flotation)
Dahibari	1.6 MTY	−50 +13 mm (Deshaling Jig)	−13 + 0.5 mm (HM Cyclone)	−0.5 mm (WO Cyclone and Spirals)
Patherdih	2.5 MTY	Not yet finalized		
Dugda	2.5 MTY			
Bhojudih	2.0 MTY			

These new washeries have been facilitated with re-washing circuit for the middling produced in their primary circuit; and MoU has already been signed for utilization of rejects produced in these washeries in FBC Power Plants and allied industries (Minutes of IMC Meeting).

With the allocation of new coal blocks for captive usage, new coal washeries in private sector are expected. S&T Mining, a joint venture between Steel Authority of India Limited (SAIL) and Tata Steel Limited has been formed to look into the Mining and Coal Preparation activities of the newly acquired coal blocks to meet the demand of coking coal for their steel plants. Moreover, some American,

European, South African and Asian Companies have formed consortiums with Indian Industries to invest in the coal washing sector in India.

With installation of upcoming coal washeries in BCCL, total designed capacity of coking coal washeries in India will be 55.77 MTY by 2020 (Ministry of Coal 2014; Chakraborty et al. 2015; Saxena 2015; Sarkar et al. 2015). The projected production of Coal in BCCL and coal washing targets have been presented in Fig. 2 (Sarkar et al. 2015).

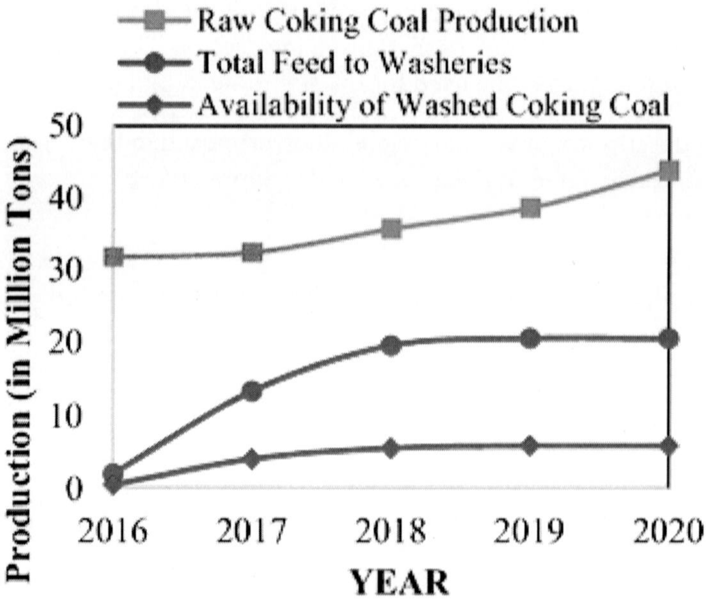

Figure 2. Projection of Coking Coal Production in BCCL and washed coal availability to steel sector

As evident from Table 4 and Fig. 2, present capacity utilization of existing washeries of BCCL is around 20 %. Study is being carried out by the authors for improving techno-economic performance of the washeries. Once the coal washing becomes a profitable proposition, BCCL management will be encouraged to revive the existing washeries for maximum capacity utilization with the objective of washing entire coking coal produced in BCCL. The other companies are expected to follow this and thus ensure utilization of Indigenous Coking Coal with no or substantial reduction in import.

6 SCOPE FOR RESEARCH AND DEVELOPMENT

As evident from Table 4 and Table 5, with deteriorating quality of coal, the complexity of washing would increase, and operators would resort to 4–5 stages of washing. Outcomes of several experiments carried out on Indian Coal and South African Coal (also of Gondwana Origin) reveal that separate washing of different size fractions have resulted in higher yield (Sapru 2015); and by integrating advanced process control, BATAC Jigs can be utilized to wash difficult-to-wash coal (Mustafi and Nandy 2015). Japan Coal has demonstrated the capability of Variwave Jigs in effectively washing Indian Coal (Koyanagi et al. 2015). FLSmidth has carried out several experiments on Indian Coal and have arrived to a conclusion that washing −13 + 1 mm fraction in Heavy Media Cyclone, −1 + 0.106 mm fraction in spiral concentrator and −0.106 mm fraction in flotation cells will yield better results compared to −13 + 0.5 mm in HM Cyclone and −0.5 mm by Flotation (Moorhead and Woodruff 2015), which is a regular practice in India. Multotech Process Equipment, SA has conducted several tests on Indian Coal for beneficiation using spiral concentrator; and the results are encouraging (Smith and Erasmus2015). Dr. Gouricharan, of CIMFR, Dhanbad has patented a novel washing scheme for LVMC Coal by adopting multi-stage, multi-stream coal washing circuit (Gouricharan 2015). However, the complexity of the operation and challenges of maintenance will increase and is likely to pose challenges in smooth operation and maintenance of the plant.

This implies that, there is scope for concerted R&D efforts to find out relatively simpler washing scheme with maximum 2–3 stage of washing, which would not only simplify the operation and maintenance, but also overcome the shortcomings of existing 2–3 stage washing circuits.

There lies scope for comparative study of Indian Coal with Gondwana Coal from other countries in the southern hemisphere. So far, studies have shown that Indian Coal differs significantly from carboniferous coal found in northern hemisphere. However, there is hardly any literature on comparative study of Indian and South African or Australian Coal. This is the reason, coal washing technology used in these countries are accepted as suitable for Indian Coal with little or no modification. Ample opportunity is there for the scientific community to develop indigenous technology designed for difficult washing characteristics of Indian Coal.

Academic & Research institution and technology providers should work in tandem with the industry, as they are ultimately responsible for implementing these technologies. Even after 60 years of Independence, Indian coal washing industry is at its nascent stage. This calls for a Academia-Research Institution and industry consortium to develop new and alternate process routes which are technically easier and superior, environmentally benign and economically sustainable.

7 CONCLUSIONS

The existing situation of Indian coal washing Industry is a resultant of gross negligence of the industries to strengthen R&D for development of indigenous technologies. The Industry has relied from the very beginning on imported technology and machinery, ignoring the fact that foreign technologies have been developed taking into account the nature of coal in those countries, which is significantly different from that of Indian Coal, which possess difficult washability characteristics because of its drift origin. Neither the industry nor the government agencies have sponsored an exhaustive R&D project dedicated to coal, despite it being the primary source of energy and infrastructural element in India.

Due to non-availability of such research facilities either at Govt. Research laboratories or Academic Institutions, Indian Scientific community has not yet been able to develop a cause—effect—solution matrix for operational and maintenance practices of Coal Washing Industry. It's one of the principal reasons behind existing condition of public sector Coal Washeries that were installed using imported technology but no indigenous intelligentsia to deal with regular operational and maintenance challenges. Had the academia–industry and research institution been working in tandem, the coal washing scenario in India would have been different.

Thus, attractiveness of Coal Washing Industry in India is high, but also full of challenges. PESTEL analysis of coal washing industry suggests that India is heading towards mandatory coal washing before end utilization. The deteriorating quality of raw coking coal poses challenges before the scientific community and the industry to find out simple and economically viable coal washing scheme for maximum recovery of coking coal for usage in metallurgical sector. There also lies scope for analyzing present condition of existing washeries, especially in public sector, to look for innovative techniques for revival of these washeries.

REFERENCES

1. Agrawal GC (1976) Public Sector Steel Industry in India, p 76
2. AICC (1965) Indian National Congress. Year of Freedom, vol 19
3. Alam M, Patel JP (2012) Coal washing in BCCL. In: Proceeding of Xstrata—ISM Flotation Workshop, Kolkata, India
4. CCL (Central Coal Fields Limited) History (2015) http://www.ccl.gov.in/comp/our_history.htm. Accessed 20 Oct 2015
5. CEMPDIL report (2016) Coal mining in India: history and perspective.
6. Chakraborty DP, Sen S, Ray PR, Kumar A , Tata Steel Limited (2015) , Augmenting supplies of metallurgical coal to steel mills by upgrading low and medium coking coals, CPSI J, 2015, 66–72
7. CIL (Coal India Limited) History (2015) http://www.coalindia.in/en-us/company/history.aspx. Accessed 20 Oct 2015
8. CMPDI (2015) Completed Projects (120 TPH Modular Washery at Lodna in 1992), Coal beneficiation section in http://scienceandtech.cmpdi.co.in/institution.htm. Accessed 20 Oct 2015
9. CCL (2012) Environmental statement, Kargali Washery, Central Coalfields Limited, 2012–2013
10. Das M (1975) Fantasy of coal nationalization, p 53

11. Ghosh AB (1964) The economics of coal washing in India, pp. 35–36

12. Gouricharan T (2015) Indian coal characteristics—emerging cost effective washing options. CPSI J, pp 125–130

13. Hindustan Steel Limited (1966) Statistics for Iron and Steel Industry in India, p 173

14. Konar BB, Banerjee SB, Choudhury SG, Das NS , Sen K (1998) Coal preparation: CFRI golden jubilee monograph, 1998, pp. 23–24

15. Koyanagi N, Endo H, Kubo Y, JCOAL and NECO (2015) Variwave jig in India, CPSI J, pp 140–144

16. Ministry of Coal (2015) Coal Reserves in India, http://coal.nic.in/content/coal-reserves. 20 Oct 2015

17. Ministry of Coal (2012) Govt. of India, Coal washeries section in http://www.coal.nic.in. Accessed 2012

18. Ministry of Coal (2014) Govt. of India, Minutes of IMC meeting to review the issues related to formulation of policy on coal rejects allocation for washery rejects based power plants held on 08 Oct 2014

19. Ministry of Coal (2014) Govt. of India, Provisional coal statistics 2014–2015

20. Moorhead RG, Woodruff D, Smidth FL (2015) Evaluation of various HMC cleaning strategies to maximize yield of Indian coals, the coal beneficiation 2015 conference, April 2015, pp 79–88

21. Mustafi G, Nandy M (2015) MBE Coal and Mineral Technology India Pvt. Ltd., beneficiation of coal with under-pulsated jigs—is there anything new happening?. CPSI J, 158–164

22. NSO (2015), Ministry of Statistics and Program Implementation, Govt. of India, energy statistics 2015

23. Office of the Chief Economist (2015), Department of Industry and Science, Government of Australia, Coal in India.

24. Rai B, Tiwari R (2009) Coking coals: availability, quality, and beneficiation, proceeding of international conference on coking coals and coke making: challenges and opportunities, January 2009, Ranchi, India

25. Sapru HL (2015) Beneficiation of low volatile medium coking NLW coal in India. CPSI J, April 2015, pp. 55–65, Brief Presentation.

26. Sarkar A, Dutta SK, Patel JP ,Bharat Coking Coal Limited (2015) Augmenting demand of metallurgical coal for Steel Industry: challenges and opportunities, Brief Presentation, the coal beneficiation 2015 conference, April 2015

27. Sarkar GG, Ash PN, Rao SRR (1972) Studies on coal preparation in India, CFRI

28. Saxena BK (2015) Coal India Limited Brief Presentation, CPSI J, Coal Quality related Challenges and CIL's Plan of Action 145–157

29. Sen K (2015) History of coal washing in India. http://www.kalyansen.com/Coal-preparation-in-India-historica-perspective.html. Accessed 20 Oct 2015

30. Singh VB, Guha B (1965) The Coal Mining Industry chapter in economic history of India. Allied Publishers, Chennai, pp 322–323

31. Sinha S (1988) Regional structure, processes, and patterns of development: a case study of Chhota Nagpur Region. Concept Publishing Company, New Delhi, p 74

32. Smith IGT, Erasmus WRJ (2015) Utilization of Spirals when Washing Coal, CPSI J, Multotec Process Equipment, SA, 170–176

33. Tata Steel Limited(2015) Environmental statement for Jamadoba, Bhelatand, West Bokaro - II and West Bokaro - III coal washeries. http://www.tatasteelindia.com/environment-statement.asp. Accessed 20 Oct 2015

Chapter 11

Feasibility Studies on Cleaning of High Sulfur Coals by Using Ionic Liquids

Binoy K. Saikia[1] • Kakoli Khound[1] • Om P. Sahu[2] • Bimala P. Baruah

[1]Coal Chemistry Division, CSIR-North East Institute of Science & Technology, Jorhat, Assam 785006, India 2 Analytical Chemistry Division, CSIR-North East Institute of Science & Technology, Jorhat, Assam 785006, India

ABSTRACT

Coal has been used as an energy resource around the world, primarily for the generation of electricity. The cleaning of coal by removing its unwanted sulfur and mineral matter components is utmost essential before their gainful utilizations. The ionic liquids (ILs) are considered as non-toxic solvents for using in different industrial processes. The effect of two ILs namely, 1-*n*-butyl, 3-methylimidazolium tetrafluoro borate (IL1) and 1-*n*-butyl, 3-methylimidazolium chloride (IL2) in oxidative de-sulfurization and de-ashing of two industrially important high sulfur coal samples from Meghalaya (India) is discussed in this paper. The maximum removal of total sulfur, pyritic sulfur, sulfate sulfur and organic sulfur are observed to be 37.36 %, 62.50 %, 83.33 % and 31.63 % respectively during this oxidative process. The quantitative diffuse reflectance Fourier transform-infrared spectroscopy analysis supports the formation of sulfoxides (S=O) and sulfones ($–SO_2$) and their subsequent removal during the oxidation of the coals in presence of ILs. The X-ray fluorescence combined with near edge X-ray absorption fine structure and scanning electron microscopic studies reveal the removal of mineral matters (ash yields) from the coal samples. The thermogravimetric analysis of the raw and clean coals indicates their high combustion efficiencies and suitability for using in thermal plants. The method is partially green and the ILs could be recovered and reused in the process.

Keywords: High sulfur coals De-sulfurization of coalIndian coalsIonic liquids De-ashing

1 INTRODUCTION

Coal has been used as the major source for electricity generation in many parts of the world including India. India is the third largest coal producing country in the world and 'Coal India Ltd.' is one of the world's largest coal mining companies (Stracher and Taylor 2004). There are over 861 billion tonnes of proven coal reserves worldwide, which is enough to last for around 100 years at current rates of production. However, cleaner approaches are vital for gainful and sustainable utilizations of coals. Coal naturally contains sulfur and mineral matters (ash yield). Sulfur in coal is generally grouped into inorganic and organic forms. The high sulfur coals create environmental problems to the community and thus, minimization of sulfur and mineral matters from coals is very much essential before their various types of utilizations. The acid mine drainage (AMD) is also caused by the oxidation of pyritic sulfur in coals (Baruah et al. 2006). The performance of a coal-based power plant is also influenced by many coal properties. The mineral matters and sulfur contents of any coal have most impact on the boiler performance. The presence of mineral matter (ash contents) can cause ash deposition and erosion damage to the boiler tube. The acid dew point of flue gas is also determined by the sulfur content of the coal. Higher coal sulfur level may promote air heater corrosion (Juniper and Pohl 1997). In steel making industry, the sulfur content of coke, sediments on the iron crystal surface make steel more brittle and decrease its plasticity property (Ehsani 2006). In recent years, chemical desulfurization of coals has attracted much attention due to the gradual reduction of the statutory sulfur contents. A large number of chemicals have been used to remove the sulfur and mineral matters from coals (Yaman et al. 1954; Mazumdar et al. 1989; Sain et al. 1991; Demirbas and

Balat 2004; Komintarachat and Trakarnpruk 2006; Baruah and Khare 2007; Ahmed et al. 2008; Zhang et al. 2009; Campos-Martin et al. 2010; Silva et al. 2011). The most common industrial process of catalytic hydro-desulfurization (HDS) needs hydrogen with additional high investments and operation cost (Schmitz et al. 2004). On the other hand, the chemicals used during the desulfurization processes also have been creating environmental hazards to the communities as well as surroundings. As a consequence, chemists are exposed to a large number of lethal chemicals during their working. Thus, the chemical desulfurization process must be compatible to the environmental restriction as well as to the chemists. So, suitable and environment friendly chemical desulfurization method is urgently needed for the coal processing plants. In the last few decades, there emerged a new class of green solvents known as "Ionic Liquid" (IL) (Rogers and Seddon 2003). They are composed of cation and anion, non-volatile, excellent solubility, great thermal/chemical stability and non-flammability and environmental friendliness. The ILs have been used potentially in different sectors including nuclear applications to the petroleum industry. The ILs have also been reported for direct extraction of S-compounds (extractive desulfurization, EDS) from liquid fuels (Bosmann et al. 2001; Esser et al. 2004; Zhao et al. 2005; He et al. 2008; Holbrey et al. 2008; Kendra-Krolik et al. 2011).

The coal resources of India are available in older Gondowana formation of Peninsular India and younger to tertiary formation of North-eastern region (NER). The NER of India has a large reservoir of high sulfur (2 %–8 %) coals with sub-bituminous in rank (Saikia et al. 2014). The majority of sulfur functionalities of these coals were found to be in the organic forms (75 %–90 %), which are retardant to remove chemically (Baruah and Khare 2007). There are five types of organic sulfur functionalities found in these NER coals including aliphatic or aromatic thiols, aliphatic or mixed sulfides, aliphatic or aromatic disulfides, heterocyclic compounds of thiophenic and thioketonic types, etc., (Barooah and Baruah 1996); moreover, the NER coals are characterized by low ash and high caking index, suitable for use in cement plant and metallurgy as coking coal. But, these coals cannot be gainfully utilized because of their high sulfur contents as mentioned above.

We present here a partially green chemical beneficiation method for de-sulfurization and de-ashing of high sulfur low grade NER coals from

Meghalaya (North-east India) by using ILs as non-toxic solvents in a mild reaction conditions with no need of hydrogen, pressure and reactor. The prime objective of this study is also to see the outcome of the oxidative desulfurization (ODS) process with those high sulfur coals in presence of the ILs. The ODS process includes oxidation of sulfur compounds followed by their extraction in solvents. The use of green solvents like ILs in this present coal cleaning process for high sulfur coals makes the process partially green and efficient. The process may also be considered in clean coal technology.

2 MATERIALS AND METHODS

The chemicals (Sigma-Aldrich and Merck) used were: hydrogen peroxide (30 %), formic acid (85 %), 1-*n*-butyl,3-methylimidazolium tetrafluoro borate (IL1) and 1-*n*-butyl,3-methylimidazolium chloride (IL2). The representative high sulfur coal samples (SCF and B) were collected from two different collieries of Bapung coalfield (N 27°13′–27°23′; E 95°35′–96°00′), Meghalaya (India) and ground to −0.211 mm sizes for other analyses by suing standard method (ASTM 2010). The areas under investigation is covered by the Barail group of rocks consist coarse sandstone, shale, minor coal lenses and carbonaceous shale. The coal seams are associated with the Lakadong Sandstone of Shella Formation of Jaintia group (Eocene) includes three persistent coal seams ranging from 1–3 m in thickness along with inter-seam sandstone and shale (Ahmed and Rahim 1996).

The proximate analysis and sulfur analysis were carried out in Proximate Analyser (TGA701, Leco) and Sulfur Analyser (LECO, 144DR) respectively by using standard method (ASTM 1991). The forms of sulfur were determined by following the standard method (ASTM 2005). The percent of organic sulfur was calculated by difference. The gross calorific values (GCV) were also determined in Bomb calorimeter (Model: Leco, USA) by using standard methods (ASTM2013).

2.1 Quantitative diffuse reflectance FT-IR spectroscopic analysis of coals

Diffuse reflectance Fourier transform infrared (FT-IR) spectra of the coal samples were recorded using FT-IR instrument (System-2000, Perkin–Elmer Inc.) in the range of 500–1500 cm^{-1}. The samples (about 6–7 mg) were accurately weighed on single pan electronic balance (Model: SI-234, Denver Instrument) and thoroughly mixed with spectroscopic grade KBr powder (about 200 mg) in an agate mortar. Pure ground KBr was used to obtain the reflectance spectrum by taking 64 numbers of scan having resolution 4.0 cm^{-1}. The FTIR spectra were processed using the software Spectrum V-3.02 (Version 3.02.01, Perkin–Elmer) in absorbance mode. At first the base line was corrected and was brought to the 0.0 reading. The spectra were normalized to 1 mg of the samples by dividing it with the amount of the sample taken for recording the drift spectra to compare the absorbance at 563 cm^{-1}.

2.2 Thermal (TG–DTG), XRF and SEM analysis of coals

Thermogravimetry (TG) and their first derivative thermogravimetric (DTG) spectra were recorded in the thermal analyzer (Model: TGA701, LECO) with heating rate of 10 °C/min in presence of oxygen environment. The combustion performance index (*S*) and energy of activation (*E*a) were calculated by using the methods as reported elsewhere (Sonibare et al.2005; Qing et al. 2011).

The high energy X-ray fluorescence (XRF) measurements of coal samples before and after the treatment were obtained from the Canadian Light Source [Saskatchewanin the Soft X-ray micro-characterization Beamline (SXRMB)]. The Si(111) crystals were used for XRF measurements with photon energy set at 8350 eV. The powder sample was mounted using double-sided, conducting carbon tape, and loaded into a vacuum chamber (base pressure of 1×10^{-8} torr). The scanning electron microscopic (SEM) images were recorded in SEM (Model: JSM-6360 JEOL).

2.3 Removal of sulfur components and mineral matters (ash yield)

Thirty grams of coal sample were refluxed with a mixture of 13.2 mL of 20 % formic acid and 75 mL of 20 % H_2O_2 for 1 h. Then, the oxidized coal samples were filtered and washed with water until the pH of the filtrate became neutral. In another two separate experiments, 30 g coal samples (0.211 mm in size) were oxidized with the mixtures of 13.2 mL of 20 % formic acid and 75 mL of 20 % H_2O_2 in the same reflux conditions for 1 h in presence of the IL1 and IL2 (0.5 g for each) respectively. The coal-IL mixtures were centrifuged and the recovered coal residues were washed with hot distilled water until neutral and oven dried for 1 h. The filtrates obtained were kept for the recovery and further use of the ILs. The oven dried oxidized coal samples were used for subsequent analysis.

The filtrates obtained in the above step were centrifuged with absolute alcohol to separate out any impurities from the mixtures. Absolute alcohol was separated by traditional evaporation and then, the filtrates were mixed with 1 % $BaCl_2$ solution followed by CCl_4 (about 50 mL) to precipitate out dissolved sulfur functionalities. White colored substances were formed upon slight heating at about 60 °C due to the precipitation of inorganic and organic sulfur compounds, which were removed from coals during oxidation. The white precipitates were then filtered through Whatmann-42 filter paper. The final filtrate containing ILs became colorless. After boiling off the CCl_4, the filtrates could be used for another desulfurization experiments with slight decrease in activity.

3 RESULTS AND DISCUSSION

3.1 Physico-chemical characteristics of coals

The physico-chemical properties of the raw and oxidized coal samples are shown in Table 1. It demonstrates that the SCF and B coals are high sulfur and high ash in nature. Moreover, the majority of sulfur is organically bound in the coal samples (up to 79.0 %). The geochemical studies of sulfur in coals provide valuable information regarding the sources of sulfur and processes of sulfur incorporation into coals which helps in the study of the depositional environments and formation history of coals (Chou 2012). The major part of inorganic sulphur in these coals consists of pyrite (FeS_2), which is usually in particulate, to agglomerated forms (Baruah et al. 2006). The high sulphur content in the Northeast Indian coals resulted from the increased availability of sulphate ions in sea water coupled with the activity of anaerobic bacteria. The presence of high organic sulfur in northeast Indian coals suggests the marine influence as sulfur compounds are quickly formed by oxidation in marine environment. The oceanic sulfate sources have high sulfate reduction rates and during the coal formation, these sulfates are incorporated to the matrix of the coal molecule giving high organic sulfur groups. The geochemistry of the high sulphur Northeast Indian coals are reported elsewhere (Nayak et al. 2009; Saikia et al. 2014).

Table 1. Physico-chemical analysis of the raw and beneficiated coal samples (as received, %wt)

Coals	M	Ash	VM	FC	TS	PS	SS	OS	%PS removal	%SS removal	%OS removal	%TS removal	%Ash removal	GCV (kcal/kg)
SCF	4.07	25.9	31.8	38.3	4.47	0.80	0.12	3.6	–	–	–	–	–	6565.60
SCF Oxd	2.58	22.3	32	43.12	3.37	0.60	0.06	2.6	25.0	50.00	27.77	24.60	13.89	7155.00
SCF +IL1	2.10	23.5	33.3	41.10	3.25	0.55	0.04	2.6	31.25	66.66	26.11	37.36	9.26	7411.79
SCF + IL2	2.25	25.6	32.4	39.7	2.80	0.30	0.02	2.48	62.50	83.33	31.11	27.29	1.15	7780.60
B	4.30	17.8	33.7	44.0	3.47	1.08	0.67	1.77	–	–	–	–	–	6687.56
B Oxd	2.34	15.9	36.9	44.9	2.84	0.75	0.45	1.65	30.55	32.83	7.34	18.15	10.67	6812.08
B + IL1	4.72	15.5	33.7	46.08	2.71	0.98	0.52	1.21	9.25	22.38	31.63	21.90	12.92	7140.12
B + IL2	2.31	16.0	37.4	44.3	2.36	0.65	0.60	1.73	39.81	10.44	2.25	14.12	10.11	7213.40

M moisture, *VM* volatile matter, *FC* fixed carbon, *TS* total sulfur, *PS* pyritic sulfur, *SS* sulfate sulfur, *OS* organic sulfur

3.2 Removal of sulfur and mineral matter (ash yield)

As observed, the oxidation these coal samples (B and SCF) in ILs (IL1 and IL2), the pyritic, organic and sulfate sulfur contents have been significantly removed. On oxidation, the maximum removal of total sulfur and mineral matters (ash yield) of the coal samples are found to be 37.36 % and 13.89 % respectively. However, the pyritic, sulfate and organic sulfur are significantly removed by 62.50 %, 83.33 % and 31.63 % respectively during the process. Moreover, the fuel values (i.e., GCV) are also found to be increased after the beneficiation process (Table 1). The Fig. 1 shows the effectiveness of the process in removal of the mineral matters (ash yield) and sulfur components from the two coal samples. It is observed that the present process of ODS is more effective for the SCF coal sample than that of B coal although they are from the same coalfield. Thus, the removal of sulfur including pyritic, sulfate and organic sulfur varies from one colliery to other in the same Bapung coalfield. It may be due to the difference in organic sulfur functional groups present in the coal structure. However, it needs geo-chemical study to understand these variations in the same coalfield. The oxidation in presence of ILs produces sulfone with other components of the product being controlled by factors such as reaction temperature and pH (Davalos et al. 1994). Some researcher found that S-removal in ODS does not linearly increases with H_2O_2 dosage in IL (Zhao et al. 2010). The results were reported that the sulfoxide functional groups decreases with coal rank, while the contents of thiophene increases (McCoy and Weiss 1954; Calkins 1994; Yperman et al. 1995; Maes et al. 1997). Thus, the stable thiophenes are formed from less stable organic sulfur compounds (thiols and sulfides) with increasing rank. On the other hand, coal of different ranks/types has a different distribution of organic-sulfur functionalities, which may affect the extent of desulfurization.

In the oxidative process, the acidic ILs decomposes H_2O_2 to form OH, which is strong oxidant for organic S-compounds present in the coals. The organic sulfur compounds were selectively oxidized into sulfone and sulfoxide in the IL phase. Generally, the divalent S-atom of the organic sulfur compounds undergoes the electrophilic addition of oxygen atom from the $H_2O_2$to form sulfone during oxidation. The present Bronsted acidic ILs is water soluble and promotes catalytic reactions; hence in presence of hydrogen peroxide they will have more oxidizing effect. However, the strength and extent of coal oxidation may vary with the rank/type of coal and nature of ILs.

3.3 X-ray diffraction analysis of coals

It is to be mentioned that the low removal of mineral matters (de-ashing) were observed during this process (Table 1). The similarities of the XRD profiles of the raw and treated coals also support the low removal of mineral matters (Figs. 2, 3). The associated minerals present in the raw and oxidized coals are also shown in Figs. 2 and 3. Meghalaya coals of the NER (India) are of Gondowana type, where the

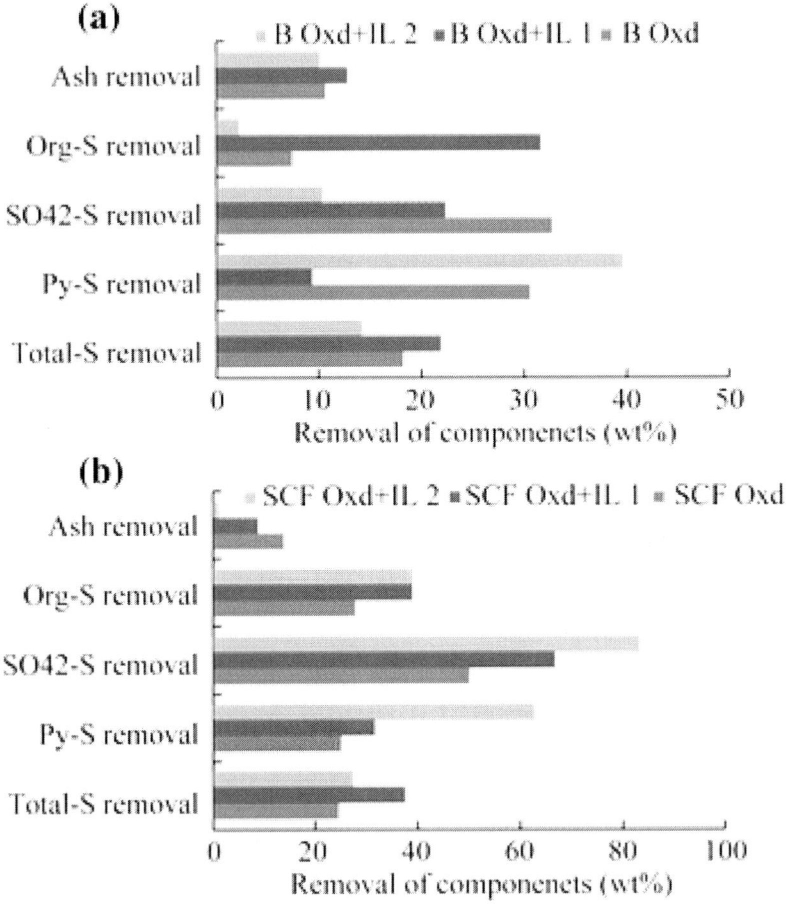

Figure 1. Extent of removal of components during oxidation of coals with ILs

majority of mineral matters are present in the inertinite parts. It has higher density and hard to remove and hence cannot be separated by mild and conventional ash removal methods (Falcon and Ham 1988).

However, the pyrite of these coals is in the vitrinite part, which is removable easily after oxidation reactions.

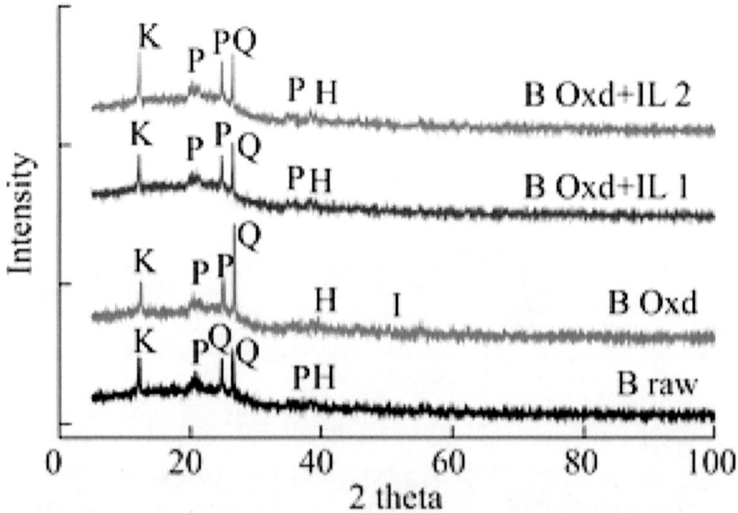

Figure 2. X-ray diffractogram of raw and beneficiated B coal sample (*K* kaolinite, *P* pyrite, *Q* quartz, *H* hematite,*I* illite)

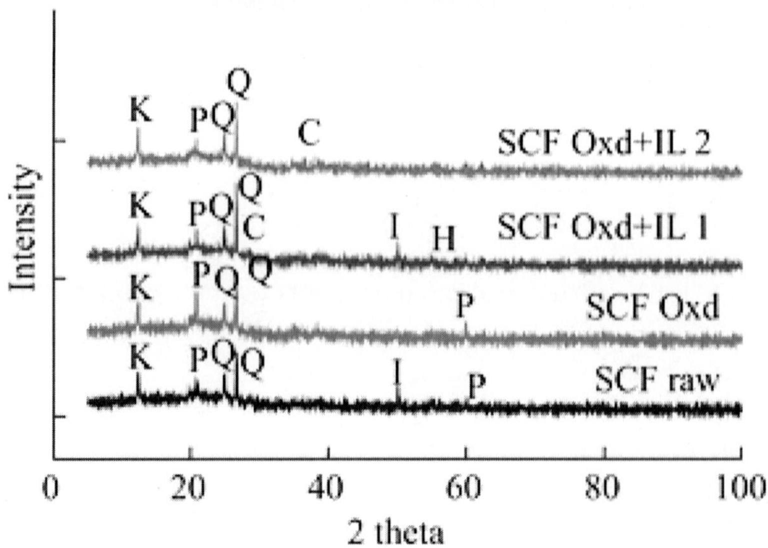

Figure 3. X-ray diffractogram of raw and IL beneficiated SCF coal sample (*K* kaolinite, *P* pyrite, *Q* quartz, *C*calcite, *H* hematite, *I* illite)

3.4 Quantitative FT-IR spectroscopic analysis of sulfur transformation

The FT-IR spectra of the raw and oxidized coals are shown in Figs. 4 and 5 and it is observed that the amounts of sulfur functionalities are different in both the coals. The absorption bands in the region of 550 cm^{-1} are assigned to the stretching vibration of S–S bonds of sulfides and C–S bonds in primary and secondary thiols (Silverstein et al. 1987). The bands in the region of 1415–1380 and 1200–1185 cm^{-1} are observed due to the presence of organic sulfates. The stretching frequencies in the range of 1070–1030 cm^{-1} are due to the presence of sulfoxide in the coal samples. The bands in the range of 1050–1020 cm^{-1} are due to the formation of S=O in the oxidized coal samples (Silverstein et al. 1987; De Filippis and Scarsella 2003). However, these regions are also seen in the raw coal sample due to the presence of sulfate ($-SO_4$). The FT-IR spectra of the raw and oxidized coals were normalized and it is interestingly observed that the absorbance of the sulfoxide, sulfones and sulfate groups were decreased appreciably during the total beneficiation process. Thus, the FT-IR analysis reveals that the sulfur has been converted to sulfoxide (S=O) and sulfone ($-SO_2$) on oxidation and subsequently removed by ILs. In this study, the extent of desulfurization differs due to the difference in sulfur contents in the coal samples.

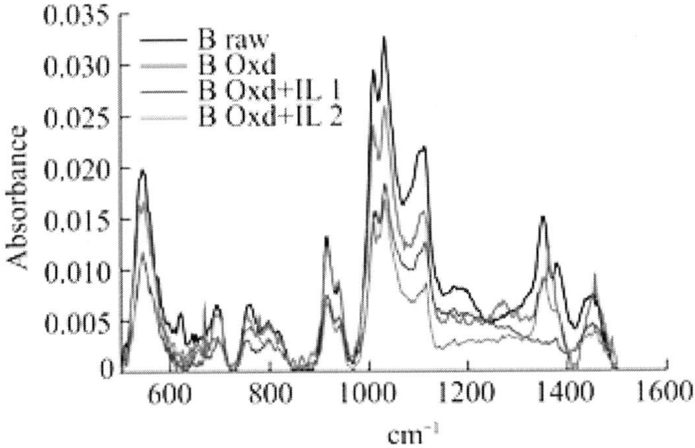

Figure 4. FT-IR spectra of the raw and beneficiated coal samples (B) in IL1 and IL2

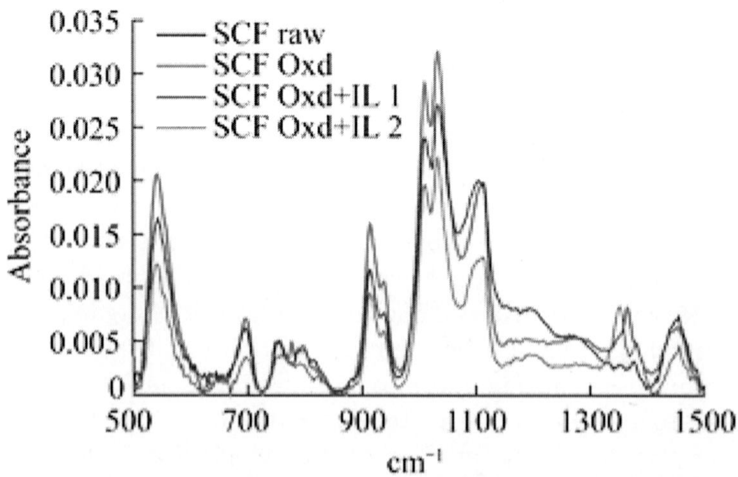

Figure 5. FT-IR spectra of the raw and beneficiated coal samples (SCF) in IL1 and IL2

3.5 X-ray fluorescence analysis on mineral matter (ash yield) removal

The XRF combined with the NEXAFS analysis of the raw and beneficiated (oxidized) coal samples are depicted in Fig. 6. The excitation energy has been chosen just above potassium. All XRF spectra were normalized to the same intensity at the excitation energy (scattered X-rays), so it is easy to say that the current beneficiation process converts the sulfur functionalities to sulfones and sulfoxide (Fig. 6f). The XRF could be useful in showing: (a) the change in sulfur intensity due to oxidation; (b) change in inorganic composition, particularly in for oxidation followed by IL extractions (Fig. 6c–f); (c) the B sample has more sulfur than SCF sample, which is in agreement with the chemical analysis. It has also a lot more Ca, Mn and Ni along with pyrite and sulfate as seen in the XAFS; (d) for SCF coal: most of the inorganic species have been removed/reduced after the treatment indicating the maximum removal of ash (13.09 %); (e) for both the coals, it seems that sulfur peak has increased after the treatment, which could be due to the conversion of sulfur forms to inorganic sulfur, such as Fe_2S and $FeSO_4$, as seen in the increase of the Fe intensity. In addition, the ILs treatment of B coal removed most of other inorganic species such as Ca, Mn and Ni.

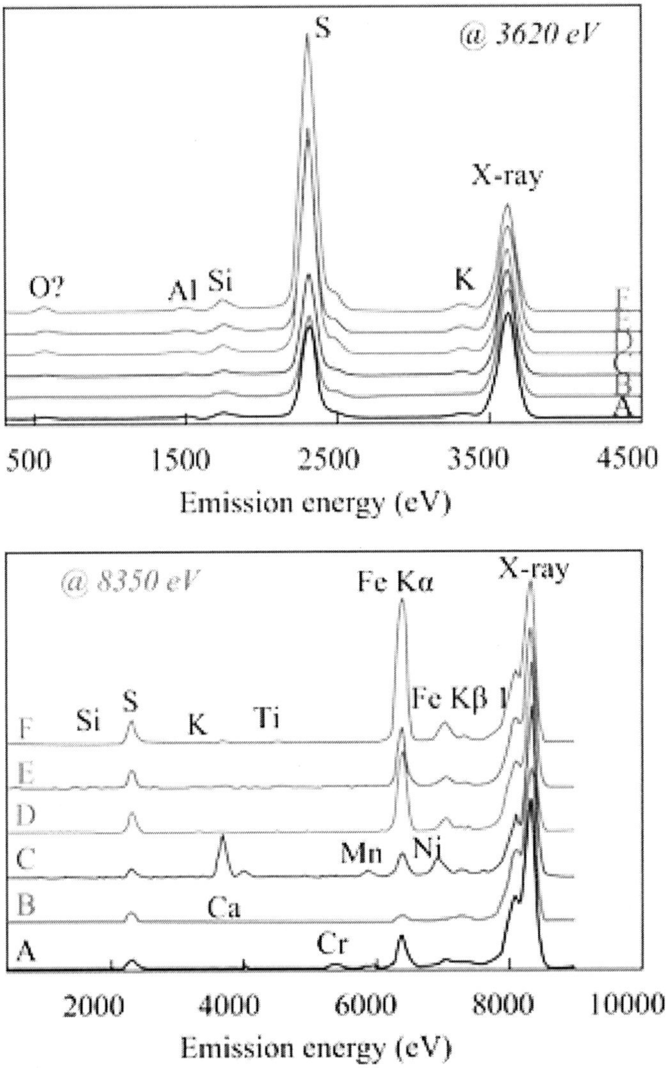

Figure 6. XRF Spectra of the raw and beneficiated coal samples at two different eV (*A* SCF raw, *B* SCF Oxd, *C* B raw, *D* B Oxd, *E* B Oxd + IL1, *F* B Oxd + IL2)

3.6 SEM analysis of raw and beneficiated coal samples

The effect of H_2O_2, HCOOH and ILs on the microstructure of the coals is displayed in Fig. 7, which compares the changes in morphologies of the raw and beneficiated coal samples. The SEM micrographs of the beneficiated coal samples (Fig. 7d–h) reveal the occurrence of leaching of the coals during oxidation. The appearance of small hollow pits on their surface might be due to the disappearance of pyritic, sulfate and mineral matters from the coal samples on oxidation. A few small and large particles (in Fig. 7d–g) are found to be dispersed in the beneficiated coals. Thus, the changes in the SEM morphologies support the present beneficiation process through removal of sulfur components and mineral matters from the coal samples (Ishak and Ismail2004).

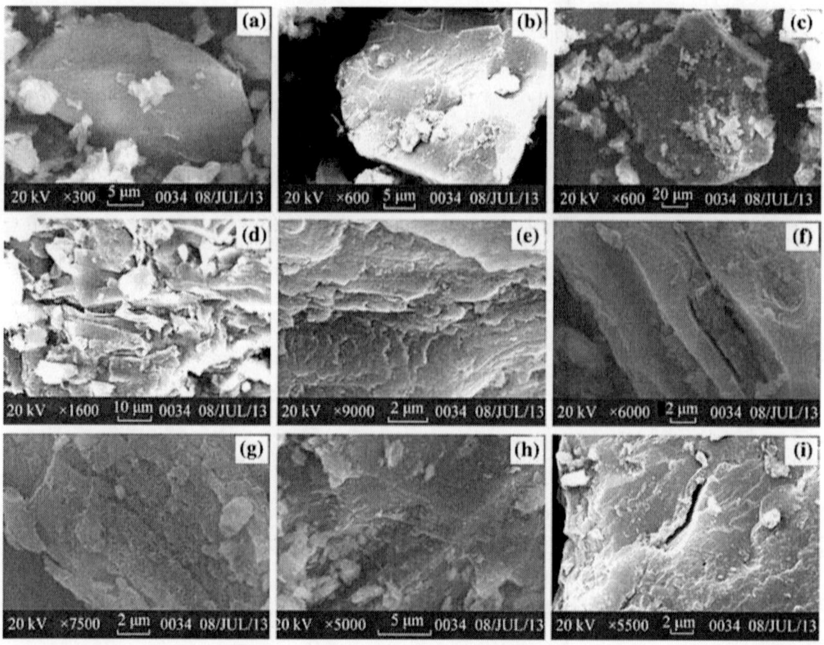

Figure 7. SEM (SE) morphology of the raw and beneficiated coal samples (**a**–**c** raw coals; **d**–**i** IL-treated coals)

3.7 Thermal properties of the coals

The TG–DTG profiles of the raw and beneficiated coals mainly indicate four temperature ranges (Fig. 8). The DTG curve express the starting and closing temperatures of the decomposition and the sharp peak gives the point of decomposition of the coal samples. The residual moisture releases at temperatures range of 25–100 °C. In this stage, the rate of weight loss increases as observed in the TG–DTG curves (Fig. 8). It is observed that the volatile matters start to be released along with the reduction of H-bonding around 223 °C. The weight loss is also increased at this stage as revealed from the TG curve of the both coal samples. However, both the beneficiated coal samples (oxidized) show significant release in volatile matters at this stage temperature. In the third stage (at higher temperature around 400 °C), the fixed carbon in coals begins to oxidize (releasing CO and H_2) and the volatile matters are completely released and the weight loss has been sharply increased. The functional groups, like R–O–R, R–OH present in beneficiated coals facilitates greater decomposition at lower temperature that are released as CO and CO_2. However, the C–H bond present in the raw coal samples are difficult to break than the C–O bond present in the oxidized coal samples (Saikia et al. 2009). Thus, the burning temperatures of the beneficiated coals are reduced indicating their better combustion efficiency. At a temperature range of 500–600 °C, the decomposition of ash occurs e.g., kaolinite → metakaoline at 550–600 °C. The combustion performance index (*S*) of the raw (B and SCF) and beneficiated coal samples (oxidized) were also determined. It is observed that the beneficiated SCF coal samples have lesser combustion performance index than that of the raw coal samples (Table 2). However, the B coal samples shows higher combustion performance index than that of raw coal samples. Thus, the beneficiated B coal samples will have higher combustion activity in the thermal application at lower temperatures. The combustion performance indexes (*S*) along with the activation energies (*E*a) of the raw and beneficiated coal samples are listed in the Table 2, which shows significant variations with different types of the treatments. The activation energies (*E*a) of the beneficiated coals are also found to be generally lesser than the raw coals that decrease with the rise in their S values. However, the activation energies of the IL oxidized coals are found to be higher, which may be due to the increase in carbon contents during the oxidation. It is to be noted that the lower the ignition temperature the easier the ignition of coal. The lower the burnout temperature, the shorter the burnout time of the coal, and the lower the combustible matter content of the ash. Thus, it is obvious that the initial ignition and burnout temperatures of coal samples not only vary with the reactants but also with the coal properties. The present partially green process can also be adopted for the development of value added products from coals.

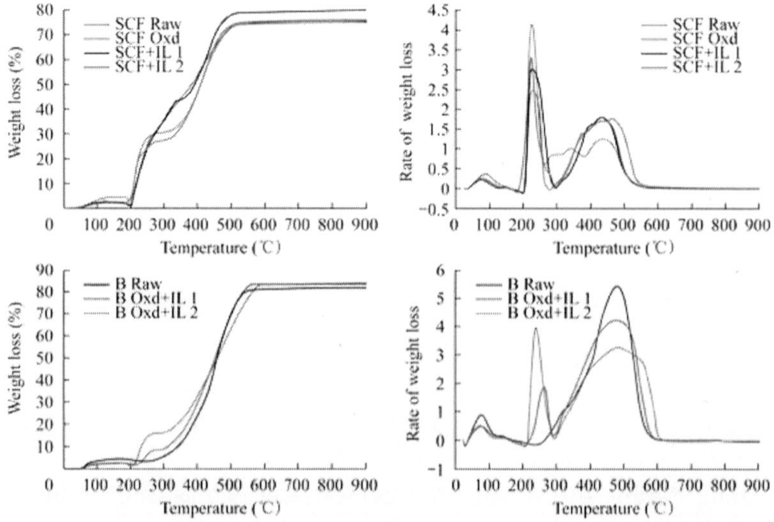

Fig. 8 TG–DTG of raw and beneficiated (B and SCF) coal samples

Table 2. Combustion characteristics of coals

Coal samples	Ignition temperature (T_i)	Burnout temperature (T_b)	The combustion performance index ($S \times 10^{-7}$)	Activation energy (Ea; kJ/mol)
SCF	222.18	586.53	3.18	164.64
SCF Oxd	240.13	564.85	1.25	155.34
SCF + IL1	219.13	564.85	1.40	192.90
SCF + IL2	219.13	546.79	1.22	192.93
B	483.61	625.55	356.8	208.00
B Oxd	492.47	641.91	323.0	241.30
B Oxd + IL1	485.74	627.43	418.0	163.70
B Oxd + IL2	487.91	652.67	450.8	181.42

4 CONCLUSIONS

The ILs can be used for the beneficiation of high sulfur coals through removal all forms of sulfur including mineral matters. In the present process, ILs are able to reduce the pyritic, sulfate and organic sulfur by maximum of 62.50 %, 83.33 % and 31.63 % from Meghalya (India) coals. The FT-IR, XRF and SEM analyses support the beneficiation process through removal of sulfur components and mineral matters from the coal samples. The GCVs of the beneficiated coals are enhanced after oxidation in ILs. The process removed some of inorganic species such as Ca, Mn and Ni. The combustion performance indexes (S) and activation energies (Ea) indicate the suitability of the beneficiated coals for thermal applications. Moreover, the chemical process finds merits for industrial application as it could be carried out in mild conditions. In summary, the present green process of chemical desulfurization of coals with ILs is observed to substitute the use lethal chemicals in coal beneficiation.

ACKNOWLEDGMENTS

The funding from the Ministry of Steel, Govt. of India (GAP-249) and CSIR (MLP-6000-WP-III) are highly acknowledged. Authors express thanks to the USIC-NEHU, Shillong (India) for the SEM images. Dr. Yongfeng Hu and Dr. Stephen G. Urquhart are highly acknowledged for the XRF spectra of coal samples at the Canadian Light Source, University of Saskatchewan. Many thanks are also given to Dr. Robert Finkelman for his comments to the draft.

REFERENCES

1. Ahmed M, Rahim A (1996) Chemical desulfurization methods of organic sulfur of Eocene coal, Jaintla Hills, Meghalaya. In: Fuel and Energy Abstracts vol 5. p 333
2. Ahmed A, Ahmad N, Shah R, Bhatti MN, Saleem M (2008) Coal desulfurization by solvent leaching methods. J Fac Eng Technol 15:47–56
3. ASTM (1991) Annual book of ASTM standards; coal and coke. Standard practice for proximate analysis of coal and coke 05:05
4. ASTM (2005) Annual book of ASTM standards; coal and coke. Standard test method for forms of sulfur in coal 05:06
5. ASTM (2010) Standard practice for collection of a gross sample of coal. ASTM International, West Conshohocken 2011
6. ASTM (2013) Annual book of ASTM standards; gaseous fuels; coal and coke; bioenergy and industrial chemicals from biomass. Standard test method for gross calorific value of coal and coke 05:06

7. Barooah PK, Baruah MK (1996) Sulphur in Assam coal. Fuel Process Technol 46:83–88

8. Baruah BP, Khare P (2007) Desulfurization of oxidized indian coals with solvent extraction and alkali treatment. Energy Fuels 21:2156–2164

9. Baruah BP, Saikia BK, Kotoky P, Rao PG (2006) Aqueous leaching on high sulfur sub-bituminous coals, in Assam, India. Energy Fuels 20:1550–1555

10. Bosmann A, Datsevich L, Jess A, Lauter A, Schmitz C, Wasserscheid P (2001) Deep desulfurization of diesel fuel by extraction with ionic liquids. Chem Commun 23:2494–2495

11. Calkins WH (1994) The chemical forms of sulfur in coal: a review. Fuel 73:475–484

12. Campos-Martin JM, Capel-Sanchez MC, Perez-Presas P, Fierro JLG (2010) Oxidative processes of desulfurization of liquid fuels. J Chem Tech Biotech 85:879–890

13. Chou CL (2012) Sulfur in coals: a review of geochemistry and origins. Int J Coal Geol 100:1–13

14. Davalos A, Pecina ET, Soria M, Carrillo FR (1994) Kinetics of coal desulfurization in an oxidative acid media. Int J Coal Prep Util 29:152–172

15. De Filippis P, Scarsella M (2003) Oxidative desulfurization: oxidation reactivity of sulfur compounds in different organic matrixes. Energy Fuel 17:1452–1455

16. Demirbas A, Balat M (2004) Coal desulfurization via different methods. Energy Sour 26:541–550

17. Ehsani MR (2006) Desulfurization of tabas coals using chemical reagents. Iran J Chem Eng 25:59–66

18. Esser J, Wasserscheid P, Jess A (2004) Deep desulfurization of oil refinery streams by extraction with ionic liquids. Green Chem 6:316–322

19. Falcon R, Ham AJ (1988) The characteristics of Southern African Coals. J S Afr Inst Min Metall 88:145–161

20. He L, Li H, Zhu W, Guo J, Jiang XLu (2008) Deep oxidative desulfurization of fuels using peroxophosphomolybdate catalysts in ionic liquids. Ind Eng Chem Res 47:6890–6895

21. Holbrey JD, Lopez-Martin I, Rothenberg G, Seddon KR, Silvero G, Zheng X (2008) Desulfurisation of oils using ionic liquids: selection of cationic and anionic components to enhance extraction efficiency. Green Chem 10:87–92

22. Ishak MAM, Ismail KB (2004) Sulfur removal from Mae Moh coal via mixture of hydrogen peroxide and formic acid. In: Proceedings of the 21st international Pittsburgh coal conference

23. Juniper LA, Pohl JU (1997) What price coal quality in electric power generation? The Australian Coal Review (7):37–42

24. Kendra-Krolik K, Fabrice M, Jaubert JN (2011) Extraction of thiophene or pyridine from n-heptane using ionic liquids gasoline and diesel desulfurization. Ind Eng Chem Res 50:2296–2306

25. Komintarachat C, Trakarnpruk W (2006) Oxidative desulfurization using polyoxometalates. Ind Eng Chem Res 45:1853–1856

26. Maes II, Gryglewicz G, Machnikowska H, Yperman J, Franco DV, Mullens J (1997) Rank dependence of organic sulfur functionalities in coal. Fuel 76:391–396

27. Mazumdar B, Saikia PC, Sain B, Baruah BP, Bordoloi CS (1989) Desulphurisation of high sulphur coals of north-east Indian Coalfields by organometallic compounds. Fuel 68:610–613

28. McCoy RN, Weiss FT (1954) Application of catalytic desulfurization to group sulfur analysis of petroleum fractions. Anal Chem 26:1928–1933

29. Nayak B, Singh AK, Upadhyay AK, September Bhattacharyya KK (2009) A note on the characters of some lower Gondwana coals

of West Siang district in the Arunachal Himalaya and their trace element content. J Geol Soc India 74:395–401

30. Qing W, Hao X, Hongpeng L, Chunxia J, Jingru B (2011) Thermogravimetric analysis of the combustion characteristics of oil shale semi-coke/biomass blends. Oil Shale 28:284–295

31. Rogers RD, Seddon KR (Eds) (2003) Ionic liquids as green solvents: progress and prospects. ACS Symposium Series

32. Saikia BK, Boruah RK, Gogoi PK, Baruah BP (2009) A thermal investigation on coals from Assam (India). Fuel Process Technol 90:196–203

33. Saikia BK, Ward CR, Oliveira MLS, Hower JC, Baruah BP, Braga M, Silva LF (2014) Geochemistry and nano-mineralogy of two medium-sulfur northeast Indian coals. Int J Coal Geol 121:26–34

34. Sain B, Saikia PC, Baruah BP, Bordoloi CS, Mazumder B (1991) Desulphurization of Assam coal by chlorinolysis and hydrothermal techniques. Fuel 70:753–755

35. Schmitz C, Datsevich L, Jess A (2004) Deep desulfurization of diesel oil: kinetic studies and process-improvement by the use of a two-phase reactor with pre-saturator. Chem Eng Sci 59:2821–2829

36. Silva G, Voth S, Szymanski P, Prokopchuk EM (2011) Oxidation of dibenzothiophene by hydrogen peroxide in the presence of bis(acetylacetonato)oxovanadium(IV). Fuel Process Technol 92:1656–1661

37. Silverstein RM, Bassler GC, Morril TC (1987) Spectrometric identification of organic compunds, 5th edn. Wiley, New York

38. Sonibare OO, Ehinola OA, Egashira R, KeanGiap L (2005) An investigation into the thermal decomposition of Nigerian coal. J Appl Sci 5:104–107

39. Stracher GB, Taylor TP (2004) Coal fires burning out of control around the world: thermodynamic recipe for environmental catastrophe. Int J Coal Geol 59:7–17

40. Yaman S, Ersoy-Mericboyu A, Küçükbayrak S (1954) Chemical coal desulfurization research in turkey. Fuel Sci Technol Int 13:49–58

41. Yperman J, Franco D, Mullens J, Van Poucke LC, Gryglewicz G, Jasienko S (1995) Determination of sulfur groups in pyrolysed low-rank coal by atmospheric-pressure T.P.R. Fuel 74:1261–1266

42. Zhang G, Yu F, Wang R (2009) Research advances in oxidative desulfurization technologies for the production of low sulfur fuel oils. Pet Coal 51:196–207

43. Zhao H, Xia S, Ma P (2005) Use of ionic liquids as 'green' solvents for extractions. J Chem Technol Biotechnol 80:1089–1096

44. Zhao D, Wang Y, Duan E, Zhang J (2010) Oxidation desulfurization of fuel using pyridinium-based ionic liquids as phase-transfer catalysts. Fuel Process Technol 91:1803–1806

Chapter 12

Benefaction and Pyrolysis of Şırnak Asphaltite and Lignite

Yildirım I. Tosun

Mining Engineering Department, Şırnak University, Şırnak, Turkey

ABSTRACT

Depending on advanced technological developments in energy production the low quality coals needed the most economical technologies and even in order to make it possible to produce coalderived products. Compliance with environmental norms of coal pyrolysis or gasification of various type of coals, feasible combustion systems and energy production facilities are needed in today's modern technology, also enable the production of liquid and gaseous coal fuels. However, raw materials and chemical nature of them requires a variety of adaptation methods. This study examined the high sulfur and ash types of Kütahya, Denizli, Aydin, Soma lignite, Sirnak asphaltite and lignite. The representative samples were taken from local areas of the lignites. Fundamentally, the conditions regarding better desulfurization way, the high quality pyrolysis lignite oil production, high value light oil, coal tar and gas products were determined at the goal of high fuel producing yield.

Keywords: Coal Pyrolysis, Desulfurization, Oil Char Production, Semi-Pyrolysis

1. INTRODUCTION

Lignite consumption in energy production is increasing in parallel with growing energy needs today. In terms of the available reserves and production amount, lignite natural resources are limited. The highest amount of coal resources are consumed in energy production in Turkey, as the most economical means [1] [2] . In Turkey approximately 80% (36 million tons) of total lignite production of almost 45 million tons was utilized in primary energy production. The clean coal fuels such as coal liquids, gases will also provide high value-added products for the benefit of the country and even will make a very positive growth and employment. Turkish Energy Ministry promotes to the highest priority researches and developments in coal cleaning and processing for evaluation low qualified coal resources in the southeastern Anatolian local region, for high coal reserves of Sirnak City and production of high quality asphaltite and lignite in Sirnak City. The most effective and cost-effective technologies are needed in today's modern technologies [3] [4] . Municipal authorities also seek to validate the feasibility of production of clean coal fuels such as coal liquids and gases and even injecting, storing and monitoring power station stack gases, CO_2 in the subsurface (geologic storage) as an approach to mitigate atmospheric emissions of CO_2. Coal industries promote technological developments required basic feasibility and the standard qualification tests[5] [6] . There are lots of signs for the co-production electricity from biomasses and lignite in various industrial fields even using irregular low capacity of cellulosic biomass wastes. Processing technologies using biomass should contribute to production of cleaner coal fuels [5] [6] . On the nature and characteristics of the based biomass and lignite, production of high value clean coal fuels should be commonly managed by pyrolysis and gasification of lignite and other cellulosic waste resources in the local region. Even, for this purpose, the country needs the further fuel cleaning technologies providing the essential clean oils and gases. For this reason, Kütahya Gediz, Denizli Çivril, Aydin, Tunçbilek, Soma, Sirnak lignites and Sirnak asphaltite were subjected to coal washing, pyrolysis and gasification and the production of coal liquids and gaseous fuels regarding energy development in the southeastern Anatolian region and further improve the diversification of fuels. Tunçbilek and Aydın washed lignite ash contents of clean coal products presented lower levels. For Tunç- bilek, Aydın lignites, various pyrolysis devices were used. Pyrolysis tests were carried out regarding the oil production ability and the efficiencies in the gasification of carbonized portion of coals, gas and liquid. The coal gas fuel production efficiencies were determined at the optimum values, and optimized pyrolysis parameters of the furnace equipment.

However, lignite-based technologies are examined on the basis of raw materials and contribute to the side of the forest waste biomass as raw material for cellulosic biomass can be processed together with the lignites. Some extent mixing was carried out in order to improve the pyrolysis quality of Sirnak asphaltite and lignite.

2. LITERATURE REVIEW

Considerable research on coal pyrolysis and gasification has been conducted over the years, but the pyrolysis results are widely dispersed because of the complex chemistry of coal [7] [8] . Time related coal-pyrolysis modelling assumes basically first-order kinetic equations, or less sensitive for heating rate. The other distributed activation model is dependent on the heating rate. The last two more advanced models need three and four constants, respectively, which basically depend on the coal properties but also cover to some extent, the effect of heat-and-mass transfer phenomena. That is the reason for the different values of the activation energy and preexponential factor cited in the literature and the lack of generally valid data. The same situation exists in the case of coal-char gasification. The reaction rate of char is influenced mainly by chemical and physical factors, which include coal type, catalytic effect of the ash and the specific surface area of char, which changes during the reaction course with the development of internal pores, and finally, their destruction [9][10] . In the case of the scaling-up procedure, the uncertainty of a complex model of the reacting system may be very high and it is reasonable under some conditions to use a methodology based on quasi-equilibrium conditions, which can be reflected at a larger scale. This approach assumes basically that the process itself, with all specific features, is a decisive factor for the path of the reactions of coal decomposition. Therefore a static model of char gasification and coal pyrolysis was developed: it is based on the assumption that the final process temperature is a decisive factor for the required volatile-matter content in the char being in a quasi-equilibrium state with respect to the gas temperature [11] [12] . To achieve this, it is necessary to create conditions of internal circulation of the transported coal and char in a riser, where the average concentration of solids amounts to 0.05 - 0.15 $m^3 \cdot m^3$, i.e. the conditions for residence time are long enough for the thermal decomposition of coal and intensive mixing so enhancing mass and heat transfers. On the other hand, the gasification of circulated char is mainly chemically controlled due to the relatively short residence time of the gas. Additionally, the characteristic feature for a given circulating fluid-bed reactor is a very narrow range of operational residence time, both for the gas and coal, which allows us to use the quasi-equilibrium approach. The model developed reflects the kinetic processes (transport phenomena of coal species, reaction rate) by constants of the developed equations, which depend mainly on the coal properties.

3. COAL PYROLYSIS IN A CIRCULATING FLUIDISED BED

A circulating fluidised-bed reactor was used as a coal pyrolysis reactor. The process was tested at a scale of 1 - 2 kg/h: collecting operational and design data to build an industrial installation. A technological diagram of the coal-pyrolysis process development unit is presented in Figure 1. As illustrated in Figure 1, where 1st unit is coal storage, 2nd unit is pyrolysis reactor, 3rd unit is coarse char cyclone, 4th unit is fine char cyclone, 5th unit is coarse char bag, 6th unit is circulating coal recycle.

In this research, representative specimens the different types of Turkish lignites; Kütahya Gediz, Denizli Çivril, Aydın, Tunçbilek, Soma, Şırnak asphaltite and lignite coals, were crushed and comminuted to minus 1 mm size by controlled screening. Air dried samples 1 - 2 kg from each different coal types were prepared and sealed in nylon bags. Proximate analysis of Turkish lignites and coals used in the experiments are given in Table 1. The qualities of processed coal products are ascertained by chemical and standard coal analysis of ASTM 3173- 317

4. RESULTS AND DISCUSSION

Screen analysis of Turkish lignite samples were made by standard Tyler Screens and particle size distributions and normal distributions of lignite samples are respectively illustrated in Figure 2. Coal particles in lignite samples were mainly distributed between 100 and 300 μm size fractions. High specific surface areas of 1160 cm^2/gr for lignite samples were found and also highly sufficient in order to react with alkali.

In the pyrolysis experiments with addition hydrated lime, reactor temperature changed between 100˚C and 400˚C and lignite samples mixed only by lime and soda and ash addition at equal rates. Produced coal specimens were subjected to chemical analysis for sulfur hold-up determination. Test results of pyrolysis by lime and addition of Turkish lignite are seen in Figure 3. Some Turkish lignite could not easily be desulfurized so that ash and sulfur contents of these lignite samples were high and further washing would improve pyrolysis results. Higher temperatures over 350˚C decreased desulfurization rate due to high volatile lost from coal samples. Lime might not be dissolved in soda mixture in the concern of nonpenetrating and lower reacting by coal surfaces so that resulted in lower desulfurization rates.

From the point of view of pyrolysis experimentation, Sirnak asphaltite calorific value was significant. That char quantity in the experimental conditions is calculated on the basis of the gas composition in the ambient

Table 1. Proximate analysis of Turkish lignites (ADB: Air dried base; DB: Dried base; DAB: Dried ashless base).

Coal Type	Ash, % ADB	Moisture, % ADB	Total S, % DB	Volatile Matter, % DAB
Şırnak Asphaltite	42.3	0.1	7.1	62.6
Şırnak Lignite	24.3	18.1	3.1	52.6
Aydın Şahinali	19.3	8.1	1.1	49.6
Kütahya Gediz	22.0	1.7	3.6	42.7
Denizli Çivril	15.2	12.7	3.0	45.2
Tunçbilek	25.7	5.7	3.9	38.2
Soma Kisrakdere	13.8	14.0	2.2	40.4

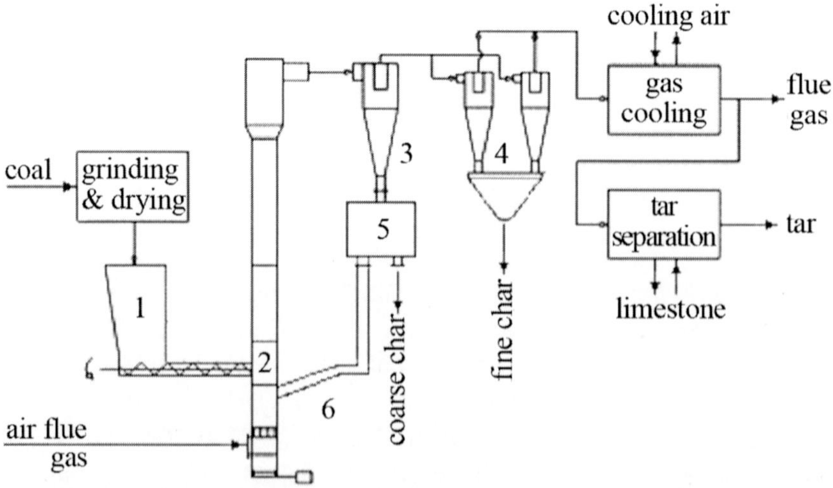

Figure 1. Fluidized coal-pyrolysis process.

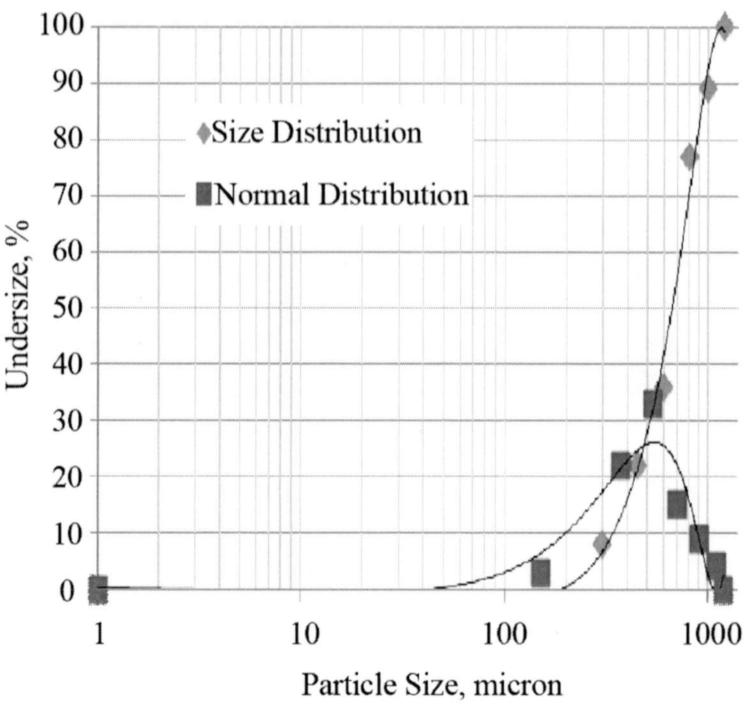

Figure 2. The particle size distributions and normal size distributions of Turkish lignites.

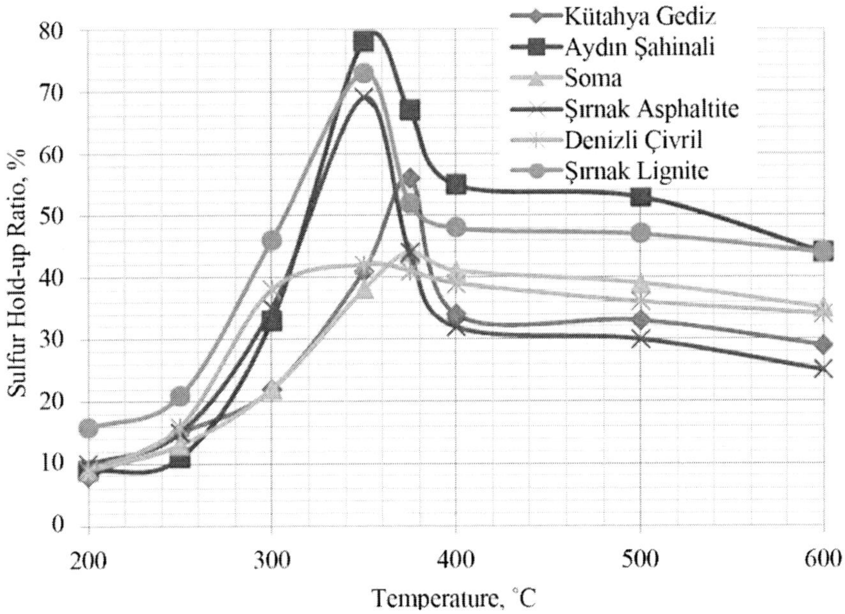

Figure 3. The sulfur hold-up rates in pyrolysis of Turkish lignites by hydrated lime and soda.

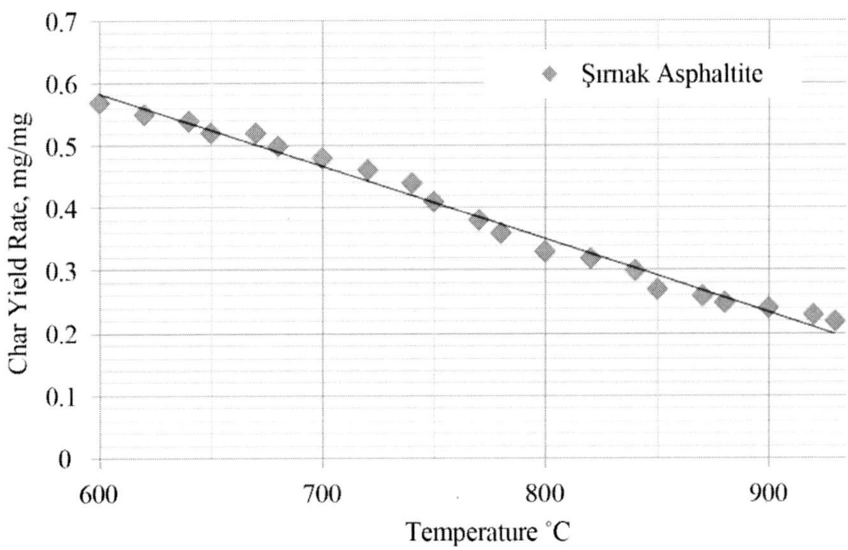

Figure 4. Comparison of experimental data and calculated results of char yield at pyrolysis temperatures.

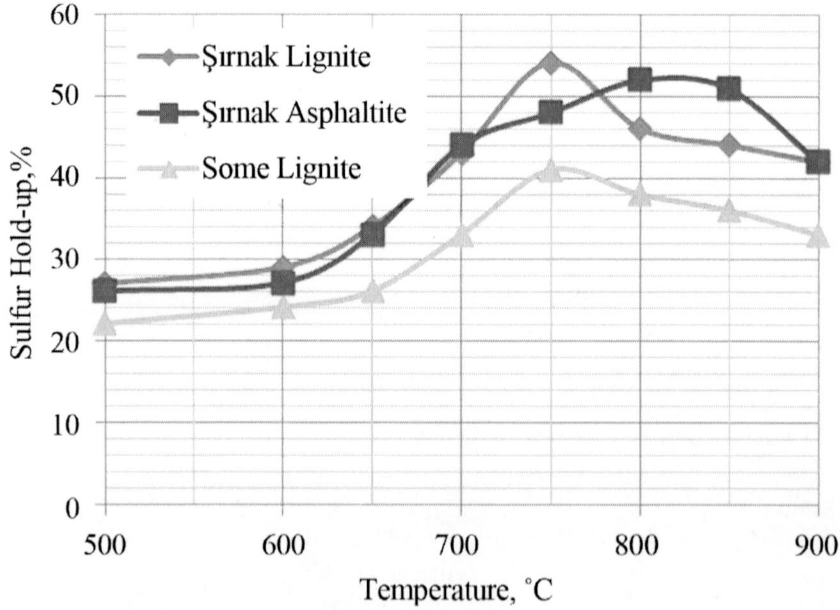

Figure 5. Sulfur hold-up rates in pyrolysis of Turkish lignites by marble waste.

state. So neither the contained water vapor nor the condensing hydrocarbons are taken into account. Pyrolysis liquid and gaseous products of Sırnak asphaltite may equal to 5 - 20 g·m^3 tar and 5 - 10 g·m^3 of benzene, toluene, xylene in unit process gas. These components increase the gas calorific value by 320 - 1200 kJ and, in total, allow one to obtain a gas of calorific value of up to approximately 6000 kJ·m^3 for Sırnak asphaltite (Figure 4). Char yield of Şırnak asfaltite and the other lignites increased at lower pyrolysis temperatures.

Test results of pyrolysis by the marble waste on desulfurization are seen in Figure 5. The specimens of Sırnak asphaltite and lignite, Soma lignite were efficiently desulfurized by marble waste (6% MgO). Semi pyrolysis and pyrolysis medium with marble waste decreased sulfur contents of chars and liquid products.

5. CONCLUSIONS

Research institutions mainly should search characteristics of lignite-to-date processes and improve technological applications and even advance processes to carry out performance tests, depending on the nature of the raw materials necessary to develop efficiencies.

Pyrolysis of different types of Turkish lignite was successfully processed in terms of deashing and desulfurization.

At higher rates of desulfurization of different types of Turkish lignites could be obtained from the tests using alkali reagents and hydrated lime mixtures. It has been clearly determined that lime and hydrated lime are much beneficial in caustic leaching of different types of Turkish lignites. It was also distinctly ascertained that marble waste fine containing approximately 6% magnesia was much efficient as a desulfurizing agent in pyrolysis. Magnesia content was efficient as hydrous lime but much more advantageous in absorption stage of gaseous medium.

Şırnak asphaltite should be cleaned and high ash content reduced char and oil yield in pyrolysis. Benefaction from Turkish lignites in the various combustion systems, in order to receive clean energy with clean coal liquid and gaseous products must be generated in lower temperature pyrolysis. It is also adviced that the high amount of formation of flue gas will be managed at higher pyrolysis temperatures over 700°C and even extracts more environmental friendly gaseous products. Also in the research works production of clean energy with the design of the addition of high-quality coal and biomass waste mixtures are processed and biomass fuels could be an alternative clean fuel sources. Clean coal fuels produced by pyrolysis may be supplied in southeastern Anatolian region in Turkey. Hence, those clean alternative resources will further enhance the industrial development in the region. For this purpose, further washing of Kütayha Gediz, Denizli Çivril, Aydın, Tunçbilek, Soma lignites, Sirnak asphaltite and lignite may improve pyrolysis products. Thus, the higher performance can be achieved by a certain mixture of solid fuel additives. A preferable advanced design in pyrolysis could produce clean coal fuels in the local site of the country so significant that needs to obtain the highest quality coal fuels.

REFERENCES

1. Anonymous (2009) TKİ Lignite Report 2009, TC Energy Ministry Reports. Ankara, Türkiye.

2. Anonymous (2009) TTK Coal Report 2009, TC Energy Ministry Reports. Ankara, Türkiye.

3. Boynton, R.S. and Gutschick, K.A. (1983) Lime, Industrial Minerals and Rocks. AIME, New York

4. Culfaz, M., Ahmet, M. and Gürkan, S. (1996) Removal of Mineral Matter and Sulfur from Lignites by Alkali Treatment. Fuel Processing Technology, 47, 99-109.

5. Reimers, G.W. and Franke, D.W. (1991) Effect of Additives on Pyrite Oxidation, RI: 9353. Bureau of Mines, Washington DC. [Paper reference:2]

6. Tosun Y.I., Rowson, N.A. and Veasey, T.J. (1994) Bio-Column Flotation of Coal for Desulfurization and Comparison with Conventional and Column Flotation. 5th International Conference of Mineral Processing, Nevşehir, 1994.

7. Wheelock T.D. (1979) Chemical Cleaning, Coal Preparation. 4th Edition, AIME, New York.

8. Yoon, R.H. (1991) Advanced Coal Cleaning, Part 2, Coal Preparation. 5th Edition, AIME, Colorado.

9. Donskoi, E. and McElwain, D.L.S. (1999) Approximate Modelling of Coal Pyrolysis. Fuel, 78, 825-835.

10. Wiktorsson, L.P. and Wanzl, W. (2000) Kinetic Parameters for Coal Pyrolysis at Low and High Heating Rates—A Comparison of Data from Different Laboratory Equipment. Fuel, 79, 701-716.

11. Fu W.-B. and Wang, Q.-H. (2001) A General Relationship Between the Kinetic Parameters for the Gasification of Coal Chars with CO_2 and Coal Type. Fuel Processing Technology, 72, 63-77.

12. Liu, G., Benyon, P., Benfell, K.E., Bryant, G.W., Tate, A.G. and Boyd R.K. (2002) The Porous Structure of Bituminous Coal Chars

and Its Influence on Combustion and Gasification under Chemically-Controlled Conditions. Fuel, 79, 617-625.

Chapter 13

Employment of Two-Stage Oxygen Feeding to Control Temperature in a Downdraft Entrained-Flow Coal Gasifier

Ting Wang[1], Xijia Lu[1], Heng-Wen Hsu[2], Cheng-Hsien Shen[2]

[1]*Energy Conversion & Conservation Center, University of New Orleans, New Orleans, LA, USA*

[2]*Green Energy and Environment Research Laboratory, Industrial Technology Research Institute, Hsinchu City, Taiwan*

ABSTRACT

The traditional practice of employing a two-stage coal-fed gasification process is to feed all of the oxygen to provide a vigorous amount of combustion in the first stage but only feed the coal without oxygen in the second stage to allow the endothermic gasification process to occur downstream of the second stage. One of the merits of this 2-stage practice is to keep the gasifier temperature low downstream from the 2nd stage. This helps to extend the life of refractory bricks, decrease gasifier shut-down frequency for scheduled maintenance, and reduce the maintenance costs. In this traditional 2-stage practice, the temperature reduction in the second stage is achieved at the expense of a higher than normal temperature in the first stage. This study investigates a concept totally opposite to the traditional two-stage coal feeding practices in which the injected oxygen is split between the two stages, while all the coal is fed into the first stage. The hypothesis of this two-stage oxygen injection is that a distributed oxygen injection scheme can also distribute the release of heat to a larger gasifier volume and, thus, reduce the peak temperature distribution in the gasifier. The increased life expectancy and reduced maintenance of the refractory bricks can prevail in the entire gasifier and not just downstream from the second stage. In this study, both experiments and computational simulations have been performed to verify the hypothesis. A series of experiments conducted at 2.5 - 3.0 bars shows that the peak temperature and temperature range in the gasifier do decrease from 600˚C - 1550˚C with one stage oxygen injection to 950˚C - 1230˚C with a 60 - 40 oxygen split-injection. The CFD results conducted at 2.5 bars show that 1) the carbon conversion ratio for different oxygen injection schemes are all above 95%; 2) H_2 (about 70% vol.) dominates the syngas composition at the exit; 3) the 80% - 20% case yields the lowest peak temperature and the most uniform temperature distribution along the gasifier; and 4) the 40% - 60% case produces the syngas with the highest HHV. Both experimental data and CFD predictions verify the hypothesis that it is feasible to reduce the peak temperature and achieve more uniform temperature in the gasifier by adequately controlling a two-stage oxygen injection with only minor changes of the composition and heating value of the syngas.

Keywords: Two-Stage Oxygen Feeding, Entrained-Flow Coal Gasification, Gasification Simulation, Syngas Composition, Clean Coal Technology, Multiphase Flow Simulation

1. INTRODUCTION

1.1. Brief Review of Fuel Feeding Scheme of Entrained Coal Gasification

Gasification is an incomplete oxidation process converting various carbon-based feedstocks into clean synthetic gas (syngas), which is primarily a mixture of hydrogen (H_2) and carbon monoxide (CO) with minor amounts of methane (CH_4) and inert nitrogen gas (N_2). There are four main classes of gasifiers: fixed bed gasifier, fluidized bed gasifier, entrained flow gasifier, and transport gasifier. The fixed bed gasifier and fluidized bed gasifier have low (425°C - 650°C) and moderate (900°C - 1050°C) outlet gas temperatures, respectively. Entrained flow gasifiers have high outlet temperatures (1250°C - 1600°C) and operate in the slagging range (the ash is fully liquid with low viscosity). The main benefits of entrained flow gasifiers can be summarized as: the ability to handle practically any coal as feed, syngas being free of oils and tars, high carbon conversion, suitable for synthesis gas products, and high throughput because of high reaction rates at elevated temperature.

According to the direction of the main flow inside the gasifiers, the entrained flow gasifiers can be classified as up-flow design such as the Shell SCGP and E-gas processes, and down-flow design such as the GE and Simmens gasifiers. For down-flow coal gasifier (GE gasifier), coal slurry and oxidant are introduced from the top of the gasifier. The hot raw gas produced during the gasification process flows toward the bottom of the gasifier. The ash forming components melt in the gasifier, flow down the walls of the reactor, and finally leave the reactor as a liquid slag.

Based on the fuel injection scheme, the entrained flow gasifiers can be categorized as single-stage injection and two-stage injection. The E-Gas process of ConocoPhillips features a two-stage up-flow gasifier design consisting of a horizontal cylindrical vessel as the first stage and a vertical cylindrical vessel as the second stage. Approximately 78% of the coal slurry and 100% of the oxygen are fed in the first stage. Additional coal-water slurry without oxygen is injected into this hot syngas in the second gasifier stage. Endothermic gasification reactions occur between the hot syngas and the second stage coal feed. This lowers the temperature of the syngas and increases the cold gas efficiency of the process. One of the most important advantages of this 2-stage practice is to keep the gasifier temperature low downstream from the 2nd stage and thus extends the life of the refractory bricks, decreases gasifier shut-down frequency for scheduled maintenance, and reduces maintenance costs. Although local stoichiometry as well as dynamic variations in the local stoichiometry will also affect corrosion of the refractory liner, this study only focuses on managing the temperature distribution.

1.2. Motivation, Hypothesis, and Objectives

Contrast to the traditional concept of two-stage coal feeding practices of injecting coal slurry into two different stages, it is hypothesized that splitting oxygen between two stages, while feeding all the coal in the first stage, i.e., a distributed oxygen injection scheme, can distribute heat release into a larger gasifier's volume and hence reduces the peak temperature and make temperature distribution more uniform in the gasifier. In this approach, the increased life expectancy and reduced maintenance of the refractory can prevail in the entire gasifier-not just downstream from the second stage as in the traditional two-stage coal slurry fed system. To verify this hypothesis, both experiments and computational simulations are conducted. Since conducting experiments are expensive, time consuming, and usually performed with limited cases, it would be more effective and meaningful to establish a computational fluid dynamics (CFD) model to simulate effects of varying operating conditions to help obtain a preliminary understanding of the fundamental thermal-flow physics inside the gasifier. However, since the coal gasification is a very complicated multi-component and multi-phase process, a trustful and fully calibrated computational model is not available. To obtain a usable computational model as a meaningful tool, the experimental data is utilized to guide interpretation of the computational result in low operating pressure (2.5 bars) cases. Afterwards, the computational model is employed to perform detailed thermal-flow and chemical reaction simulation for more parametric studies that have not been performed by experiments. To this end, the objectives of this paper are:

1) Conduct experiments to investigate the effect of a controlled two-stage distributed oxygen approach on a gasifier's inner wall temperature distribution.

2) Interpret the CFD results and determine how the results can be effectively used by comparing them with the baseline experimental data.

3) Use the CFD model to simulate experimental cases with different oxygen split fed from 100% to 40% under the condition of 2.5 bars and 0.6 stoichiometric ratio.

4) Investigate computationally the effect of distributed oxygen fed on gasification performance and temperature distribution under a higher operating pressure (25 bars) and 0.4 stoichiometric ratio.

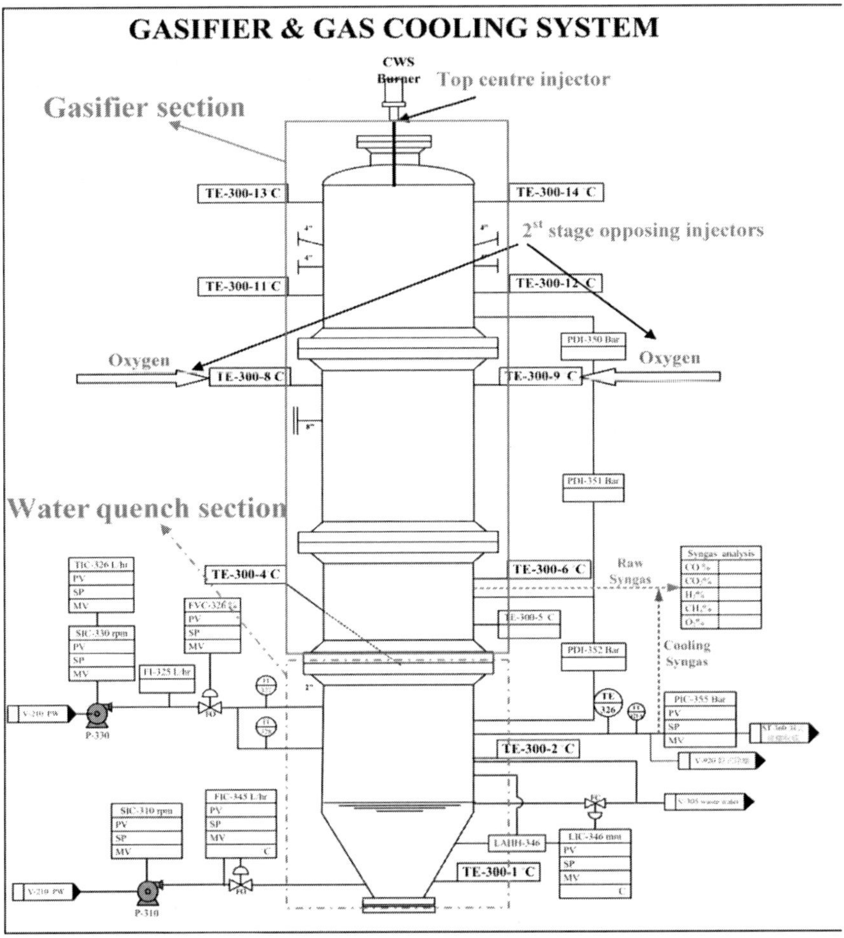

Figure 1. ITRI coal-slurry fed, oxygen blown, downdraft entrained-flow gasifier. The outline of the gasifier is based on the outside diameter of the refractory.

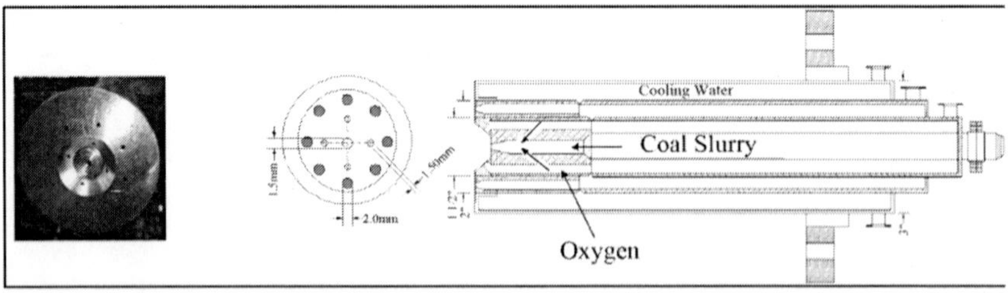

Figure 2. The internally premixed slurry-coal/oxygen injector with the coal-slurry fed in the center with oxygen fed through the co-centric outer annular pipe.

2. EXPERIMENTAL FACILITY

Industrial Technology Research Institute (ITRI) has designed a 2 ton/day, two-stage coal slurry fed, down draft entrained-flow gasifier for this study, as shown in Figure 1. This experimental facility is located in the southern Taiwanese city of Kaohsiung. The height of the gasification section above the quench section is 4.831 m, the inner diameter is 0.27 m, and the thickness of the refractory is 0.35 m. The coal powder is first mixed with water at a mass ratio of approximately 43% (coal)/57% (water). The coal slurry is then transported through the center tube of the injector as shown inFigure 2. All the oxygen is fed through the outer concentric annular pipe and is mixed with the slurry coal before injection. The total oxygen fed to the gasifier is about 60% of the stoichiometric amount needed for complete coal combustion (i.e., with a stoichiometric ratio of 0.6). The injector is water cooled through an annular water jacket wrapping around the fuel and oxygen passages. The blunt injector tip with multiple holes is derived from a previous study which investigated the performance and life expectancy of two injectors' designs [1] . It was discovered that the injector with a blunt tip geometry was able to reduce temperature gradient along the injector and significantly increase the life of the fuel injector more than the injector with the conical tip. The injector used in this study is designed and manufactured for experimental use only with three options for feeding oxygen-via the outer eight holes or inner four holes for external mixing, or using premixing scheme as shown in Figure 2. In this study, premixing scheme is used and other holes are plugged.

The gas temperatures are sampled with thermocouples being placed in cavities recessing one-inch in the wall. The gasification process occurs in a down-drafted mode with the syngas exiting near the bottom of the gasifier and continuously entering the quenching section. A detailed discussion about the quench section has been documented in Wang et al. [2] .

With the purpose of finding an optimum oxygen feeding ratio between injections from the top and a second location about 1/3 from the top of the gasifier, a series of experiments were conducted. The experiment is first conducted with 100% oxygen injected from the top, but it is discovered that the temperature is too high on the top and too cold in the middle section of the gasifier. Moreover, the carbon conversion ratio is poor, at around 72%. To improve the temperature distribution and carbon conversion ratio, part of the oxygen up to 60% (Figure 1), is diverted from the top injector to the side injectors in the middle section. This is done to hopefully provide more effective oxidation and hence more energy for feeding the gasification process in the middle section. The results are encouraging and show that the previous low temperatures in the middle and lower sections of the gasifier increased while the high temperature on the top reduced, resulting in a more uniform temperature distribution along the entire gasifier. More detailed analysis of the experimental results will be presented together with the computational results later.

3. COMPUTATIONAL MODEL

3.1. Global Gasification and Chemical Reactions

This study deals with the global chemical reactions of coal gasification [3] that can be generalized in reactions (R1.1) through (R1.10) in Table1

In this study, the methanation reactions are not considered since the production of methane is negligible under the studied operating conditions. The volatiles are modeled to go through a two-step thermal cracking process (R7) and gasification processes (R8, R9) with CH_4 or C_2H_2 as the intermediate products. The coal used in this study is Chinese Shanxi Pinghsuo coal, and its compositions are given in Table2 The compositions of volatiles are derived from the coal's heating value, proximate analysis, and ultimate analysis.

For the reaction rate of water-gas shift (WGS) reaction (R5), Lu and Wang [8] [9] tested Jones' rate by comparing the syngas composition results with the experimental data and, due to the difference in operating pressure and other boundary conditions, they purposely reduce the rate constant to A = 2.75 to match the experimental data. Consequently, A = 2.75, E = 8.38 ´ 10^7 is used for the WGS reaction rate in this paper.

Table 1. Summary of reaction rate constants referenced in this study.

	Reactions	Reaction Type	Reaction heat, ΔH°_R (MJ/kmol)	$k = AT^n \exp(-E/RT)$ (n = 0)		Reference
				A (s^{-1})	E (J/kmol)	
			Heterogeneous Reactions			
R 1	$C(s) + ½ O_2 \rightarrow CO$	Partial combustion	−110.5	0.052	6.1×10^7	Chen *et al.* [4]
R 2	$C(s) + CO_2 \rightarrow 2CO$	Gasification, Boudouard reaction	+172.0	0.0732	1.125×10^8	Chen *et al.* [4]
R 3	$C(s) + H_2O \rightarrow CO + H_2$	Gasification	+131.4	0.0782	1.15×10^8	Chen *et al.* [4]
			Homogeneous Reactions			
R 4	$CO + ½ O_2 \rightarrow CO_2$	Combustion	−283.1	2.2×10^{12}	1.67×10^8	Westbrook and Dryer [5]
R 5	$CO + H_2O(g) \leftrightarrow CO_2 + H_2$	Water Gas shift	−41.0	2.75×10^{10} #	8.38×10^7	Jones and Lindstedt [6]
R 6	$CO + 3H_2 \leftrightarrow CH_4 + H_2O$	Methanation	−205.7	$k_f = 4.4 \times 10^{11}$	1.68×10^8	
				$k_b = 5.12 \times 10^{-14}$	2.73×10^4	Benyon [7]
R 7	$CH_{2.997}O_{0.3456} \rightarrow 0.3456CO + 0.1897H_2 + 0.6544CH_4$ (Two-step Volatiles Cracking)		+12.088	Eddy dissipation		
R 8	$CH_4 + ½ O_2 \rightarrow CO + 2H_2$ Volatiles gasification via CH_4		−35.71	Eddy dissipation		N/A
R 9	$C_2H_2 + O_2 \rightarrow 2CO + H_2$ Volatiles gasification via C_2H_2		−447.83	Eddy dissipation		
R 10	$H_2 + ½ O_2 \rightarrow H_2O$	Oxidation	−242	6.8×10^{15}	1.68×10^8	Jones and Lindstedt [6]

1) All ΔH°_R at 298 K and 1 atm. 2) "+" Endothermic (absorbing heat), "−" Exothermic (releasing heat). # The value of A in [6] has been modified as 2.75 in this study.

3.2. Computational Methodology

The computational model and submodels (devolatilization, reactions, particle dynamics, gasification) used in the study are the same as developed by Silaen and Wang (2010) [10] , as well as Lu and Wang (2013) [8] [9] . Therefore, the governing and associated equations and detailed modeling intricacies are not repeated here, but they are briefly summarized below.

Table 2. The proximate and ultimate analyses of Chinese Shanxi Pinghsuo coal.

Coal	Pingshuo Coal
Proximate Analysis, wt%	
Moisture	4.31
VM	32.60
Ash	13.52
Fixed Carbon	49.57
Higher Heating Value (HHV) (kcal/kg)	6612
Ultimate Analysis, wt%	
Moisture	2.04
Ash	12.69
C	70.94
H	4.78
N	0.78
S	0.58
O	8.19

The time-averaged steady-state 3-D Navier-Stokes equations as well as the mass and energy conservation equations are solved. Species transport equations are solved for all gas species involved. It is well known that turbulence modeling has always been a sensitive issue for flow simulations because each turbulence model usually works only for some specific flow behaviors and none of any turbulence model can predict all the flows with high fidelity. Since investigation of the effect of turbulence models on gasification process is not an objective of this study, selection of an appropriate turbulence model is guided by a previous research performed by Silaen and Wang [10] . They applied five turbulence models (standard k-ε, k-ω, RSM, k-ω SST, and k-ε RNG) and reported that the RSM model predicted the best results. However, the standard k-ε turbulence model yields reasonable results without requiring significant computational time when compared to the more elegant RSM model. Therefore, the standard k-ε turbulence model is used in this study to provide closure. Enhanced wall function and variable material property are used. The P1 model is used for computing the radiation effect.

The flow (continuous phase) is solved in Eulerian form as a continuum while the particles (dispersed phase) are solved in Lagrangian form as a discrete phase. Stochastic tracking scheme is employed to model the effects of turbulence on the particles. The continuous phase and discrete phase are communicated through drag forces, lift forces, heat transfer, mass transfer, and species transfer. The finite-rate combustion model is used for the heterogeneous reactions, but both the finite-rate and eddy-dissipation models are used for the homogeneous reactions, and the smaller of the two is used as the reaction rate. The finite-rate model calculates the reaction rates based on the kinetics while the eddy-dissipation model calculates based on the turbulent mixing rate of the flow. Gasification or combustion of coal particles undergoes the following global processes: 1) evaporation of moisture; 2) devolatilization; 3) gasification to CO and 4) combustion of volatiles, CO, and char. The Chemical Percolation Devolatilization (CPD) model [11] - [13] is chosen as the devolatilization model based on the finding by Silaen and Wang (2010) [10] that the Kobayashi two-competing rates devolatilization model [14] is very slow while the CPD model gives a reasonable result.

For solid particles, the rate of depletion of the solid, due to a surface reaction, is expressed as a function of kinetic rate, solid species mass fraction on the surface, and particle surface area. The reaction rates are all global net rates, i.e., the backward reaction, calculated by equilibrium constants, are included in the global rate. Therefore, the finite rate employed in this study implicitly applies to the local equilibrium approach. Reaction rate constants used in this study are summarized in Table1

For liquid droplets, water evaporates from the particle's surface when the temperature is higher than the saturation temperature (based on local water vapor concentration). The evaporation rate is controlled by the water vapor diffusion from the droplet surface to the surroundings until 100% relative humidity is achieved. When the boiling temperature (determined by the air-water mixture pressure) is reached, water evaporation rate is controlled by the heat transfer rate at the droplet surface. After the moisture is evaporated, due to either high temperature or low moisture partial pressure, the vapor diffuses into the main flow and is transported away. Droplet breakup model and coalescence model have been applied for tracking the droplet movement. Please refer to Silaen and Wang [10] for details, which are not repeated here. Since this paper focuses on using an established existing CFD model and submodels to study the effect of distributed oxygen injection rather than develop a new model, no review of other research group's modeling efforts is conducted here.

3.3. Computational Domains

The computational domain and elements on the gasifier wall are shown in Figure 3. The computational domain contains roughly 1.1 million elements. A grid sensitivity study of 3-D geometries is conducted. Three different grids are used including a coarse grid (0.4 million cells), a medium grid (1.1 million cells), and a fine grid (1.6 million cells). The results of the temperature distribution along the gasifier show a difference of 7% between the coarse and medium grids and 2% between the medium and the fine grids. In order to save computational time, the medium grid of 1.1 million cells is chosen.

FLUENT 12.0.16 from ANSYS, Inc. is used as the CFD solver. The simulation is steady-state and uses the segregated solver, which employs an implicit pressure-correction scheme and decouples the momentum and energy equations. The SIMPLE algorithm is used to couple the pressure and velocity. Second order upwind scheme is selected for spatial discretization of the convective terms and species. For the finite rate model where the Eulerian-Lagrangian approach is used, the iterations are conducted alternatively between the continuous and the dispersed phases. Initially, two iterations in the continuous phase are conducted followed by one iteration in the discrete phase to avoid the flame from dying out. Once the flame is stably established, five iterations are performed in the continuous phase followed by one iteration in the dispersed phase. The drag, particle surface reaction, and mass transfer between the dispersed and the continuous phases are calculated. Based on the dispersed phase calculation results, the continuous phase is updated in the next iteration, and the process is repeated. Converged results are obtained when the residuals satisfy mass residual of 10^{-3}, energy residual of 10^{-5}, and momentum and turbulence kinetic energy residuals of 10^{-4}. These residuals are the summation of the imbalance in each cell, scaled by a representative for the flow rate.

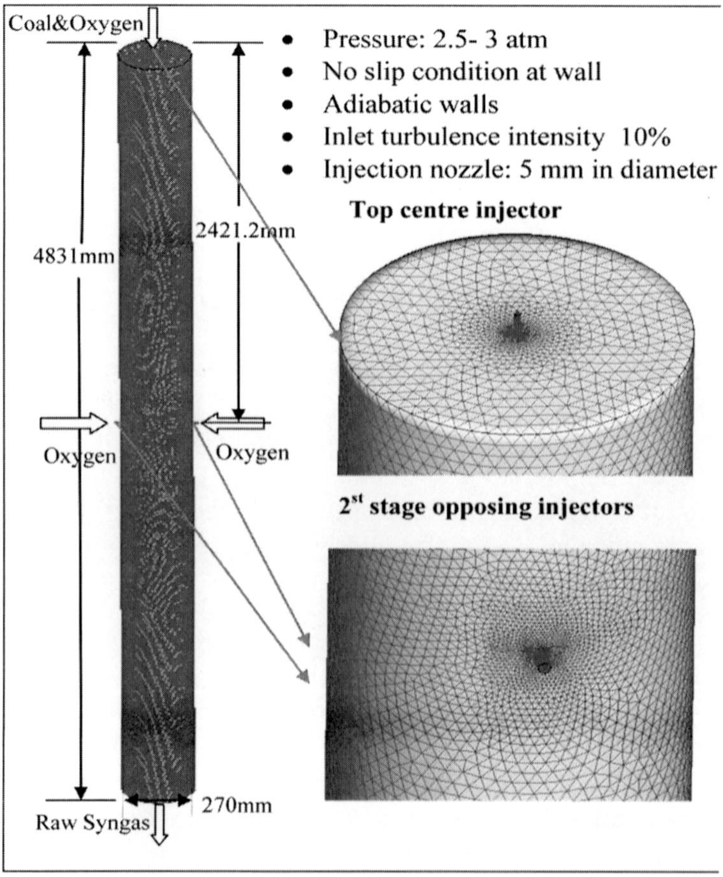

Figure 3. Meshed computational domain of the two-stage entrainedflow gasifier.

3.4. Boundary and Inlet Conditions

The summary of the studied cases is listed in Table3 In Case 1, a two-stage configuration is used with oxygen distribution of 100% - 0% between the top and the second stages. The coal is fed as slurry. The total mass flow rate of the coal slurry is 92 kg/hr (0.026 kg/s), and the total volume flow rate of oxidant is 36.8 Nm^3/s (0.0146 kg/s). The top and the second injection nozzles are both 5 mm in diameter. To match the experimental conditions, the operating pressure is set as 3 bars for Case 1. From Case 2 to Case 8, the oxygen distribution is 100% - 0%, 90% - 10%, 80% - 20%, 70% - 30%, 60% - 40%, 50% - 50%, 40% - 60%

between the top and the second injection stages respectively. Total mass flow rate of the coal slurry is 85 kg/hr (0.024 kg/s), and the total volume flow rate of oxidant is 38.6 Nm^3/s (0.0153 kg/s). The operating pressure is 2.5 bars for Cases 2 to 8.

The above eight cases from 1 - 8 are set up with the same operating conditions as the experiments. The following five cases from Cases 9 - 12 increase the operating pressure to 25 bars at which no experiments have been conducted in the current facility. From Case 9 to Case 12, the oxygen distribution between the top and the second injection stages is 100% - 0%, 80% - 20%, 60% - 40%, 40% - 60% respectively. Total mass flow rate of the coal slurry is 85 kg/hr (0.024 kg/s), and the total volume flow rate of the oxidant is 17.4 Nm^3/s (0.006912 kg/s), which is 30% of the stoichiometric amount for complete combustion.

Table 3 . Parameters and operating conditions of the studied cases.

	Case 1	Cases 2 - 8	Cases 9 - 12 (CFD only)
O_2 distribution (top-second)	100% - 0%	(100% - 0%) - (40% - 60%)	(100% - 0%) - (40% - 60%)
Coal slurry injection (kg/s)	0.026	0.024	0.024
O_2 injection (kg/s)	0.0146	0.0153	0.009216
Operating pressure (bars)	3.0	2.5	25

The oxidant is considered to be a continuous flow, and the coal particles are considered to be the discrete phase. The discrete phase includes the fixed carbon and liquid water droplets from both the inherent moisture content of coal (4.31% wt) and the injected water content form the coal slurry (57.1% wt). Other components of the coal, such as N, H, S, O, and ash, are injected as gas together with the oxidant in the continuous flow to keep the mass flow the same as in the experiments. N is treated as N_2, H as H_2, and O as O_2. S and ash are not modeled, and their masses are lumped into N_2. Since studying the effect of coal particle sizes on gasification performance is not in the objectives of this study, the coal slurry size is uniformly given as 50 mm for the purpose of conveniently tracking the change of particle sizes. The effect of coal particle size on gasification process has been discussed in [5] and is not investigated in this study.

The walls are assigned as adiabatic with internal emissivity of 0.8. In case of dry wall, interactions between the droplets and wall usually manifest three major phenomena, including reflect, break-up and trap. The actual phenomenon depends on the incoming Weber number of the droplet. Here, the Weber number is the ratio of kinetic energy of a droplet to the surface energy of a droplet ($We = \rho d_d v_d^2 / \sigma$). If the Weber number ($We_{in}$) is less than 10, the droplet will reflects elastically after it hits the wall. As the incoming We increases further to $We_{in} > 80$, the droplet falls into disintegration region which leads breaking-up the droplet to several small droplets. In the transition region of $30 < We < 80$, the droplet has the chance to either reflect or breakup. In the case of this study, since the Weber number is less than 10, the boundary condition of the discrete phase at walls is assigned as "reflect", which means the discrete phase elastically rebound off once reaching the wall. At the outlet, the discrete phase simply escapes/exits the computational domain. An area near the coal injection locations is initially patched with temperature of 1500 K to simulate the ignition process of a real operation. The limit of the highest temperature is assigned as 5000 K, and the limit of the lowest temperature is assigned as 400 K. This will remove potential runaway conditions caused by erratic, unreasonably high, or low temperatures during the iteration process.

4. RESULTS AND DISCUSSIONS

4.1. Comparison between CFD Results and Experimental Data

The computational cases 1 - 8 are conducted with the conditions matching those in the experiments. Among these eight cases, the detailed temperature and species distributions of Case 1 and Case 6 are exhibited to gain fundamental understanding of the thermal-flow behavior inside the gasifier. The conditions of these two cases are listed below.

- Case 1: Oxygen distribution (100% - 0%), coal slurry (92 kg/hr), oxidant (36.8 Nm^3/s or 0.0146 kg/s), and operating pressure (3 bars).
- Case 6: Oxygen distribution (60% - 40%), coal slurry (85 kg/hr), oxidant (38.6 Nm^3/s or 0.0153 kg/s), and operating pressure (2.5 bars).

4.1.1. Case 1 (100% - 0% for Oxygen Distribution)

Case 1 is one stage oxygen-blown operation with the oxygen distribution being 100% - 0% between the top and the second stages. Gas temperature and species mole fraction distribution on the horizontal and selected center vertical planes in the gasifier are shown in Figure 4. Since all the fuel is injected from the top, the gas temperature is higher in the top injection region than at the second stage location. Maximum gas temperature in the top injection stage reaches 2300 K. The dominant reaction in the top injection stage is the intense char combustion ($C + ½ O_2 \rightarrow CO$ and $CO + ½ O_2 \rightarrow CO_2$) as well as volatiles combustion via our 2-stage model in the first stage. Near the second stage, gasification reactions (mainly $C + ½ CO_2 \rightarrow CO$, $C + H_2O \rightarrow CO + H_2$) dominate. Due to the exothermic property of combustion process and endothermic property of the gasification reactions, the temperature drops gradually from top injection to the exit of the gasifier. Oxygen is completely depleted through the char combustion in the top injection region. The exit syngas composition calculated by CFD is shown in Table4

The mass-flow-rate weighted exit syngas temperature across over the entire exit section calculated by CFD is 1044 K and the near-wall average temperature is 986 K. They are 90 K and 32 K higher than the measured data, respectively. Note that the gas temperature was too hot to be measured by thermocouple, so the experimental temperature data was measured by thermocouples placed in cavities in the wall. Thus, it is not intended to compare the measured temperatures with the simulated core gas temperatures, rather the measured data are used to provide a reference for comparing the local near-wall syngas temperature at 5 mm from the wall.

Therefore, the comparison indicates that the average cross-sectional gas temperature is about 90 K hotter than the wall temperature and the near-wall temperature is about 32 K higher than the wall temperature.

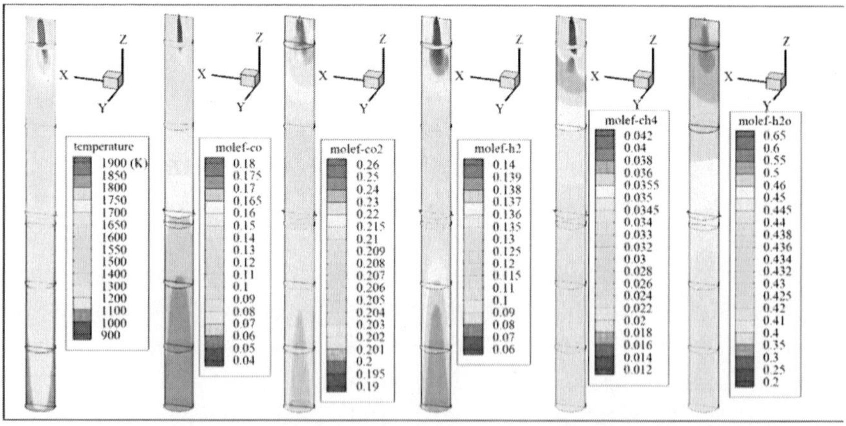

Figure 4. CFD simulated gas temperature and species mole fraction distributions for Case 1 (100% - 0% for oxygen distribution).

Table 4. Exit syngas composition of CFD results for Cases 1 and 6 and comparison of near-wall syngas temperature with measured temperatures in a wall cavity.

CFD syngas composition (Vol.)	Case 1	Case 6
CO	17%	18%
CO_2	20%	22%
H_2	14%	14%
H_2O	46%	44%
CH_4	3%	2%
CFD Temp (K)		
Cross-section average	1044	1550
Near-wall average	986	1172
Experi. Temp (K) in wall cavity	954	1150

4.1.2. Case 6 (60% - 40% for Oxygen Distribution)

Case 6 is the two-stage oxygen-blown operation with oxygen distribution of 60% - 40% between the first and the second stages. Gas temperature and species mole fraction distribution on the horizontal and center vertical planes in the gasifier are shown in Figure 5. The gas temperature is higher at the second stage injection region than at the top injection region. Maximum gas temperature reaches 2600 K at the second stage injection location and 1900 K at the top injection stage. The dominant reaction near the top injection location is the intense char combustion ($C + ½ O_2 \rightarrow CO$ and $CO + ½ O_2 \rightarrow CO_2$) as well as the volatiles oxidation via our two-stage model; then the endothermic gasification reactions occur in the region between the first and second stage injection locations as can be seen by increased CO and H_2 but reduced temperature. WGS also occurs, evident by reduced H_2O. At the second stage injection location, unburned char, CO and H_2quickly react with injected oxygen from two sides and release the second wave of combustion/oxidation heat ($C + ½ O_2 \rightarrow CO$, $CO + ½ O_2 \rightarrow CO_2$ and $H_2 + ½ O_2 \rightarrow H_2O$). The gasification reactions dominate the process after the oxygen is completely consumed in the region between the second stage injection location and the gasifier's exit.

Figure 6 and Figure 7 show CFD cross-sectional and mid-plane plots of the velocity vector, contours of temperature and mole fraction of CO_2 at the second injection stage for both Case 6 (60% - 40% for oxygen distribution) and Case 3 (90% - 10% for oxygen distribution) respectively. In Case 6, 40% oxygen, which is injected from two opposing injectors at the second stage, squeezes the main flow laterally to both sides perpendicular to the injection direction (top and bottom of Figure 6(a)and Figure 6(b)). The contours of temperature and mole fraction of CO_2 show that the chemical reactions are affected by the local flow behavior, which is demonstrated by the high temperature and CO_2 concentration presented towards top and bottom in Figure 6(a) and Figure 6(b). In Case 3, only 10% oxygen is injected at the second stage. The opposing injecting effect is not strong enough to block the main flow, so the combustion reaction only takes place below the second stage, which is shown by the contours of temperature and mole fraction of CO_2 in Figure 7.

The CFD calculated temperature is shown in Table 4 by two values. The first value (1550 K) is the mass-flow-rate weighted average temperature over the cross-section on the exit plane, and the second value (1172 K) is the mass-flow-rate weighted average temperature in the near-wall region (5 mm to the wall). The near-wall temperature is comparable to the experimental data (1150 K) within 22 K. Since the thermocouples were placed in cavities recessing in the wall, the measured data adequately reflects the relatively cooler syngas temperature near the wall. Figure 8 shows the temperature contour on the cross-section of the exit plane. It clearly shows that the temperature changes significantly from the center at around 2000 K to around 1150 K in the near-wall region.

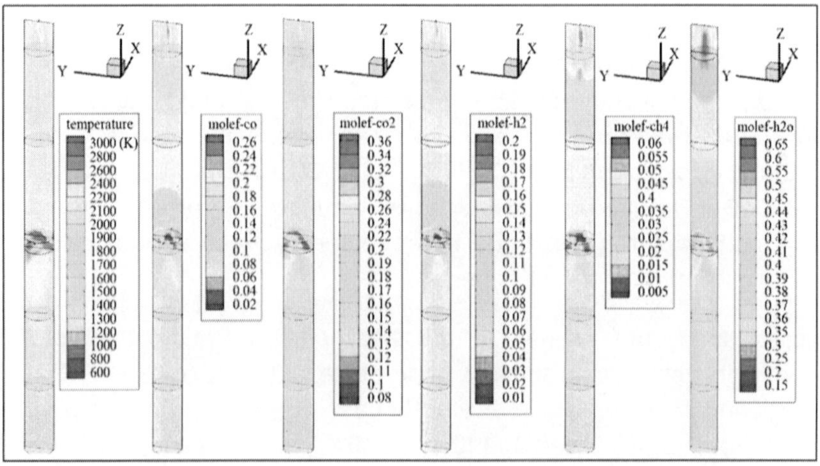

Figure 5. CFD simulated gas temperature and species mole fraction distributions for Case 6 (60% - 40% for oxygen distribution).

4.2. Effect of Oxygen Injection Distribution under the Experimental Condition (Operating Pressure: 2.5 Bars, O_2: 0.6 of Stoichiometric Amount)

To investigate the effect of different oxygen-distribution feeding schemes on the gasification performance under the low operating pressure condition, seven computational simulations cases corresponding to the experimental conditions (operating pressure: 2.5 bars, O_2: 0.6 of stoichiometric amount for complete combustion) are conducted with the oxygen fed from top at 100%, 90%, 80%, 70%, 60%, 50%, and 40% respectively. These cases are numbered from Case 2 to Case 8 as shown in Table3 The CFD results of these 7 cases are shown in Table 5, including the syngas composition at the exit, the syngas temperature at four locations (T_{exit} = TE-300-4, T2 = TE-300-8, T3 = TE-

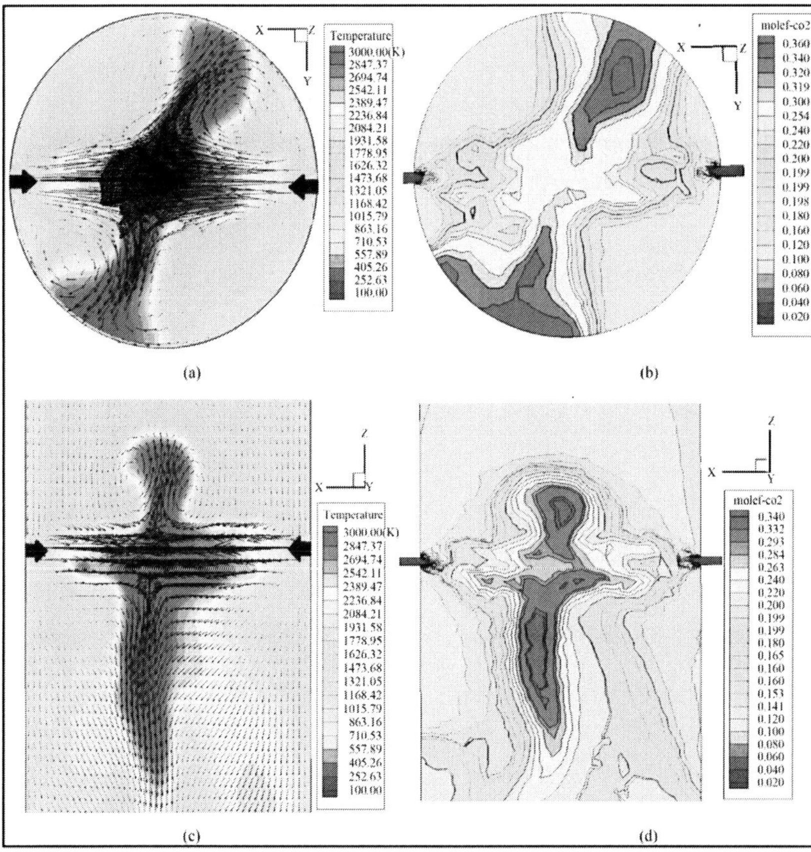

Figure 6. CFD cross-sectional and mid-plane plots of velocity vector, contours of temperature and mole fraction of CO_2 for Case 6 (60% - 40% for oxygen distribution). (a) Cross-sectional temperature contour and velocity vector at the second injection. (b) Cross-sectional mole fraction of CO_2 at the second injection. (c) Mid-plane temperature contour and velocity vector near second injection. (d) Mid-plane mole fraction of CO_2near second injection.

300-12, T4 = TE-300-13), carbon conversion ratio (CCR), and the higher heating value (HHV). The comparisons between CFD results and experimental data are shown graphically in Figure 9. The variations of simulated syngas compositions for different cases are shown in Figure 10.

Figure 9 shows that the experimental data and the CFD results yield the same trend of change for temperature distribution at the exit as the oxygen feeding distribution changes between two stages. When the percentage of oxygen injected from top reduces from 100% to 40%, the temperature decreases at the first stage (T4 and T3)

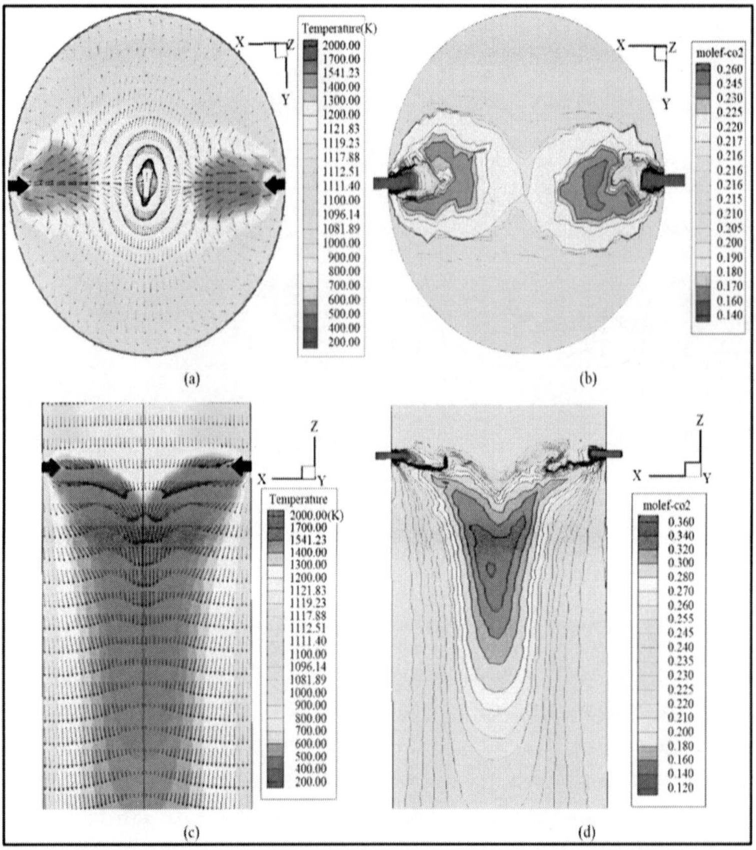

Figure 7. Cross-sectional and mid-plane plots of velocity vector, contours of temperature and mole fraction of CO_2 at second injection for Case 3 (90% - 10% for oxygen distribution). (a) Cross-sectional temperature contour and velocity vector at the second injection. (b) Cross-sectional mole fraction of CO_2 at the second injection. (c) mid-Plane temperature contour and velocity vector second injection. (d) Mole fraction of CO_2 at the second injection.

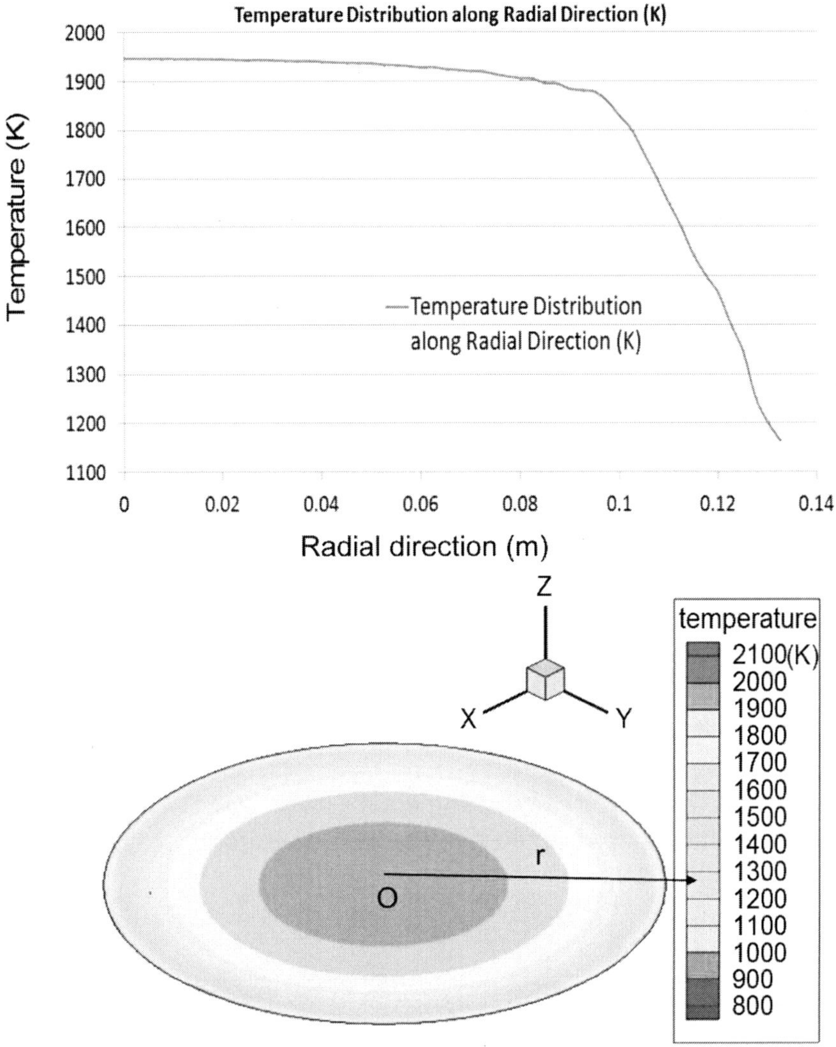

Figure 8. Gas temperature distribution at exit cross-sectional plane of the gasifier for Case 6 (60% - 40% oxygen distribution).

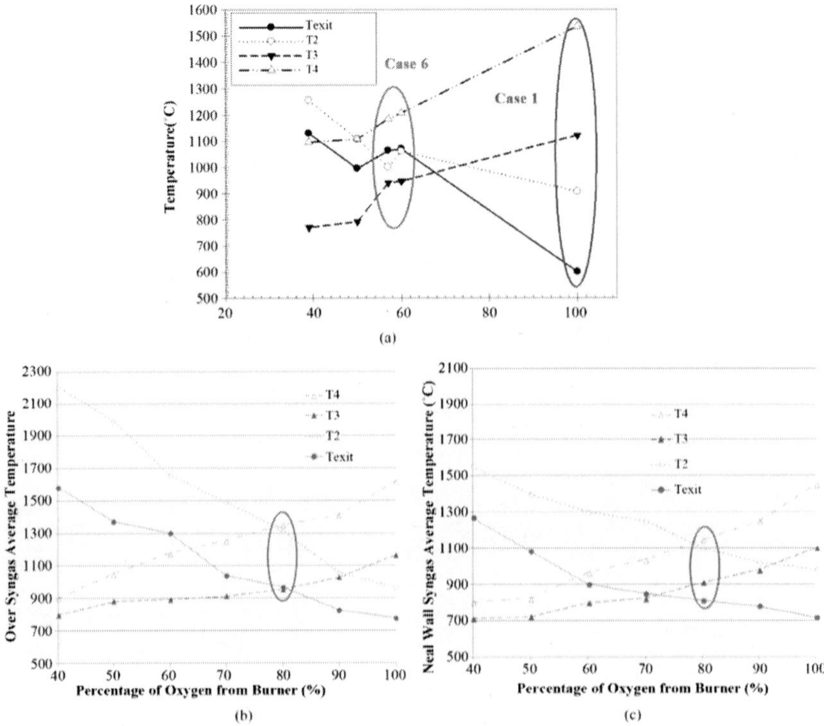

Figure 9. The effect of two-stage oxygen injection distribution on the gasifier wall temperature (a) experimental data (in wall cavities); (b) CFD results of mass-flow weighted syngas exit temperature; (c) CFD results for near-wall temperature (5 mm to the wall) for Cases 2 to 8 (Note: T_{exit}, T2, T3, and T4 in this figure correspond to TE-300-4, 8, 12, and 13 in Figure 1).

and increases at the second stage (T2 and T_{exit}). This is because if less oxygen is injected from the top, then less char will go through the exothermic combustion process and less combustion heat is released. At the first glance, the variation trends appear to reverse around 60% - 50%. Therefore, the analysis and explanations are separated into two parts based on the oxygen distribution on the top: 100% - 60% vs. 60% - 40%. The CFD simulated syngas compositions are used to provide explanation of the variations of near-wall temperature distributions.

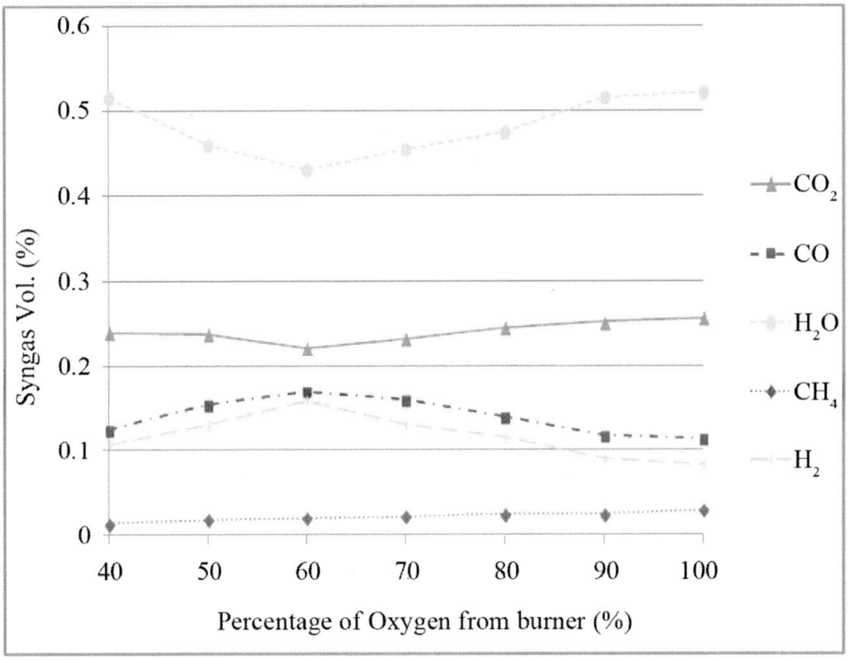

Figure 10. CFD simulated effect of two-stage oxygen injection distribution on the syngas composition for Cases 2 to 8.

Table 5. CFD results of syngas compositions, syngas temperatures at different locations, carbon conversion ratio, and HHV for Cases 2 to 8. $T_{exit,\ plane}$ is the mass flow weighted average over the cross-section of the exit plane. $T_{exit,\ near\ wall}$ is the circumferential average temperature at 5 mm from the wall.

O_2 distribution	Case 2 100% - 0%	Case 3 90% - 10%	Case 4 80% - 20%	Case 5 70% - 30%	Case 6 60% - 40%	Case 7 50% - 50%	Case 8 40% - 60%
CO_2 (Vol)	0.256	0.252	0.245	0.233	0.222	0.238	0.240
CO (Vol)	0.113	0.117	0.140	0.160	0.170	0.154	0.125
H_2O (Vol)	0.520	0.516	0.475	0.455	0.430	0.460	0.515
H_2 (Vol)	0.082	0.090	0.115	0.130	0.158	0.130	0.107
CH_4 (Vol)	0.029	0.025	0.025	0.022	0.020	0.018	0.013
T4 (K)	1892	1689	1627	1530	1450	1325	1174
T3 (K)	1439	1301	1228	1186	1164	1154	1071
T2 (K)	1237	1329	1590	1769	1934	2267	2484
$T_{exit\ plane}$ (K)	1052	1101	1242	1314	1550	1647	1859
$T_{exit,\ near\ wall}$ (K)	992	1052	1084	1124	1172	1356	1542
CCR	72.90%	74.62%	85.86%	93.02	98.88%	99.22%	99.94%
HHV (kJ/kmol)	73,180	78,455	97,559	110,776	127,713	109,078	89,284

100% - 60% top O_2 injection-Figure 10 shows that CFD simulated mole fractions of CO and H_2 at the exit increase when the top injected oxygen reduces from 100% to 60%, but they reduce when the top injected oxygen decreases from 60% to 40%. The mole fraction of CO_2 has an opposite changing trend.

60% - 40% top O_2 injection-If the mass fraction of top injected oxygen continues to reduce below 60%, there is more unburned char going though the gasification process (mainly $C + ½ CO_2 \rightarrow CO$, $C + H_2O \rightarrow CO + H_2$), resulting in more production of CO and H_2 at the first stage. At the second injection stage, some of the injected oxygen reacts with CH_4 to produce CO and H_2. Therefore, the total amount of CO and H_2 increases and CO_2 decreases at the exit of gasifier. However, if more than 40% - 50% of the total oxygen is injected from the second injection location, more CO and H_2, which were produced through the gasification process at the first stage, react with the oxygen through the combustion process at the second stage and result in more CO_2 and H_2O. Therefore, as indicated in Figure 10, CO and H_2 decrease and CO_2 and H_2O increase at the exit of the gasifier when the portion of oxygen injected from the top is reduced from 60% to 40% based on the CFD results Meanwhile, the carbon conversion ratio increases, and the exit mole fraction of CH_4 reduces when the percentage of oxygen injected from the top reduces from 100% to 40%. The reason for this trend is that the unburned char and CH_4 react with oxygen through $C + ½ O_2 \rightarrow CO$ and $CH_4 + ½ O_2 \rightarrow CO + 2H_2$ at the second injection stage. When oxygen injected from the second stage increases beyond 40%, the carbon conversion ratio reaches 99%.

One of the advantages of employing the two-stage oxygen control is to reduce the wall temperature gradient, and hence, the thermal stress along the gasifier can be reduced. The experimental results in Figure 9(a) show that the overall near-wall temperature difference in the gasifier is reduced to a narrower range of 900˚C - 1200˚C (with a 60% - 40% two-stage oxygen injection split) from a wider range of 600˚C - 1500˚C (with 100% top injection). In contrast, the CFD result in Figure 9(b) shows that the case of 80% - 20% yields the most uniform syngas temperature within a range from 955˚C to 1354˚C, while the case of 60% - 40% in Table 5 yields the highest syngas HHV (127,713 kJ/kmole). The case of 80% - 20% yields the lowest peak temperature and the most uniform syngas temperature within a range from 810˚C to 1150˚C. No experimental data are available for comparing the 80% - 20% CFD case.

4.3. Effect of Oxygen Injection Distribution under High Operating Pressure Condition (25 Bars with 40% of Stoichiometric O_2)

Since most of the gasifiers are operated under the condition with high pressures above 25 bars and low sub-stochiometric oxygen feeding below 40%, the rest of the analysis will be focused on presenting the CFD simulation at a high pressure (25 bars) and low oxygen stoichiometric ratio (0.4). In total, four cases (Cases 9 to 12 as listed in Table 3) have been conducted.

Simila to the discussion in section 4.2, the syngas composition and temperature at different locations, CCR and HHV for Cases 9 - 12 are provided in Table 6, Figure 11 and Figure 12respectively. The results are summarized below.

1) Compared to the results of the low operating pressure cases, the carbon conversion rates for different oxygen injection distribution are all above 95% in high operating pressure cases. Two mechanisms are considered to contribute to this higher carbon conversion rate: a) The partial pressure of the CO_2, H_2O, and even O_2 increases by a factor of 10. The char reaction rate expression contains the partial pressure of these gases at the coal particle surface, so the char conversion rate (mass per time) increases notably as the operating pressure increases. b) The longer particle residence time could also contribute to achieving a higher carbon conversion ratio. The average residence time is about 6 - 7 seconds for the high pressure cases while it is 3 - 4 seconds for the low-pressure cases.

2) Both the high operating pressure and low pressure cases follow the same trend of change of exit average temperature as the oxygen feeding distribution changes between two stages.

3) Consistent with the low pressure cases, the changing trend of syngas composition reverses at the 60% - 40% case. The 80% - 20% case produces the lowest peak temperature and most uniform temperature in the gasifier, but with the lowest CCR.

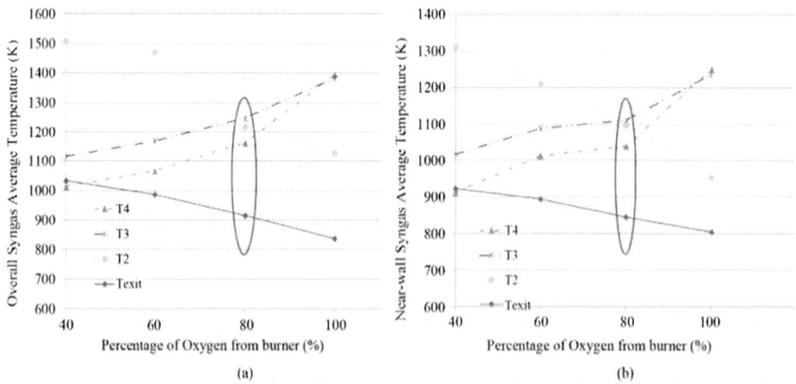

Figure 11. CFD simulated effect of two-stage oxygen injection distribution on the mass flow weighted syngas temperature at high pressure (25 bars) for Cases 9 to 12. (Note: The temperature locations T_{exit}, T2, T3, T4 in this figure correspond to TE-300-4, 8, 12, and 13 in Figure 1). (a) Average cross-sectional temperature. (b) Average near-wall temperature.

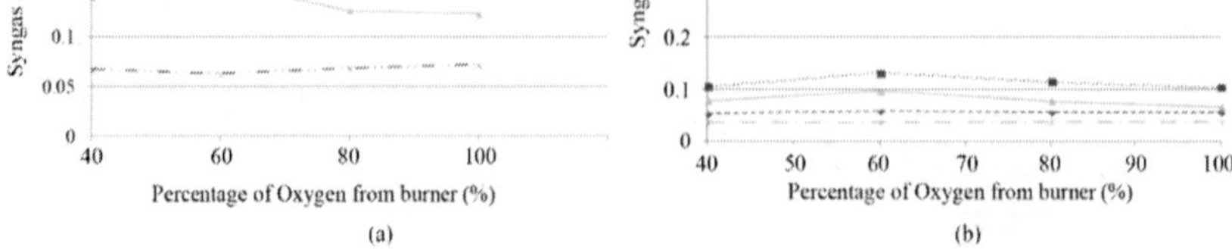

Figure 12. CFD simulated effect of two-stage oxygen injection distribution on the syngas composition at high pressure (25 bars) for Cases 9 to 12. (a) Mass fraction. (b) Volume fraction.

Table 6. CFD predicted syngas conditions at 25 bars for Cases 9 to 12.

O_2 distribution	Case 9 100% - 0%	Case 10 80% - 20%	Case 11 60% - 40%	Case 12 40% - 60%
CO_2 Vol (Mass)	0.056 (0.285)	0.057 (0.275)	0.059 (0.263)	0.053 (0.268)
CO Vol (Mass)	0.103 (0.328)	0.115 (0.343)	0.131 (0.359)	0.105 (0.332)
H_2O Vol (Mass)	0.067 (0.123)	0.078 (0.126)	0.097 (0.155)	0.078 (0.140)
H_2 Vol (Mass)	0.734 (0.186)	0.712 (0.180)	0.674 (0.153)	0.726 (0.182)
CH_4 Vol (Mass)	0.037 (0.072)	0.037 (0.069)	0.036 (0.063)	0.035 (0.068)
T4 (K)	1395	1162	1067	1014
T3 (K)	1383	1247	1169	1116
T2 (K)	1129	1220	1472	1506
$T_{exit\ Plane}$ (K)	838	915	986	1033
$T_{exit,\ near\ wall}$ (K)	804	845	894	922
CCR	99.99%	95.34%	98.88%	99.97%
HHV (kJ/kmol)	392,295	391,271	376,653	395,495

4) As more oxygen is injected at the second stage, CO and CO_2 have the same trend of change as in the low pressure condition; however, H_2O and H_2 change in the opposite direction.

5) The major differences between the high and the low pressure cases are: a) the average temperature is much lower at each corresponding location in the 25 bars condition; b) H_2dominates the syngas composition (about 70% vol.) at 25 bars case.

6) The 40% - 60% case produces the syngas with the highest HHV in the 25 bar case vs. 60% - 40% case in the 2.5 bars cases, but with the highest peak wall temperature.

5. CONCLUSIONS

1) The experiments at 2.5 bars show the 60% - 40% oxygen distribution yields the most uniform wall temperature along the gasifier.

2) In cases of split two-stage oxygen injections from 100% to 40% with 0.6 stoichiometric oxygen at 2.5 bars, the CFD predicts a) the similar trend of changing temperature as in the experimental data; b) the 80% - 20% case yields the lowest peak temperature and the most uniform gas temperature distribution along the gasifier; and c) the 60% - 40% case produces the syngas with the highest HHV.

3) The CFD's prediction of the effect of oxygen distribution under 25 bars operating pressure case shows that a) the carbon conversion ratio for different oxygen injection schemes are all above 95%; b) H_2 (about 70% vol.) dominates the syngas composition at exit; c) same as in the 2.5 bar cases, the 80% - 20% case yields the lowest peak temperature and most uniform gas temperature distribution along the gasifier; and d) the 40% - 60% case produces the syngas with the highest HHV.

4) Chemical reactions are affected by the local flow behavior. Strong opposing injection of oxygen at second stage squeezes the main flow laterally on the vertical planes perpendicular to the injection direction, making combustion spread accordingly in the vertically plane at the second stage.

Both experimental data and CFD predictions verify the hypothesis that it is feasible to reduce the peak temperature and achieve more uniform temperature in the gasifier by adequately controlling two-stage oxygen injection with only minor changes of CCR and HHV under the high-pressure (25 bars) condition. However, the variations of CCR and HHV are notable under the low pressure (2.5 bars) condition.

After this study was completed, we realized that Tsinghua University in China has already successfully developed an oxygen-staged gasifier [15] . One of the merits claimed by the Tsinghua research group was that the oxygen-staged scheme helped to reduce the syngas temperature and thus significantly extends the fuel injector's life expectancy. It appears that the success of Tsinghua's oxygen-stage scheme supports the hypothesis of this study. Since no detailed technical information has been released by Tsinghua, the information generated by this study from both experiments and CFD simulation are considered to be useful reference to the gasification community.

ACKNOWLEDGEMENTS

This study was partially supported by the Louisiana Governor's Energy Initiative via the Clean Power and Energy Research Consortium (CPERC) administered by the Louisiana Board of Regents. The ITRI gasifier was funded by the Taiwan Bureau of Energy.

REFERENCES

1. Wang, T., Silaen, A., Hsu, H.W. and Shen, C.H. (2010) Investigation of Heat Transfer and Gasification of Two Different Fuel Injectors in an Entrained-Flow Gasifier. ASME Journal of Thermal Science and Engineering Applications, 2, 011001/1-10.

2. Wang, T., Lu, X., Hsu, H.W. and Shen, C.H. (2011) Investigation of the Performance of Syngas Quench Cooling Design in a Downdraft Entrained Flow Gasifier. Proceedings of the 28th International Pittsburgh Coal Conference, Pittsburgh, 12-15 September 2011, 1-18.

3. Smoot, L.D. and Smith, PJ. (1985) Coal Combustion and Gasification. Plenum Press, New York.

4. Chen, C., Horio, M. and Kojima, T. (2000) Numerical Simulation of Entrained Flow Coal Gasifiers. Chemical Engineering Science, 55, 3861-3833.

5. Westbrook, C.K. and Dryer, F.L. (1981) Simplified Reaction Mechanisms for the Oxidation of Hydrocarbon Fuels in Flames. Combustion Science and Technology, 27, 31-43.

6. Jones, W.P. and Lindstedt, R.P. (1988) Global Reaction Schemes for Hydrocarbon Combustion. Combustion and Flame, 73, 233.

7. Benyon P. (2002) Computational Modelling of Entrained Flow Slagging Gasifiers. PhD Thesis, School of Aerospace, Mechanical & Mechatronic Engineering. University of Sydney, Sydney.

8. Lu, X. and Wang, T. (2013) Water-Gas Shift Modeling in Coal Gasification in an Entrained-Flow Gasifier Part 1: Development of Methodology and Model Calibration. Fuel, 108, 629-638. [Paper reference:2]

9. Lu, X. and Wang, T. (2013) Water-Gas Shift Modeling in Coal Gasification in an Entrained-Flow Gasifier Part 2: Gasification Application. Fuel, 108, 620-628. [Paper reference:2]

10. Silaen, A. and Wang, T. (2010) Effect of Turbulence and Devolatilization Models on Gasification Simulation. International Journal of Heat and Mass Transfer, 53, 2074-2091.

11. Fletcher, T.H., Kerstein, A.R., Pugmire, R.J. and Grant, D.M. (1990) Chemical Percolation Model for Devolatilization: 2. Temperature and Heating Rate Effects on Product Yields. Energy and Fuels, 4, 54-60.

12. Fletcher, T.H. and Kerstein, A.R. (1992) Chemical Percolation Model for Devolatilization: 3. Direct Use of 13C NMR Date to Predict Effects of Coal Type. Energy and Fuels, 6, 414-431.

13. Grant, D.M., Pugmire, R.J., Fletcher, T.H. and Kerstein, A.R. (1989) Chemical Percolation of Coal Devolatilization Using Percolation Lattice Statistics. Energy and Fuels, 3, 175-186.

14. Kobayashi, H., Howard, J.B. and Sarofim, A.F. (1976) Coal Devolatilization at High Temperatures. Symposium (International) on Combustion, 16, 411-425.

15. Zhang, J. (2013) Innovativeness and Development of Tsinghua Gasification Technology. Gasification Technology Conference, Colorado Spring, 13-16 October 2013, 17 p.

Chapter 14

Integrated Utilization of Sewage Sludge and Coal Gangue for Cement Clinker Products: Promoting Tricalcium Silicate Formation and Trace Elements Immobilization

Zhenzhou Yang [1], Yingyi Zhang [1], Lili Liu [1], Seshadri Seetharaman [2], Xidong Wang [1] and Zuotai Zhang[1,3,*]

[1]*Beijing Key Laboratory for Solid Waste Utilization and Management and Department of Energy and Resource Engineering, College of Engineering, Peking University, Beijing 100871, China*

[2]*Department of Materials Science Engineering, Royal Institute of Technology, Stockholm SE 100-4, Sweden*

[3]*School of Environmental Science and Engineering, South University of Science and Technology of China, Shenzhen 518055, China*

ABSTRACT

The present study firstly proposed a method of integrated utilization of sewage sludge (SS) and coal gangue (CG), two waste products, for cement clinker products with the aim of heat recovery and environment protection. The results demonstrated that the incremental amounts of SS and CG addition was favorable for the formation of tricalcium silicate (C_3S) during the calcinations, but excess amount of SS addition could cause the impediment effect on C_3S formation. Furthermore, it was also observed that the C_3S

polymorphs showed the transition from rhombohedral to monoclinic structure as SS addition was increased to 15 wt %. During the calcinations, most of trace elements could be immobilized especially Zn and cannot be easily leached out. Given the encouraging results in the present study, the co-process of sewage sludge and coal gangue in the cement kiln can be expected with a higher quality of cement products and minimum pollution to the environment.

Keywords: waste; sewage sludge; coal gangue; eco-cement; clinker; trace element

1. INTRODUCTION

With the rapid development of urbanization and industrialization, various solid wastes accumulate year by year, not only occupying land but also causing serious environment problems. Traditional waste disposal methods, such as pile up or landfill, have proven to be of low efficiency, a waste of resources and even environmentally harmful. The increasing shortage of resources and the negative impact on the environment, seen from a global perspective motivated the present work with respect to finding efficient process route toward waste utilization.

Among these solid wastes, sewage sludge (SS) and coal gangue (CG) are two typical common wastes generated. Sewage sludge is a problematic waste generated from wastewater treatment. The production of dried SS in China is currently eight million tons per year and is still increasing at a rate between 8% and 10% annually [1]. Due to the high content of hazardous substances such as organic, non-biodegradable compounds, bacteria and viruses, *etc.*, the improper disposal of SS could pollute the environment severely. The most common treatment methods in China are agricultural utilization, landfilling, and incineration [2,3,4]. However, the reduced availability of land and the presence of pathogens and contaminants such as heavy metals, micro-pollutants, and antimicrobial agents limit applications of these methods. Coal gangue, as one of the common industrial wastes, is one of the main byproduct from coalmining and beneficiation and accounts for about 15–20 wt % of coal production [5,6,7]. CG stockpiles have reached 4.5 billion tons in China and are increasing at the speed of 750 million tons per year. The traditional disposal method of accumulation is also infeasible due to increasingly stringent environmental regulations. Thus, an environment-friendly and economically viable process to jointly dispose of these two wastes is urgently required.

Cement kiln is well known as an ideal device to dispose wastes, which can completely decompose the hazardous substance such as dioxins and furans due to the high temperature and long residual time [8]. Based on the characteristics of SS and CG, these two wastes may be utilized in the cement kiln for clinker production, where the organic matter can serve as an alternative fuel and the residual inorganic ash can be incorporated into clinker as part of raw materials. Thus, the use of SS or CG in cement kiln offers multiple advantages such as reduced use of primary raw materials, lower fuel consumption and significant reduction in CO_2 emission. Besides, this integrated activity can also produce additional revenue to subsidize government waste disposal.

A number of investigations have been carried out with a view to utilize SS as a replacement for some raw materials used in cement manufacturing [9,10,11,12]. The main focus in the case of most of these studies has been the impact of SS on the properties of cement such as setting time and compressive strength. However, very few investigations have been conducted on how to combine cement plant with CG and SS disposal and their effect on the clinker properties, such as crystalline structure and crystal formation mechanism when they substitute a part of the raw materials for manufacturing of clinkers. Furthermore, due to the increased public concern with environment risk, it is important to evaluate the effect on the environment during the process of utilization of SS and CG, especially the trace element problem. It is well known that the interphase in the clinker has a high capacity to immobilize trace elements due to its specific structure and the silicate phases can also stabilize trace elements through solid solution [13,14]. In addition, the clay minerals in CG can help stabilize trace elements during calcinations. Therefore, the co-process of SS and CG in cement kiln can be expected to be a promising method for environmental protection. To the best of the knowledge of the present authors, the quantitative analysis of trace elements immobilizing into clinkers during the co-process of SS and CG has not been systematically investigated.

In this regard, the aim of these studies is to compare the characteristics of eco-cement clinker (made with SS and CG substitution) with ordinary cement clinkers synthesized in laboratory environment. The formation mechanism, crystalline phase components and structure, and crystalline micrographs were investigated in this study. The distribution behavior of trace elements during eco-clinker production was also studied. Further, the energy and raw materials consumptions were also evaluated. The present studies are expected to provide a method of integrated utilization of SS and CG in cement industry, which can save energy and resources, and improve trace elements immobilization at the same time.

2. EXPERIMENTAL SECTION

2.1. Materials Characterization

The dried sewage sludge pellet and coal gangue samples were collected from a municipal wastewater treatment plant located in Beijing, China. Coal gangue was supplied by power generation plant of Pingshuo in Shanxi Province, China. The results of proximate and ultimate analysis of SS and CG are shown in Table 1. As can be seen, SS had lower ash yield and fixed carbon content than that of CG, whereas it had higher volatile matter contents. The contents of sulfur in CG and SS are at low level, which can effectively increase the amount of liquid phase, resulting in decreasing the melting temperature point and its viscosity. The formation of C_3S might therefore be accelerated. Furthermore, the calorific value of SS was around 9.35 MJ/kg, which was higher than that of CG (4.82 MJ/kg).

The chemical compositions were determined by an X-ray fluorescence spectrometer (XRF, S4-Explorer, Bruker, Karlsruhe, Germany). The XRF analysis of SS and CG are also present in Table 1. It can be seen that the major components of SS and CG are SiO_2, Al_2O_3, Fe_2O_3 and CaO, which is the major components to prepare cement clinker. Besides, SS contains high amounts of alkalis and alkaline earth metal elements. It also should be noted that phosphorus levels in SS are much higher than that in CG. The concentrations of trace elements in SS and CG are also shown in Table 1. It can be seen that the trace elements in SS are notably higher than those in CG, especially Zn and Cu.

Table 1. Chemical compositions of coal gangue and sewage sludge.

	Proximate Analysis (wt %)					Ultimate Analysis (wt %)				Calorific Value (MJ/kg)	
-	moisture, ad [a]	ash, ad [a]	volatile matter, ad [a]	fixed carbon, ad [a]	Total	C, ad	H, ad	N, ad	S, ad	-	
CG	0.8	67.9	15	16.3	100	17.5	12.6	0.6	1.2	4.82	
SS	2	53.3	35.8	8.9	100	23.2	4.0	3.0	1.3	9.35	
				Chemical Composition (wt %)							
-	SiO_2	Al_2O_3	Fe_2O_3	CaO	MgO	K_2O	Na_2O	P_2O_5	Others	LOI	Total
CG	32.8	28.8	1.2	0.1	<0.1	0.1	<0.1	0.1	2.5	34.4	100
SS	25.2	5.6	4.2	6.4	1.5	1.2	0.6	4.9	1.4	48.6	100
				Trace Elements Contents (mg/kg)							
-	As	Ba	Cd	Co	Cr	Cu	Mn	Ni	Pb	Zn	
CG	19	ND	ND	8	25	15	6	4	9	14	
SS	50	256	2	11	124	358	543	41	48	1084	

[a] ad, air dried; LOI, loss on ignition at 1000 °C; Others, other elements; ND, not detected.

To understand the structure of solid waste, the SS and CG were analyzed by X-ray diffraction (XRD, D/Max 2500, Rigaku, Tokyo, Japan) and Fourier transformation infrared (FT-IR) and the results are shown inFigure 1. It can be seen from Figure 1a that the major mineral phases in SS are quartz and calcite. In addition, anorthite was also identified in the SS. The XRD patterns of CG show the distinctive peaks belonging to the characteristics of kaolinite. Besides, the mineral phases of clinochlore was also identified in the CG. The FT-IR spectra of SS and CG are presented in Figure 1b. As for SS, the peaks at 3404, 2923, 2854, 1439, 2353, and 1652 cm^{-1} are related to the organic function groups. This indicated the presence of high content of organic material such as carboxylic, alcoholic and amide compounds [15]. The high content of organic material likely lead to a high calorific value of SS compared to that of CG. Besides, the peaks at 1420 and 873 cm^{-1} indicate the existence calcite, while the peaks at 1010, 920, 750, 660, 530 and 462 cm^{-1} are the characteristic bands of anorthite [16]. As for CG, the peaks at 3694, 3669, 3653, 3620, 1115, 1100, 1034, 1009, 934, 912, 792, 754, 540 and 431 cm^{-1} are the characteristic bands of kaolinite while the peak at 3620, 1085, 1630, 1009, 655 and 460 cm^{-1} belong to the characteristics of clinochlore [16,17]. Furthremore, the peaks at 1080, 798, 690 and 462 cm^{-1} indicate the existence of quartz in CG and SS [16]. It can be seen that a strong band at 1033 cm^{-1} can be found in the spectra of CG and SS. For CG, this band belong to the Si–O stretching of Si–O–Si while for SS, it might be attributed to the O–H in mineral components or the existence of C–O function group [18,19].

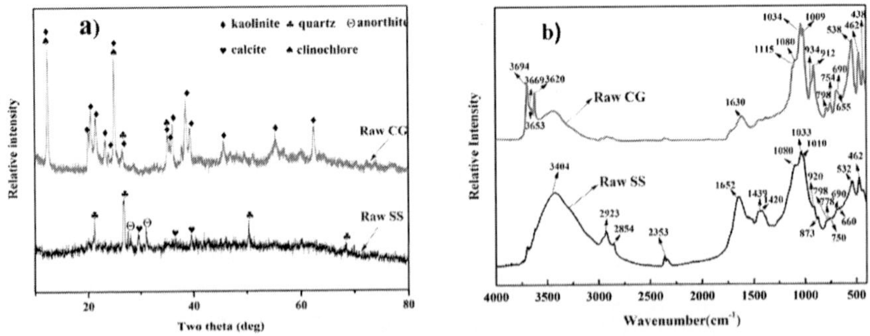

Figure 1. The X-ray diffraction (XRD) patterns and Fourier transformation infrared (FT-IR) spectra of raw coal gangue (CG) and sewage sludge (SS): (**a**) XRD patterns; and (**b**) FTIR spectra.

The thermal analyzer (Q600SDT, TA Instuments, New Castle, PA, USA) was used to study the thermal behavior of SS and CG. The measurements were conducted from ambient temperature to 1200 °C at a linear heating rate of 10 °C/min under air as atmosphere and the results are shown in Figure 2. It could be seen that the most rapid mass loss occurred at around 200 °C and two distinct peaks appeared in the derivative thermogravimetric (DTG) and differential scanning calorimetric (DSC) curves of SS. The former peak (200 to 370 °C) was attributed to the combustion of volatiles and the second peak (370 to 500 °C) corresponds to the char oxidation. On the contrary, the CG showed a different thermal behavior: rapid mass loss occurred in the temperature range between 400 and 600 °C. The DTG and DSC curves showed only one distinct peak. It could also be noted that the spontaneous combustion did not occur until the temperature exceeded 200 °C. This suggested that the SS and CG can easily be stored and handled, without any risk of potential explosion from spontaneous combustion during the manufacturing process when using SS and CG as raw materials [20].

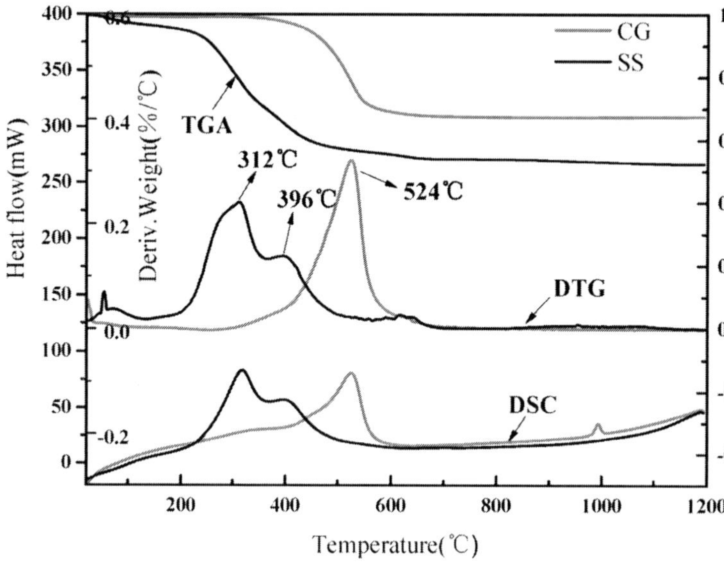

Figure 2. The thermogravimetric (TG), derivative thermogravimetric (DTG) and differential scanning calorimetric (DSC) curves of raw materials.

2.2. Eco-Cement Clinkers Preparation

The raw meals (raw materials to prepare for clinker) were prepared by mixing $CaCO_3$, SiO_2, Al_2O_3, $Fe_2O_3$and various amounts of SS or CG. All the reagents used here were of analytical reagent grade. The compositional parameters in cement chemistry are listed as Equations (S1)–(S4).

The raw meals were dried in air at 105 °C for 24 h and ground in a ball mill until the particles can pass through #200 mesh metallic sieve. Then, they were prepared under different blending ratios with SS additions of 5, 10, 15 and 30 wt %, respectively. In consideration of cement clinker parameters, the mixing ratio of CG was chosen as 5 and 10 wt %, as shown in Table 2. All the mixture were pressed to ϕ20 mm × 5 mm slices by applying a pressure of 10 Mpa and calcined in the programmable electrically heated tube furnace. The furnace temperature was raised at the rate of 10 °C/min from room temperature to 1450 °C. The temperature was maintained at 950 °C for 30 min to ensure the complete decomposition of $CaCO_3$ and held at 1450 °C for 2 h. After calcination, the eco-cement clinkers were cooled rapidly in air and pulverized to pass through #200 mesh metallic sieve. The samples were collected for XRD, FTIR, and scanning electron microscope (SEM, S4800, Hitachi Ltd., Tokyo, Japan) analyses and the particle size distribution of raw meals are shown in Figure S1.

Table 2. The addition amount of sewage sludge and coal gangue in raw meal.

Sample	Addition Amount (%)	
	SS	CG
S1(Blank sample)	0	0
S2	5	0
S3	10	0
S4	15	0
S5	30	0
S6	0	5
S7	0	10
S8	15	5

2.3 Trace Element Immobilized Characterization

In order to investigate the transformation behavior of trace elements during calcinations, the samples were digested in microwave assisted digestion system (EXCEL, TOPEX, PreeKem, Shanghai, China) in triple acid (HNO_3:HF:H_3PO_4 = 4:2:1). The concentrations of selected trace elements As, Cd, Co, Cr, Cu, Ni, Pb and Zn were measured by inductively coupled plasma-atomic emission spectroscopy (ICP-AES, Prodigy XP, Leeman, Hudson, NY, USA). Standard reference materials GBW08401 (coal fly ash) [17] and GBW07406 (GSS-6, soil) [21] were used for calibration. The accuracy for the trace elements was within ±10%.

In order to determine the potential leachability of trace elements in clinkers, Toxicity Characteristic Leaching Procedure (TCLP) method of the US Environmental Protection Agency (US-EPA) was applied to the clinker. The extracting solution used was 0.1 M acetic acid at pH 4.93 and the trace elements in leachate were determined by ICP-AES.

3. RESULTS AND DISCUSSION

3.1. Mineralogical Characterization of the Eco-Cement Clinker

The XRD patterns of eco-cement clinker with different amounts of SS addition are shown in Figure 3a. It can be seen that the major components of eco-cement clinkers are C_3S (Ca_3SiO_5), C_2S (Ca_2SiO_4), C_3A ($Ca_3Al_2O_6$ and $Ca_3(Al,Fe)_2O_6$) and C_4AF ($Ca_4Fe_2Al_2O_{10}$). Thus, the synthesized eco-cement clinkers exhibit the characteristics similar to those of ordinary Portland cement. Amongst these components, C_3S is the major component of Portland cement clinker that determines its quality. The most important characteristic peaks of C_3S appeared at 2θ of about 32° and the intensity of peak that is indicative of the relative contents of C_3S in the clinker [22,23] (see Figure 3b). It should be noted that the characteristic peak of C_3S shows no significant changes with SS addition up to 10 wt %. With further increasing SS addition, the intensity of diffraction peak of C_3S is found to increase significantly, indicating the positive effect on the formation of C_3S. However, the C_3S diffraction peak dropped sharply with higher SS additions of up to 30 wt % and the phases of f-CaO and C_2S-$0.05C_3P_2(Ca_3(PO_4)_2)$ are formed. These results indicate that appropriate amount of SS addition is favorable for the formation of C_3S, while excessive SS addition shows the opposite effect on the formation of C_3S.

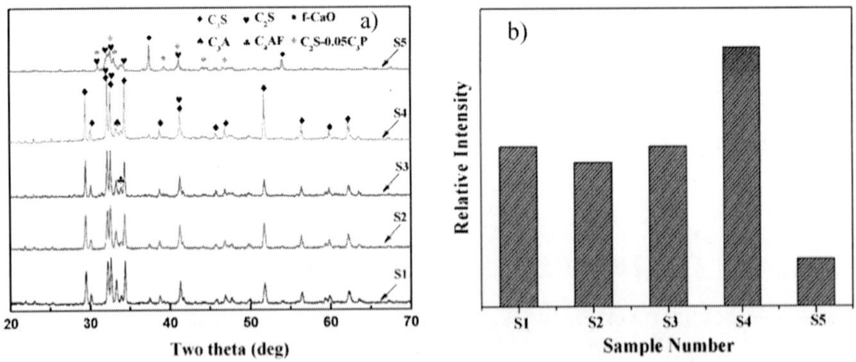

Figure 3. The XRD patterns of the clinker adding with different amount of SS and the peak intensity of C_3S: (**a**) XRD patterns; and (**b**) comparison of XRD peak intensity at 2θ range from 31.8° to 32.4°.

It is also noted that SS addition plays a significant impact on the C_3S crystal structure, which is illustrated in Figure S2. For the blank sample S1 without SS addition, the only diffraction peak observed with regard to the crystal structure of C_3S is between 51.2° and 52.2°, indicating a rhombohedral (R) structure [24,25]. The split blip between 31.5° and 33° indicates the existence of triclinic or monoclinic crystal structure of C_3S in blank sample. With the amount of SS additions increasing, the split blip became insignificant and the C_3S showed the total rhombohedral structure. With the SS additions up to 15 wt %, the shoulder peak appeared between 31.5°~33° and 51°~52.5°, which is characteristic of monoclinic (M) structure. Thus, the C_3S polymorphs show transition from rhombohedral to monoclinic structure with increasing amount of SS addition.

To study the differences of crystal characteristics between clinkers with various amounts of SS and CG additions, a comparison of the crystalline phases formed in the clinkers with various additions of SS and CG is presented in Figure 4. In view of the high content of alumina in CG (28.1%), the maximum amount of CG to be added cannot exceed 10 wt %. It can be seen from Figure 4a that the major crystalline phases of the clinkers produced by the addition of CG are very similar to that of commercially available Portland cement. With a higher amount CG addition, a more intense peak of C_3S was observed, as shown in Figure 4b, indicating that addition of CG favor C_3S phase formation. Moreover, it is also noteworthy that the intensity of C_3S characteristic peak with CG addition was higher than that with SS addition. The results might indicate that addition of CG is more favorable for the formation of C_3S in the clinkers compared with SS addition. In addition, as shown in Figure S3, the characteristic diffraction peak between 31.5°~33° and 51°~52.5° show clearly that the structure of C_3S was rhombohedral (R), which indicates that addition of CG has very little impact on the C_3S crystal structure.

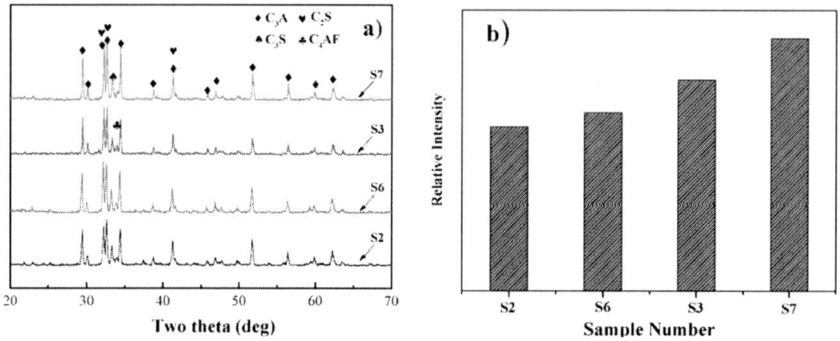

Figure 4. The comparison XRD patterns of the clinker adding with different amount of SS, CG and the peak intensity of C_3S: (**a**) XRD patterns; and (**b**) comparison of XRD peak intensity at 2θ from 31.8° to 32.4°.

Compared to blank sample S1, the positive effect on the C_3S formation with CG addition may be attributed to the special structures of CG. The mineral phase of CG consists mainly of kaolinite, which is weakly bonded with hydroxy group. During the process of calcinations, CG is likely to generate a large amount of active silica and alumina, which can then react with calcium oxide to promote the formation of C_3S. Meanwhile, the hydroxyl group is also broken, and the released oxygen may attribute to depolymerizing Si–O within the raw meals [26,27].

3.2. Structure Analysis of the Eco-Cement Clinker

In order to further identify the mechanism of phase generation and the structure transformation, clinkers with different amounts of SS and CG additions were analyzed by FT-IR spectra. The results obtained are shown in Figure 5 and the detail peak assignments are presented in Supplemental Table S1.

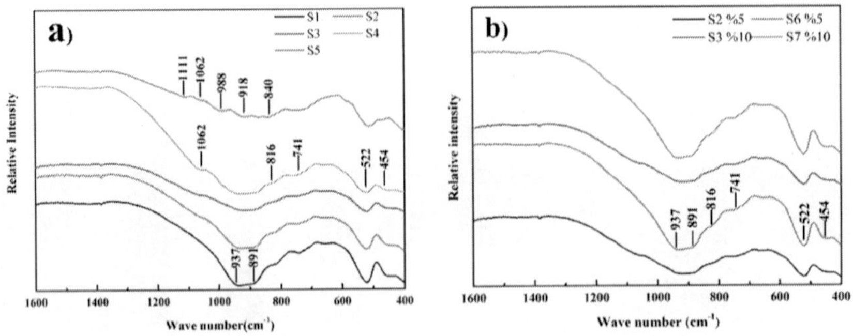

Figure 5. FT-IR spectra for the eco-cement clinkers: (**a**) clinkers with different amounts of SS additions; and (**b**) clinkers with different amounts of SS and CG additions.

The most intense peaks at 937 and 891 cm^{-1} and the shoulder at 816 cm^{-1} could be attributed to the characteristic of C_3S [28]. Meanwhile, the peak at 741 cm^{-1} can be attributed to the Al–O vibrations in C_4AF while the peak at 454 cm^{-1} correspond to the Al–O vibration at C_3A [29] is indicative of the existence of C_3A and C_4AF in the clinker. For the sample S1 without SS addition, the peaks at 937 and 891 cm^{-1} are separated. With SS addition up to 10 wt %, the two break peaks at 937 and 891 cm^{-1} merged into a single peak, and the peak at 816 cm^{-1} became less profound. This indicates that the crystal structure of C_3S transformed from a relatively poor symmetry structure to a more symmetrical total rhombohedral structure. The peaks at 937 and 891 cm^{-1} became more distinct again and the peak at 816 cm^{-1} reappeared as the SS addition was increased to 15 wt %, indicating the appearance of monoclinic structure in C_3S. Furthermore, the peaks of 937 and 891 cm^{-1} became more profound, indicating the higher contend of C_3S to some extent. With the SS addition at 30 wt %, the characteristic peaks of the C_3S became less intense and the peaks at 988, 918 and 840 cm^{-1} appeared. The three peaks correspond to the Si–O vibration of C_2S [30]. Besides, the peaks at 1062 cm^{-1}indicated the existence of (PO_4) group. The results demonstrate that the decomposition of C_3S and the formation of C_2S solid solution, which conform to the XRD results (Figure 3a).

Figure 5b shows the differences between the eco-cement clinkers with simultaneous addition of SS and CG. It could be seen that the clinkers kept the C_3S structure with CG addition but there were some differences. The intensity of C_3S characteristic peaks with CG addition is more profound than that with SS addition, indicating the formation of C_3S with CG addition. Moreover, the results also show that there has no obvious modification effect on the C_3S polymorphs with different amounts of CG addition and its crystal structure is more asymmetric compared to the one with SS addition.

3.3. Liquid Phase Formation during Clinker Calcination Process

DSC experiments were conducted to examine the influence of SS and CG additions on the thermal behavior during the clinker formation and the results are shown in Figure 6. It can be seen that the first peak occurred at approximately 785 °C corresponding to the decomposition of $CaCO_3$ and the second peak appeared at temperature between 1350 and 1400 °C indicating the formation of liquid phase. The results reveal the fact that addition of SS and CG has no significant effect on the decomposition of $CaCO_3$ but can lower the temperature of liquid phase formation, which can be attributed to the existence of trace elements, such as Cu and Zn in the SS and CG (Table 1). The impurities would reduce the melting point of local area melt and favor for the formation of liquid phase during the process of calcinations [31,32]. The chemical reaction between C_2S and CaO could be accelerated with increasing liquid phase. Besides, the elements of magnesium and sulfur introduced by SS or CG can also lower the viscosity of liquid [33]. Therefore, the additions of SS and CG are likely to boost C_3S formation and save energy consumption to some extent.

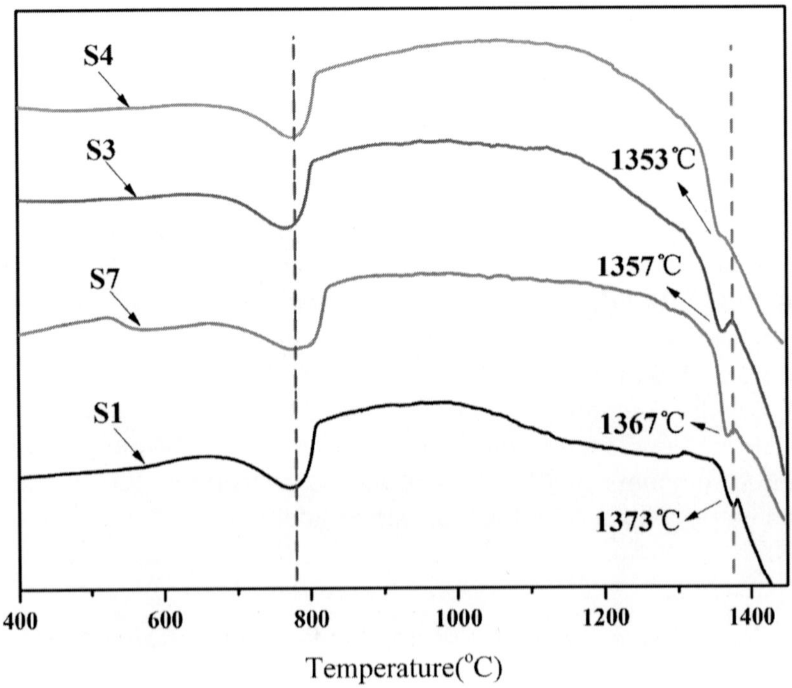

Figure 6. DSC curves for raw meals during heating process.

3.4. The Immobilization Behavior of Traces Elements

Trace elements have received more and more attention due to their hazardous effect on human health and ecosystem [34,35]. The concentration of selected trace elements in the ash of SS and the eco-cement clinkers were analyzed and the following equation can be used to describe the immobilized effect of the trace elements [36].

$$G = \frac{K}{\frac{S}{1 - L}}$$

where *G* is the immobilization ratio of trace elements; *K* represents the trace elements concentration in clinker (μg/g); *S* stands for the trace elements concentration (μg/g) in raw meals; and *L* is the ignition loss (%) after burning the raw meals at 1450 °C. Figure

7 illustrates the trace elements immobilization ratio in ash/clinker during the calcinations process of raw SS (samples S4 and S8). Due to the low concentration of Cd in SS and CG, this element could barely be detected in both raw meals as well as clinker. On the other hand, Pb showed high concentration in raw meals. However, the concentration of lead was found to be below the detection limits in the ash and the clinker, which could be due to the high volatility. Hence, only distribution of As, Ba, Co, Cr, Cu, Ni and Zn are discussed in this part.

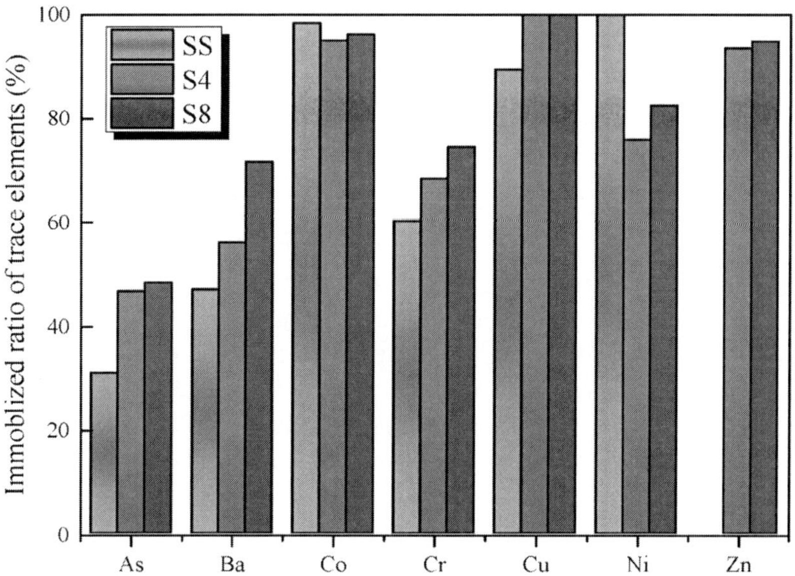

Figure 7. The immobilized ratio of trace elements in ash/clinker during the calcinations process of raw SS, sample S4 and sample S8.

For raw SS, it can be seen that Co, Ni and Cu are non-volatile and are mainly enriched in the ash, while As, Ba and Cr are enriched in the ash to different degrees and are classified as semi-volatile elements. Furthermore, it should be noted that Zn is recognized as a highly volatile element in SS and ended up in the gas phase during the calcination process at 1450 °C. This is consistent with previous study showing that Zn showed high volatility due to the high content of chlorine in the SS [37]. For sample S4 with 15 wt % SS addition, it can be seen that the raw meals do have the immobilization effect with respect to most of the trace elements in SS during the burning process, especially Zn and over 90% of Zn can be trapped in the clinker. It is known that the mineral phases in the clinker has a high capacity to immobilize trace elements. Zn, As, Co, Cu and Ni can exist in both of interphase and silicate phase while Cr is preferentially found in the silicate phase [14,38]. Thus, the high retention of Zn can be ascribed to the formation of oxidic solid solution in the mineral phase of clinker during calcination. This phenomenon was also observed by Barros *et al.* [39].

In order to immobilize trace elements of the SS in the raw meals to a greater extend and avoid vaporization of volatile elements, SS should be blended with CG as additive into raw meal. The maximum amount of CG addition was recommended as 5 wt % in consideration of cement clinker modulus, as shown in sample S8. It could be seen that, with the CG addition, immobilization ratio of almost all the trace elements increased as compared with sample S4. The elevated retention effect of trace elements can be attributed to the specific structure of kaolinite. During the calcinations, the crystalline structure of kaolinite is transformed to semi-crystalline metakaolinite and further transformed to mullite. During the reaction series, the decomposition and transition of the crystal structure results in a charge imbalance and the elements may be chemically bonded to the aluminosilicate structure [40]. Therefore, the addition of CG can facilitate the prevention of trace elements emission.

The trace elements incorporated in the clinkers might be leached out under acidic conditions and cause damage to the environment. To investigate this possibility, the toxicity characteristic leaching procedure (TLCP) was conducted to identify the toxicity of the clinker. The regulatory threshold values in china are in accordance with GB 5085.3-2007 (identification standards for hazardous wastes-identification for extraction toxicity of China) [9]. Table 3 shows the leaching concentration of toxic elements and the limit of the extraction procedure toxicity standards in china. For the cement clinker with SS addition (sample S4), it can be seen that only Ba and Zn are present in the leachate and none of them exceeded the regulatory thresholds. This indicates the immense immobilization effect of clinker mineral for trace elements. In addition, no big differences on the leaching ability of trace elements could be noticed with the addition of CG into the clinker (sample S8) except for Zn and

Ba. The addition of CG caused a little higher concentration of Ba but lower concentration of Zn in the leachate. The leaching experiment results show that the eco-cement clinkers are safe to the environment.

Table 3. Toxicity characteristic leaching procedure (TCLP) leaching concentrations (mg/L) of eco-cement clinkers.

Sample	Elements (mg/L)								
	As	Ba	Cd	Co	Cr	Cu	Ni	Pb	Zn
S4	ND	1	ND	ND	ND	ND	ND	ND	0.06
S8	ND	2.13	ND	ND	ND	ND	ND	ND	ND
GB 5085.3-2007	5	100	1	~	15	100	5	5	100

ND: not detected, ~: not mentioned in the regulation.

3.5. Energy and Material Balances with SS and CG Addition

The previous study indicated that one tonne of Portland cement clinker production need 3–4 GJ primary energy consumptions and 1.6 tonne of raw meals supplement, respectively [41,42]. It could be seen fromTable 1 that the calorific value of SS is equivalent to 9.35 GJ/t. According to the present results, 15 wt % the raw meals could be replaced by SS (0.24 t), *i.e.*, SS can supply about 2.25 Gt energy during the process of calcinations. Under this condition, the fuel consumption in the manufacture of cement could be reduced by more than 50% with 15 wt % SS addition. Further, for the purpose of immobilizing more trace elements, the blend of SS and CG could be added into raw meals as mentioned above (sample S8). According to the present calculation, about 2.5 Gt energy can be saved, which could reduce the consumption of energy above 60%.

Furthermore, it could also be calculated that incorporation of SS at 15 wt % in raw meal would result in reducing raw materials requirement of CaO by approximately 8.1 kg/t raw meals, Al_2O_3 by 1.1 kg/t raw meals, SiO_2 by 1.1 kg/t raw meals and Fe_2O_3 by 0.8 kg/t raw meals. The raw meals are traditionally extracted from natural materials including clay, limestone, sand and iron powder. Assuming that using the same nature materials in reference to produce clinker [21], the chemical analyses of the raw materials are shown in the Table S2. Figure 8 illustrates the change of compositions of raw meals made by the nature materials by SS and CG additions. For the raw meals with only SS addition, it can be seen that there will be a significant decrease in the raw materials requirement with 15 wt % SS addition, especially for clay. Due to the incorporation of SS, the raw material requirement of limestone would be reduced by 59.2 kg/t clinker, clay by 163.2 kg/t clinker, sand by 12.8 kg/t clinker and $Fe_2O_{3(ref)}$ by 4.8 kg/t clinker, *i.e.*, the addition of SS can replace approximately 4.7 wt % of limestone, 73.4 wt % of clay, 14.3 wt % of sand and 23.1 wt % of $Fe_2O_{3(ref)}$. For the raw meals with the blend of SS and CG addition, due to the limit by the modulus of raw meals, the maximum amount of CG addition could only be up to 1.5 wt % when mixed with 15 wt % SS. It can be seen that, with the CG addition, there is no significant decrease in the raw materials requirements except the clay, which can be totally replaced. It is admitted that this would lead to a minor increase in the demand for limestone and sand. Therefore, it is important to take into account the costs of raw materials in the cement manufacturing process and determine the most cost-effective raw material proportion.

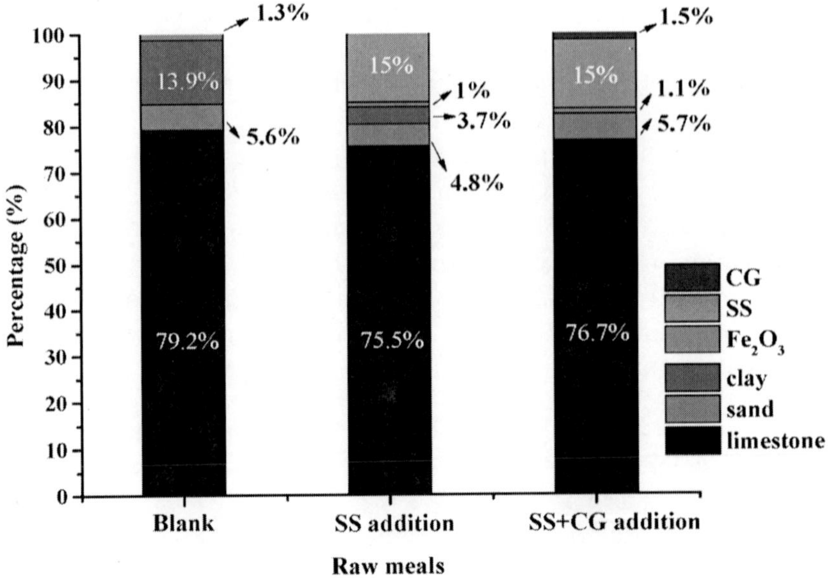

Figure 8. The impact on raw meals compositions with SS or the blend of SS and CG addition.

4. CONCLUSIONS

The present results clearly demonstrate that sewage sludge and coal gangue can be used successfully as alternative materials and fuel in clinker manufacturing at the laboratory scale. The main conclusions of the work are summarized below:

1. Appropriate SS and CG addition could be beneficial to C_3S formation while excess SS addition had the negative effect. The decomposition of C_3S into C_2S and f-CaO was found to occur as the SS addition was increased to 30 wt % and the amount of SS blending in the raw meals should be strictly controlled. The positive effect of SS and CG additions for C_3S formation can be attributed to different mechanisms. The former could be described by introducing a mass of impurities with low melting point, which could lead to decreasing the temperature of liquid phase formation while the special structure of CG could make it easier for the reaction between f-CaO and C_2S. In addition, the raw meals with CG addition could boost more C_3S appearance compared with SS.

2. SS could lead to the modification of C_3S polymorphism from the relatively asymmetric structure to more symmetrical structure (R) with increasing SS addition. As SS addition is increased to 15 wt %, the crystal structure of C_3S becomes monoclinic (M).

3. It was found that clinkers had a good effect on immobilizing most of the trace elements in SS, especially Zn. CG can also help stabilize trace elements during the calcinations process. The TCLP results show that the eco-cement clinkers met the standards of the current Chinese regulatory thresholds.

4. Integrated utilization of SS and CG as alternative raw materials and fuel in cement manufacture could effectively reduce raw material and energy consumption. They are expected to reduce energy consumption by more than 60% as well as reduce a fair amount of the natural raw material consumption according to the present estimation.

ACKNOWLEDGMENTS

This study was sponsored by the National Natural Science Foundation of China (51522401, 51472007, 51272005), and the Ministry of public welfare industry scientific research projects (201511062).

AUTHOR CONTRIBUTIONS

Zuotai Zhang and Zhenzhou Yang designed the experiments; Zhenzhou Yang conducted the experiments, analyzed the data, and wrote the manuscript; and Yingyi Zhang, Lili Liu, Seshadri Seetharaman and Xidong Wang advised the experiments and helped editing the manuscript.

CONFLICTS OF INTEREST

The authors declare no conflict of interest.

REFERENCES

1. Hong, J.M.; Xu, C.Q.; Hong, J.L.; Tan, X.F.; Chen, W. Life cycle assessment of sewage sludge co-incineration in a coal-based power station. Waste Manag. 2013, 33, 1843–1852.
2. Folgueras, M.B.; Alonso, M.; Diaz, R.M. Influence of sewage sludge treatment on pyrolysis and combustion of dry sludge. Energy 2013, 55, 426–435.
3. Murakami, T.; Suzuki, Y.; Nagasawa, H.; Yamamoto, T.; Koseki, T.; Hirose, H.; Okamoto, S. Combustion characteristics of sewage sludge in an incineration plant for energy recovery. Fuel Process Technol. 2009, 90, 778–783.
4. Zhou, H.B.; Ma, C.; Gao, D.; Chen, T.B.; Zheng, G.D.; Chen, J.; Pan, T.H. Application of a recyclable plastic bulking agent for sewage sludge composting. Bioresour. Technol. 2014, 152, 329–336.

5. Querol, X.; Izquierdo, M.; Monfort, E.; Alvarez, E.; Font, O.; Moreno, T.; Alastuey, A.; Zhuang, X.; Lu, W.; Wang, Y. Environmental characterization of burnt coal gangue banks at Yangquan, Shanxi Province, China. Int. J. Coal Geol. 2008, 75, 93–104.

6. Zhou, C.C.; Liu, G.J.; Wu, D.; Fang, T.; Wang, R.W.; Fan, X. Mobility behavior and environmental implications of trace elements associated with coal gangue: A case study at the Huainan Coalfield in China. Chemosphere 2014, 95, 193–199.

7. Zhou, C.C.; Liu, G.J.; Yan, Z.C.; Fang, T.; Wang, R.W. Transformation behavior of mineral composition and trace elements during coal gangue combustion. Fuel 2012, 97, 644–650.

8. Conesa, J.A.; Rey, L.; Egea, S.; Rey, M.D. Pollutant formation and emissions from cement kiln stack using a solid recovered fuel from municipal solid waste. Environ. Sci. Technol. 2011, 45, 5878–5884.

9. Lin, Y.M.; Zhou, S.Q.; Li, F.Z.; Lin, Y.X. Utilization of municipal sewage sludge as additives for the production of eco-cement. J. Hazard. Mater. 2012, 213, 457–465.

10. Pavsic, P.; Mladenovic, A.; Mauko, A.; Kramar, S.; Dolenec, M.; Voncina, E.; Vrtac, K.P.; Bukovec, P. Sewage sludge/biomass ash based products for sustainable construction. J. Clean. Prod. 2014, 67, 117–124.

11. Saikia, N.; Kato, S.; Kojima, T. Production of cement clinkers from municipal solid waste incineration (MSWI) fly ash. Waste Manag. 2007, 27, 1178–1189.

12. Yen, C.L.; Tseng, D.H.; Lin, T.T. Characterization of eco-cement paste produced from waste sludges.Chemosphere 2011, 84, 220–226.

13. Taylor, H.F.W. Cement chemistry. In Chemistry for Engineers; Academic Press: London, UK, 1990.

14. Cui, S.P.; Lan, M.Z.; Zhang, J.; Wang, C.Y. Effect and incorporation mechanism of heavy metal elements in hazardous

industrial wastes during clinker formation. J. Chin. Ceram. Soc. 2004, 32, 1264–1270.

15. Fan, H.J.; Zhou, H.; Wang, J. Pyrolysis of municipal sewage sludges in a slowly heating and gas sweeping fixed-bed reactor. Energy Convers. Manag. 2014, 88, 1151–1158.

16. Yang, N.R.; Yue, W.H. The Handbook of Inorganic Matalloid Materials Atlas; Wuhan University of Technology Press: Wuhan, China, 2000. (In Chinese)

17. Zhang, Y.Y.; Nakano, J.; Liu, L.L.; Wang, X.D.; Zhang, Z.T. Trace element partitioning behavior of coal gangue-fired CFB plant: Experimental and equilibrium calculation. Environ. Sci. Pollut. Res. 2015, 22, 15469–15478.

18. Percival, H.J.; Duncan, J.F.; Foster, P.K. Interpretation of the kaolinite-mullite reaction sequence from infrared absorption spectra. J. Am. Ceram. Soc. 2006, 57, 57–61.

19. Liu, G.R.; Song, H.J.; Wu, J.H. Thermogravimetric study and kinetic analysis of dried industrial sludge pyrolysis. Waste Manag. 2015, 41, 128–133.

20. Rodriguez, N.H.; Martinez-Ramirez, S.; Blanco-Varela, M.T.; Donatello, S.; Guillem, M.; Puig, J.; Fos, C.; Larrotcha, E.; Flores, J. The effect of using thermally dried sewage sludge as an alternative fuel on Portland cement clinker production. J. Clean. Prod. 2013, 52, 94–102.

21. El-Rjoob, A.W.O.; Massadeh, A.M.; Omari, M.N. Evaluation of Pb, Cu, Zn, Cd, Ni and Fe levels in rosmarinus officinalis labaiatae (rosemary) medicinal plant and soils in selected zones in Jordan.Environ. Monit. Assess. 2008, 140, 61–68.

22. Li, H.X.; Xu, W.; Yang, X.J.; Wu, J.G. Preparation of Portland cement with sugar filter mud as lime-based raw material. J. Clean. Prod. 2014, 66, 107–112.

23. Xu, W.; Xu, J.C.; Liu, J.; Li, H.X.; Cao, B.; Huang, X.F.; Li, G.M. The utilization of lime-dried sludge as resource for producing cement. J. Clean. Prod. 2014, 83, 286–293.

24. Bigare, M.; Guinier, A.; Mazieres, C.; Regourd, M.; Yannaqui, N.; Eysel, W.; Hahn, T.; Woermann, E. Polymorphism of tricalcium silicate and its solid solutions. J. Am. Ceram. Soc. 1967, 50, 609–619.

25. Urabe, K.; Nakano, H.; Morita, H. Structural modulations in monoclinic tricalcium silicate solid solutions doped with zinc oxide, M(i), M(ii), and M(iii). J. Am. Ceram. Soc. 2002, 85, 423–429.

26. Horvath, E.; Frost, R.L.; Mako, E.; Kristof, J.; Cseh, T. Thermal treatment of mechanochemically activated kaolinite. Thermochim. Acta 2003, 404, 227–234.

27. Qiu, G.H.; Luo, Z.Y.; Shi, Z.L.; Ni, M.J. Utilization of coal gangue and copper tailings as clay for cement clinker calcinations. J. Wuhan Univ. Technol. 2011, 26, 1205–1210.

28. Diouri, A.; Boukhari, A.; Aride, J.; Puertas, F.; Vazquez, T. Stable Ca3SiO5 solid solution containing manganese and phosphorus. Cem. Concr. Res. 1997, 27, 1203–1212.

29. Rodriguez, N.H.; Martinez-Ramirez, S.; Blanco-Varela, M.T.; Guillem, M.; Puig, J.; Larrotcha, E.; Flores, J. Evaluation of spray-dried sludge from drinking water treatment plants as a prime material for clinker manufacture. Cem. Concr. Comp. 2011, 33, 267–275.

30. Elkhadiri, I.; Diouri, A.; Boukhari, A.; Puertas, F.; Vazquez, T. Obtaining a sulfoaluminate belite cement by industrial waste. Mater. Construcc. 2003, 53, 57–69.

31. Kakali, G.; Parissakis, G. Investigation of the effect of Zn oxide on the formation of Portland-cement clinker. Cem. Concr. Res. 1995, 25, 79–85.

32. Kakali, G.; Parissakis, G.; Bouras, D. A study on the burnability and the phase formation of PC clinker containing Cu oxide. Cem. Concr. Res. 1996, 26, 1473–1478.

33. Masaki, K.; Maki, I. Effect of prolonged heating at elevated temperatures on the phase composition and textures of Portland cement clinker. Cem. Concr. Res. 2002, 32, 931–934.

34. Tian, H.Z.; Liu, K.Y.; Zhou, J.R.; Lu, L.; Hao, J.M.; Qiu, P.P.; Gao, J.J.; Zhu, C.Y.; Wang, K.; Hua, S.B. Atmospheric emission inventory of hazardous trace elements from China's coal-fired power plants-temporal trends and spatial variation characteristics. Environ. Sci. Technol. 2014, 48, 3575–3582.

35. Swaine, D.J. Why trace elements are important. Fuel Process. Technol. 2000, 65, 21–33.

36. Liu, Y.; Wang, Y.; Wang, L.; Wang, X. Incorporation features of Cu~(2+),Zn~(2+),Cr~(6+),Cd~(2+) and Pb~(4+) ions in cement clinker. In Proceeding of the 8th International Symposium on Cement & Concrete (ISCC2013), Nanjing, China, 20 September 2013. (In Chinese).

37. Folgueras, A.B.; Diaz, R.A.; Xiberta, J.; Prieto, I. Volatilisation of trace elements for coal-sewage sludge blends during their combustion. Fuel 2003, 82, 1939–1948.

38. Stephan, D.; Mallmann, R.; Knofel, D.; Hardtl, R. High intakes of Cr, Ni, and Zn in clinker Part I. Influence on burning process and formation of phases. Cem. Concr. Res. 1999, 29, 1949–1957.

39. Barros, A.M.; Tenorio, J.A.S.; Espinosa, D.C.R. Evaluation of the incorporation ratio of ZnO, PbO and CdO into cement clinker. J. Hazard. Mater. 2004, 112, 71–78.

40. Li, X.; Li, H.; Yang, G. Promoting the adsorption of metal ions on kaolinite by defect sites: A molecular dynamics study. Sci. Rep. 2015, 5, 14377–14389.

41. Huntzinger, D.N.; Eatmon, T.D. A life-cycle assessment of Portland cement manufacturing: Comparing the traditional process with alternative technologies. J. Clean. Prod. 2009, 17, 668–675.

42. Habert, G.; Billard, C.; Rossi, P.; Chen, C.; Roussel, N. Cement production technology improvement compared to factor 4 objectives. Cem. Concr. Res. 2010, 40, 820–826.

Chapter 15

Recovery of Metals and Other Beneficial Products From Coal Fly Ash: A Sustainable Approach For Fly Ash Management

Prafulla Kumar Sahoo[1,2] • Kangjoo Kim[2] • M. A. Powell[3] • Sk Md Equeenuddin[4]

[1] *Vale Institute of Technology, Baventura da Silva, Belem, Para 66055-090, Brazil*

[2] *Department of Environmental Engineering, Kunsan National University, Jeonbuk 573-701, Republic of Korea*

[3] *Prometeo, SDIC, Secretary of Higher Education, Science, Technology and Innovation (SENESCYT), Quito, Ecuador 4 Department of Mining Engineering, National Institute of Technology, Rourkela, Orissa 769008, India*

ABSTRACT

Increasing production and disposal of coal fly ash (CFA) is a matter of serious environment concern. However, CFA contains various beneficial metals and mineral matters whose demand is increasing in the industrialized world, while natural supplies are diminishing. Therefore, recovery of these potential resources from CFA can be an alternative way to save mineral resources, as well as to reduce the environmental burden of CFA disposal. There are numerous methods developed for the recovery of beneficial products from CFA. Based on the US patents and journal literatures, the present review describes the recovery status and technologies of major elements such as Al, Si, Fe and Ti, and trace elements such as V, Ga, Ge, Se, Li, Mo, U, Au, Ag, Pt groups and rare earth elements (REEs) and other beneficial products such as magnetic materials, cenospheres, and unburned carbon from CFA. It also highlights the recovery efficiency and drawbacks for their extraction, and suggests future research to develop satisfactory results in terms of selective recovery and purification.

Keywords : Coal fly ash Beneficial elements Recovery Hydrometallurgical methods Extraction efficiency

1 INTRODUCTION

Coal based thermal power plants are one of the largest sources of electricity, contributing around 39 % of the total electricity generated throughout the world (Fig. 1) (US EIA 2014). This is due the fact that coal is the most abundant fossil fuel, and it is comparatively much lower priced than other fuels such as oil and natural gas. However, growing use of coal in power plants is generating huge amounts "waste", known as coal fly ash (CFA), with China, the US and India being the largest producers (Heidrich et al. 2013) (Table 1). The current global annual production of CFA is approximately 750 million tonnes (Yao et al. 2015), and this is anticipated to increase in the near future. Thus, increases in the global production of CFA is a matter of serious environmental concern because of issues related not only to its disposal but also the threat to public health and potential damage to soil, agriculture and natural beauty. These problems are more critical in areas experiencing increasing land use competition. Fly ash can even reach the sub-soil and contaminate the groundwater with heavy metals (Prasad and Mondal 2009; Ramya et al. 2013). Thus, there is a pressing need of its recycling and utilization (Ilic et al. 2003; Yao et al. 2015). More efforts have been undertaken on concrete production, paving roads, mine fillings, building material production, land reclamation, soil stabilization, toxic element immobilization, and synthesis of polymers and agriculture etc. (Kruger 1997; Basu et al. 2009; Ahmaruzzaman 2010; Alam and Akhtar 2011; Loya and Rawani 2014). Taking this into account, on average only 25 % of CFA is utilized (Sommerville et al. 2013), wherein 38 % for US, 47 % for the EU and 15 % for India (Yao et al. 2014a, b; reference therein). The rest is considered as pollutant and disposed waste. Hence, there is an urgent need for developing of its novel applications (Blissett et al. 2014).

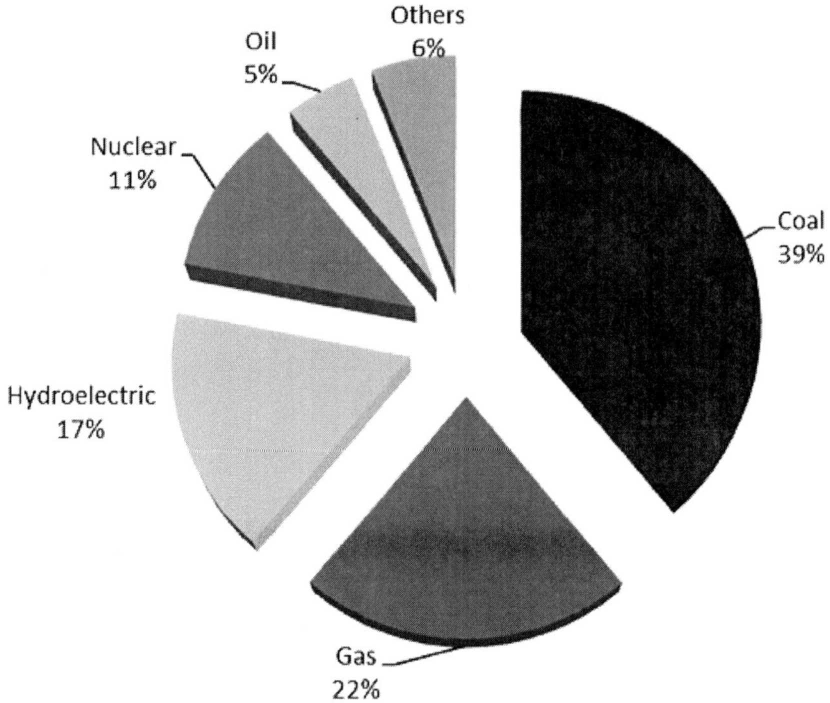

Figure 1. World electricity production from all energy in 2014 (US EIA 2014)

Currently, CFA is being explored as an alternative source of valuable metals, since it mainly composes of a variety of oxygenated inorganic compounds, such as SiO_2, Al_2O_3, Fe_2O_3 and TiO_2, with smaller amounts of trace elements (Lai-shi et al. 2011; Blissett and Rowson 2012; Seredin et al. 2013; Mayfield and Lewis 2013). With diminishing metal resources as well as ever-increasing their demand, recovery of these products from CFA would be highly worthwhile and a challenging topic for current international research (Cao et al. 2008; Mayfield and Lewis 2013; Tolhurst 2015). Nevertheless, recovery of metals from CFA can reduce environmental risks associated with disposal and leaching

Table 1. Worldwide annual production of coal fly ash (Heidrich et al. 2013)

Country	Ash production (Mt)
China	385
USA	118
India	105
Europe	52.6
Middle East and Africa	32.6
Russian Federation	26.6
Other Asia	16.7
Australia	13.1
Japan	11.1
Canada	6.8

of metals to surface and subsurface waters. There have been several reviews based on journal articles that focused on the recovery of beneficial elements and products from CFA (Jha et al. 2008; Cao et al. 2008; Meawad et al. 2010; Blissett and Rowson 2012; Yao et al. 2014a, b; Franus et al. 2015). However, the present review focused on US Patents, as well as journal articles to present an overview of the recovery status and technologies of metals and other beneficial products from CFA using several metallurgical and bio-metallurgical methods with the aim of evaluating their extraction efficiency, impact and applicability.

2 PROPERTIES OF COAL FLY ASH

The physico-chemical properties of CFA are highly variable depending on coal type, combustion temperature, combustion technique, air/fuel ratio, and particle size of the feed coal (Dhadse et al. 2008). The major oxides for CFA from different coal types are presented in Table 2 (Meawad et al. 2010). The principal components of all ash samples are SiO_2, Al_2O_3, and Fe_2O_3, with varying amounts of carbon, reported as loss on ignition (LOI), CaO, and MgO. Higher concentrations of CaO and MgO, and lower concentrations of SiO_2, Fe_2O_3 and carbon are found in lignite and sub-bituminous CFA compared to bituminous CFA. According to the American Society of Testing Materials (ASTM C618 2000), two classes of CFA have been defined based on the coal usage. Fly ash produced from anthracite or bituminous coal is classified as class F, while class C fly ash is produced from lignite or sub-bituminous coal. This classification is based on differences in contents SiO_2, Al_2O_3, and Fe_2O_3, and LOI in the ash. In classes C and F, the $SiO_2 + Al_2O_3 + Fe_2O_3$ contents are >50 % and >70 % respectively, and LOI is <6 % and <12 %, respectively. In addition, class C fly ash corresponds to high-lime (10 % up to 40 %), while Class F fly ash has lime concentrations <10 % CaO. Coal fly ash is considerably rich in trace elements (Karayigit et al. 2001; Vassilev et al. 2005; Dhadse et al. 2008). The normal range of those elements is given in Table 3 (National Research Council 2006; USEPA 1999; Meawad et al.2010). Elements in CFA are enriched mainly based on their organic/inorganic affinity and host mineralogy, followed by volatility (Xu et al. 2003; Vassilev et al. 2005). During fly ash gasification, the majority of mineral fractions hosting important trace elements is volatilized and then condensed on the surfaces of the fly ash (Clark and Sloss 1992; Querol et al. 1995). US Patent No. 0274886A1 (2007) illustrates the volatility classes of elements during combustion (Fig. 2). This indicated that elements such as As, B, Bi, Cd, Ge, Ga, Mo, Pb, S, Se, Sb, Sn, Te, and Ti are classified into Class II category. These elements have tendency to more concentrate in the finer particles of the ashes during the condense-volatilization process (US Patent No. 46860311987; Gieré et al. 2003).

3 RECOVERY OF MAJOR ELEMENT

There are a number of methods have been developed for the recovery of major elements from CFA (Table 4), which are discussed below.

Table 2. Major oxide concentrations in fly ash from different coal types (%) (Meawad et al. 2010; references therein)

Component	Bituminous	Sub-bituminous	Lignite
SiO_2	20–60	40–60	15–45
Al_2O_3	5–35	20–30	10–25
FeO_3	10–40	4–10	4–15
CaO	1.12	5–30	15–40
MgO	0–5	1–6	3–10
Na_2O	0–4	0–2	0–6
K_2O	0–3	0–4	0–4
LOI	0–15	0–3	0–5

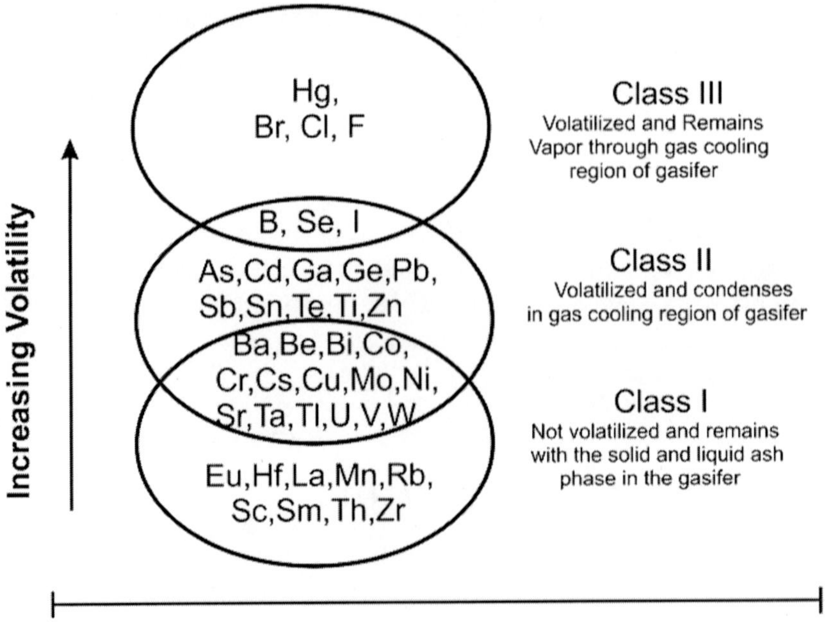

Figure 2. Relative volatility of trace elements in coal gasification systems (US Patent 0274886A1 2007)

Table 3. Trace element concentrations in fly ash (mg/kg) (National Research Council 2006; USEPA 1999; Meawad et al. 2010)

Element	Median	Range
As	43.4	0.0003–391
Ba	806.5	0.02–10,850
Be	5	0.2–2105
B	311	2.98–2050
Cd	3.4	0.01–79
Cr	136	3.6–437
Co	35.9	4.9–79
Cu	112	0.2–655
F	29	0.04–320
Pb	56.8	0.02–273
Mn	250	24.5–750
Hg	0.1	0.013–49.5
Ni	77.7	0.1–1270
Se	7.7	0.0003–46.5
Sb	4.6	0.2–205
Ag	3.2	0.01–49.5
Sr	775	30–3885
Tl	9	0.15–85
V	252	43.5–5015
Zn	148	0.28–2200

Table 4. Recovery of major and trace elements, their extraction methods and recovery efficiency from coal fly ash

References	Element	Method	Chemical reagent/others	Recovery efficiency
Burnet et al. (1977)	Al	High-temperature chlorination	Carbon, chlorine	Not known
Mehrotra (1982a, b)	Al, Ti	High-temperature chlorination	Carbon, carbon monoxide, chlorine	25 % Al, 25 % of Ti
US Pat. No. 4159310 (1979)	Al, Ti	High-temperature chlorination	Carbon monoxide, chlorine, H_2SO_4	70 %–80 % Al, 80 % Ti
US Pat. No. 4243640 (1981)	Al	Magnetic separation, acid leaching, calcination	HNO_3	Up to 90 % Al
US Patent No. 4567026 (1986)	Al, Ti	Magnetic separation, leaching, calcination	H_2SO_4, NAOH, CO_2	Not known
US Patent No. 4254088 (1981)	Al, Ti	Salt-soda sintering, leaching	NaCl, Na_2CO_3, HNO_3/H_2SO_4	90 % Al
Matjie et al. (2005)	Al, Ti	Calcination, leaching, solvent extraction	CaO, H_2SO_4	99.4 % Al_2O_3, 97 % TiO_2
US Pat. No. 4539187 (1985)	Al, Ti	Acid leaching	Fluorosilicic acid, HF	Not known
Wu et al. (2012)	Al	Pressure acid-leaching	H_2SO_4	82.4 % Al
US Pat. No. 4130627 (1978)	Al, Fe	Alkaline leaching, electrolysis	NaOH, HCl/HF	65 % Fe, 89 % Al
Padil and Sohn (1985)	Al	Lime-soda sintering	$CaCO_3$, $NaCO_3$	50 %–60 % alumina
Lu et al. (2003)	Al	Lime-soda sintering	Limestone, $NaHCO_3$	70 % alumina
Bai et al. (2010)	Al	Lime-soda sintering	Lime, $NaCO_3$, NaOH	90 % alumina
US Pat No. 4397822A (1983)		Lime-soda sintering	$CaCO_3$, $NaCO_3$, Sulfur, carbon	90 % alumina
US Pat No. 0119426A1 (2010)	Si, Al	Leaching, calcination	NaOH, CO_2, CaO, H_2SO_4	Not known
US Pat No. 7871583B2 (2011)	Si, Al	Leaching, calcination	NaOH, CO_2, CaO, H_2SO_4	Not known
US Pat No. 5231018 (1993)	Al, Ti	Bioleaching	C and N source, GIN-1, growth medium	Not known
US Pat No. 5278069 (1994)	Al, Ti	Bioleaching, acid precipitation	*Thiobacillus thiooxidans*	Not known
US. Pat No. 4539187 (1985)	Si	Acid leaching	SiF_4, HF	Not known
US Pat. No 0287827A1 (2010)	Si	Acid leaching	HF, coal, S	Not known
US Pat. No. 0119426A1 (2010)	Si	Alkaline leaching, carbonation	NaOH, H_2SO_4, CaO, CO_2	>98.5 % SiO_2 (purity)
US Pat. No. 7871583B2 (2011)	Si	Alkaline leaching, carbonation	NaOH, H_2SO_4, CaO, CO_2	>98.5 % SiO_2 (purity)
US Pat. No 4814150 (1989)	V	Leaching, precipitation	NaOH, H_2SO_4, ammonium salt	80 %–90 %
US Pat. No. 4966761 (1990)	V	Leaching, solvent extraction	NaOH, resin, HCl, NH4Cl, quaternary amine	Up to 93.4 % V
US Pat. No. 7498007B2 (2009)	V	Acidic leaching	Calcium hydroxide, quick lime, acid	>98 %
Xiang-yang et al. (2010)	V	Acid leaching	H_2SO_4	95 %
Vitolo et al. (2000)	V	Combustion, acid leaching	H_2SO_4, $NaClO_3$	83 %
US Pat. No. 6652819B2 (2003)	V	Combustion, acid leaching	H_2SO_4	Not known
US Pat No. 4475993 (1984)	Ga, others	Alkali halite melt, electrolysis	Aluminium trichloride	69 % of Ga
US Pat No. 4643110 (1987)	Ga, Ge	Sublimation		Not known
US Pat No. 4678647 (1986)	Ga, Ge	Sublimation		Up to 90 % Ga
US Pat No. 4757772 (1988)	Ge, Ga	Sublimation		Not known
US Pat No. 4678647 (1986)	Ge, Ga	Sublimation		Up to 90 % Ga

References	Element	Method	Chemical reagent/others	Recovery efficiency
Arroyo et al. (2014)	Ge, Ga	Acid, alkaline, complexant leaching	Oxalic acid, sulfuric acid	Up to 90 % Ge; up to 82 % Ga
Gutiérrez et al. (1997)	Ga	Acid leaching	HCl, Amberlite LA-2	83 % of Ga
Fang and Gesser (1996)	Ga	Acid leaching, flotation	HCl	Up to 80 % of Ga
Arroyo et al. (2009)	Ge	Water leaching, solvent extraction	Water, catechol, NaOH	95 % Ge
Font et al. (2005)	Ge	Water leaching		86 % of Ge
Mok et al. (1984)	Mo	Acid digestion, solvent extraction	HCl, HNO_3, dithiocarbamate	Not known
Ogata et al. (2011)	Mo	Acid/alkaline leaching	HCl, NaOH, gibbsite	90 % of Mo
US Pat. No. 3857920 (1974)	Li	Leaching	Water, CO_2	80 %–90 % of Li_2CO_3
Wang et al. (2007)	Se	Leaching	HNO_3, NaOH	55 %–69 % at pH 12
Sparton Res Inc. (2007)	U	Acid leaching, ion-exchange	H_2SO_4, resin	Not known
Maslov et al. (2010)	U, Ra	Acid leaching	HNO_3, HF	U (99 %), Ra (97 %)
Us Pat No. 6827837B2 (2004)	Au, Pt group	Combustion, acid digestion, precipitation	HNO_3, HCl	Not known
US Pat No. 0056548A1 (2005)	Au, Pt group	Combustion, acid digestion, precipitation	HNO_3, HCl	Not known
US Pat No. 4649031 (1987)	REEs	Sieving, aqueous alkali/acid digestion	Na_2CO_3, HCl, HNO_3	Not known
US Pat No. 0287653A1 (2013)	REEs	Acid leaching, Ion-exchange	HNO_3, tributyl phosphate, kerosene	Not known
US Pat No. 8968688B2 (2015)	REEs	Acid leaching, Ion-exchange	HNO_3, tributyl phosphate, kerosene	Not known
Narukawa et al. (2007)	Cr	Alkaline leaching	NaOH	Not known
Huang et al. (2011)	Pb, Zn	Alkaline leaching	NH4OH/NaOH	Not known
Okada et al. (2007)	Zn, Pb	Acid leaching	CH_3COOH	97 % Zn, 98 % Pb
Hong et al. (2000)	Cr, Cu, Pb, Zn	Leaching	Chelating agents (EDTA)	20 %–50 % Cr, 60 %–95 % Cu 60 %–100 % Pb, 50 %–100 % Zn
US Pat. No. 5278069 (1994)	Co, others	Bioextraction	*Thiobacillus thiooxidans*	Not known

3.1 Aluminum and titanium

Aluminum (Al) is an industrially important metal; it is light, corrosion resistant, and malleable and is mainly used in the aircraft industry. It is also used in the production of aluminum chemicals and in refractories (Matjie et al. 2005). Titanium (Ti) is used for various purposes in industry. In particular, it has been used in the aerospace industry. Titanium oxide is used as a pigment in the paint industry. On larger-scale, Ti and titanium alloy products are used for making automotive parts (Fontana et al. 2005). Coal and coal residues such as CFA are enriched with these elements (Halina et al. 2007; Seredin 2012; Dai et al. 2012a).

Several methods have been reported for recovering Al from CFA. Burnet et al. (1977) introduced a chlorinating process, where the fly ash is first separated into magnetic and non-magnetic fractions. The non-magnetic fraction of CFA was mixed with carbon and chlorinated in a fixed bed. The iron remaining in this fraction was removed by lowering the temperatures between 400 and 600 °C. The chlorinated aluminum fraction was recovered by increasing the temperature to 850–950 °C. Following Burnet et al. (1977), Mehrotra et al. (1982a,b) studied a high-temperature chlorinating process, and found that only 25 % of Al and 25 % of Ti in ash could be chlorinated at temperature above 900 °C. However, this process left about one-third of the iron with the nonmagnetic fraction. US Patent No. 4159310 (1979) gives a more thorough discussion of the chlorinating process for the recovery of Al and Ti. In this Patent the nonmagnetic fraction of ash was chlorinated in an oxidizing atmosphere and selectively separated Fe from other metals. The remaining residue was then chlorinated in a reducing atmosphere in the absence of solid carbon to suppress the chlorination of Si and vaporized the chlorides of other metals. Aluminum and Ti were selectively separated and recovered by condensation from the vapors. Overall, this method recovered 70 %–80 % Al and 80 % Ti. However, this method can become very expensive if the chlorination of unwanted metals is not suppressed.

Another method for the recovery of Al and Ti involves leaching CFA by acid/alkali leaching followed by precipitation, solvent extraction, or re-crystallization (Matjie et al. 2005; Halina et al. 2007; Li et al. 2011). US Patent No. 4243640 (1981) investigated a process for the recovery of Al from CFA where the ash was subjected to a magnetic separation and subsequently leaching was carried out on the non-magnetic fraction using nitric acid (strength about 30 %–56 %). The leachate was then evaporated and calcinated at 1000 °C for the recovery of Al. The calcinated product contained 90 % alumina along with other impurities like Fe oxides. US Patent No. 4567026 (1986) reported another acid leaching method for the recovery of Al and Ti from CFA. The process involved separation of magnetic fractions from the ash, followed by leaching the non-magnetic fraction with sulfuric acid, precipitating and removing Ti and Fe oxides from the leach solution by adding a strong base (NaOH), and precipitating and removing Al by contacting the reclaiming solution with CO_2. The precipitated aluminium hydroxide is calcinated to get aluminum oxide. However, direct sulfuric acid leaching yielded a low aluminum extraction (Matjie et al. 2005). Nayak and Panda (2010) also showed that direct sulphuric acid leaching at low concentration did not yield high aluminum recovery.

In contrast, CFA was treated pelletized with lime and calcinated, leaching with sulfuric acid that yielded better recovery. US Patent No. 4254088 (1981) discusses a process where the CFA was sintered with a mixture of NaCl and Na_2CO_3 at a temperature 700–900 °C. After sintering, 90 % of the aluminum CFA was in the acid soluble fraction. Recently, Matjie et al. (2005) discussed this calcinated technique at high temperature to decompose mullite phases, and then the calcinated CFA was leached in sulfuric acid to produce a solution containing Fe, Al and Ti species. The selective recovery of Al and Ti was carried out using various extraction methods, solvent extraction was more efficient than others and recovered 99.4 % of Al_2O_3 and 97 % of TiO_2. US Patent No. 4539187 (1985) also used an acid treatment where CFA was reacted with concentrated fluorosilicic acid and hydrogen fluoride at high temperature to form aqueous silicon fluoride vapor and fluoride and fluorosilicate of Al and Fe. Separated the Al and Fe bearing fluorides and fluorosilicate from aqueous silicon vapor, hydrolyzed and then recovered the substantial amount of pure Fe and Al following electroplating process. Recently, Wu et al. (2012) demonstrated pressure acid-leaching method, wherein the high extraction (82.4 %) of Al was obtained by diminishing CFA size (74 μm) and increasing acid concentrations (50 %). However in this method, the non-target metals readily leached into the acid, resulting in impurities. In addition, acid leaching had limited application due to its highly corrosive nature, besides this process required high cost, acid resistant equipment, and air-tight processing to protect the work environment.

Recovery of alumina by alkaline leaching is less problematic than techniques using acid leaching. US Patent No. 4130627 (1978) describes the alkaline leaching method where in the process takes place in an aqueous alkali solution of approximately pH 13, using NaOH, to form an aqueous CFA blend. The blend was agitated on a hotplate at temperatures of about 90 °C, and filtered to recover an alkaline solid. The filtrate was then contacted with hydrochloric or hydrofluoric acid to form soluble salts of various metals. Iron and aluminum were then selectively precipitated and recovered. This process recovered 65 % by weight of Fe and about 89 % of Al. However, this process has some disadvantages, such as being complex, time consuming, requiring larger quantities of processing materials, and high facility costs. This problem can be dealt with lime-soda sinter process, where a mixture of CFA, calcium carbonate, and sodium carbonate was sintered at high temperature to form a soluble sodium and calcium aluminate (Padil and Sohn 1985). Lu et al. (2003) reported this process where the CFA was mixed with limestone, sintered at 1260 °C, and then dissolved in sodium carbonate solution at 80 °C. This process obtained 99.9 % purity of alumina, and its recovery efficiency was 70 %. Recently, Bai et al. (2010) also used a lime-soda sintering method for the recovery of aluminum from an alkali desilicated CFA. The final Al extraction efficiency was 90 %. This process extracted Al from nearly 88 %–92 % at temperatures below 85 °C, while Al was not fully released from the sintered ash when the temperature was greater than 85 °C. US Patent No. 4397822A (1983) demonstrated an improvement method which added a small amount of sulfur and carbon to the lime-CFA sinter mixture before sintering to lower the optimum sinter temperature. This method recovered almost 90 % of alumina from CFA, and the results indicate that the addition of a combination of S and C, either elementally or as coal refuse, produces a higher recovery of alumina from CFA and at lower sintering temperatures, resulting in an energy savings. Recently, two US Patent Nos. 0119426A1 (2010) and 7871583B2 (2011) released a method for the recovery of alumina from high Si containing CFA, where CFA is leached with highly concentrated NaOH (>40 %). Silica is first removed from the CFA in the form of Na_2SiO_3, followed by separation of the residues having Al:Si ≥ 2. Subsequently the product was calcinated at 900–1100 °C to get pure alumina.

Besides acid/alkaline leaching, bioleaching also is used for the recovery of metals from CFA. This is discussed by in US Patent No. 5231018 (1993), where the CFA is suspended in an aqueous saline solution having pH ranging from 5 to 8, and then forming a reactive mixture by adding a pure culture a gram-positive bacteria (GIN-1), genus Rhodococcus. The mixture is then incubated for a period of time sufficient for the formation of microorganism cells/metal oxide adsorbate agglomerates, which separate from the reaction mixture. In this process, incubation reaction time was adjusted to selectively recover TiO_2 and Al_2O_3 from CFA. In another biochemical study, US Patent No. 5278069 (1994) used *Thiobacilus thiooxidans* strain in an acidic media for the recovery of metals from CFA. The metals were selectively recovered by precipitating the solution by adjusting the pH. Aluminum was mostly precipitated at pH 3.1–4.7, while Ti was precipitated at pH 1.5–3.1. These results indicate that bioleaching is one of the most cost-effective methods for the removal of metals from CFA; however, the percentage of metal recovery and their purity have not been reported.

3.2 Silica

Silica is used for various purposes, such as semi-conductors, soda-lime glass, high temperature thermal protection fabrics, glasses (including borosilicate glass and lead glass), and contact lenses and medical devices (Lukkassen and Meidell 2007). Coal fly ash is enriched with silica.

Several methods have been developed for the recovery of silica such as acid or alkali leaching, followed by separation and purification using different techniques. US Patent No. 4539187 (1985) describes a method where CFA is reacted with fluorosilicic acid and hydrogen fluoride at a high temperature to form fluorides of Fe, Al and Si. The fluoride mixture is distilled, separating the aqueous and gaseous silicon fluoride from the mixtures, hydrolyzing the silicon fluorides vapor, and recovering silicon dioxide in high purity form. Recently, US Patent No. 0287827A1 (2010) released a method where a mixture of coal and CFA is reacted with hydrogen fluoride in water to produce a liquid stream containing silicon fluoride and metal fluorides, and a solid stream containing unreacted coal and S compounds. Sulfur compounds are reacted with metallic nitrates dissolved in the water, which separated the aqueous solution of nitrate, S and metallic ions from the solid coal, and the solid coal is then washed with water. Subsequently, silicon fluoride and metal fluorides are contacted with metallic nitrates in an aqueous mixture to form solid silicon dioxide, and solid silicon dioxide is separated from the aqueous mixture. This process has recovered high purity silica from CFA. However, acid leaching is not always appropriate because it needs better acid resistant and air-tight processing equipment and it causes environmental pollution. This problem can be solved by using alkaline leaching methods. US Patent Nos. 0119426A1 (2010) and 7871583B2 (2011) released an alkaline leaching method for the recovery of silica from CFA, where Si is leached using NaOH in the form of

Na_2SiO_3solution. The solution is vaporized to obtain Na_2SiO_3, and pure silica (>98.5 %) is produced by carbonation.

4 RECOVERY OF TRACE ELEMENTS

There are a number of methods have been developed for the recovery of trace elements from CFA (Table 4), which are discussed below.

4.1 Vanadium

Vanadium (V) is mainly used in ferrous and non-ferrous alloys due to its tensile strength, hardness, and fatigue resistance (Moskalyk and Alfantazi 2003; Navarro et al., 2007). In addition, V is used in vanadium redox batteries. Although, V is extracted from various sources, V-enriched coal and carbonaceous residues such as CFA are one of the most important sources (Dai et al. 2015a).

Numerous methods have been used to extract V from carbonaceous residues. The most common method is wet treatment where the residue is subjected to multi-stage leaching, followed by precipitation via pH adjustment and oxidation/reduction (Europian Patent No. JP-A-H08-325651 1996; H10-251025 1998). US Patent No. 4814150 (1989) created a process where the combustion residue is reacted with sodium hydroxide in the presence of an oxidizer, dried at 100–250 °C, and leached with water to selectively extract V. Vanadium is precipitated from the alkaline extract by adding H_2SO_4 (up to pH 2) and an ammonium salt. The precipitated ammonium polyvanadate is calcinated to get vanadium pentoxide, which yields 80 %–90 % of V. US Patent No. 4966761 (1990) discusses a process in which CFA is mixed with water to prepare a slurry, carbon is skimmed off followed by alkaline pressure leaching and solvent extraction for the recovery of V (up to 93.4 %). US Patent No. 7498007B2 (2009) uses an acid solution to recover V. The extraction is carried out by precipitating vanadium pentoxide as a result of increasing pH by adding calcium hydroxide, quick lime, or calcium carbonate to the acid solution. After V is precipitated, the solid is separated by filtration or centrifugation. The percentage of V recovery from the acidic solution was ≥98 %. Recently, Xiang-yang et al. (2010) also described an acid leaching process using large volumes of sulfuric acid, extracting V up to 95.8 % from the leachate; however, no purification of V was attempted. Wet treatment techniques using acid/alkali leaching still have problems since they need a relatively large amount of expensive reagents, and they deal with complicated, troublesome repetition of steps, including conversion of solid CFA to a solution, isolating V from other metals and reconversion of V the solution into a solid. In addition, this technique may be unsuitable for CFA samples containing a high carbonaceous fraction and a low concentration of V because leaching V may not be efficient. To cope with this problem, dry treatment techniques have been developed (US Patent Nos. 4420464 1983, 4816236 1989). In these methods, the carbon is removed by combusting the V-containing residue at a controlled temperature, and then the residue is leached with a strong alkali/acid to recover the vanadium. Vitolo et al. (2000) used this method for the recovery of V from a fly ash at temperatures of 650–1150 °C, followed by an acid leaching and oxidative precipitation of vanadium pentoxide. Overall, this process yielded 83 % V. US Patent No. 6652819B2 (2003) also employed this method for the recovery of V from carbonaceous residues. The method involves a preliminary combustion of the residue at temperatures of 500–690 °C, followed by leaching with H_2SO_4 at a pH range of 1.5–4, and selective precipitation V by adjusting the pH in the range of 4.5–7.5, and separating the V ion as a tetravalent vanadium compound.

4.2 Gallium and germanium

Gallium (Ga) is used for the manufacture of high technology optical devices such as semiconductors, DVD's, laser diodes, and other electronic devices (Moskaly 2004). Germanium (Ge) is used in the manufacture of photovoltaic cells, fiber optics and as a catalyst for polyethylene terephthalate production (Butterman and Jogerson 2005). Germanium and gallium in the Earth's crust are mainly associated with Zn and its mineral phases. Thus, they are typically recovered from by-products of Zn mineral processing. Germanium and Ga are also enriched in coals (Gluskoter et al. 1977; Harvey et al. 1983; Dai et al. 2012a, b; Seredin 2012; Dai et al. 2014b; Qin et al. 2015a, b; Dai et al. 2015b). Thus, coal fly ash is a promising alternative resource of those elements (Fang and Gesser 1996; Zhang et al. 1999; Arroyo et al.2014; Dai et al. 2014a, b). After combustion in a coal burning power plant, both elements condense from the vapor phase onto the surface of the CFA, thus are enriched in CFA relative to the feed coal (Font et al. 2005; Querol et al. 1995).

Several techniques have been developed for the recovery of Ge and Ga from CFA. Alkali halite melt is one of the methods, and is discussed in US Patent No. 4475993 (1984). In this method, the CFA is reacted with aluminium trichloride to form a soluble compound of trace metals in the melt. Gallium was recovered from the melt by electrolysis. However, this process is uneconomical and recovered a moderate percentage (69 %) of Ga. Another method of Ga and Ge extraction is sublimation, which is discussed in several US Patents (Nos. 4757772 1988, 4757770 1988, and 4643110 1987). The steps include heating the CFA in an oxidizing atmosphere to vaporize the Ge and Ga, then treating the oxides in a reducing atmosphere to reduce oxides to suboxides, which subsequently precipitate on the surface of sand particles, and recovery of the products from the sand particles using a cold liquid solution. This method has problems involving other volatile compounds such as As and S. To cope with this problem, US Patent No. 4678647 (1986) introduced a method where CFA is pelletized and treated in the presence of an oxidizing gas at a temperature of about 900 °C. This treatment removes As and S from the CFA, so that they will not remain in the concentrated Ga and Ge fractions. Then, the oxides are treated in the presence of a reducing gas at the same temperature range to reduce the oxides to suboxides. Recover of Ga suboxides is done by condensation at a lower temperature (nearly 700–800 °C) and dissolving the condensed material using an acidic solution. The recovery rate of Ga for this method is up to 90 %. This method can be more economical and more efficient than the alkali halite method.

Other techniques for the recovery of Ga and Ge include using water/acid/alkali leaching of CFA, followed by separation/purification by using precipitation, distillation, ion flotation, adsorption into activated carbon, and solvent extraction (Meawad et al. 2010). Arroyo (2014) optimized Ge and Ga recovery from CFA using different extractants in a wide range of extraction conditions (acid, alkaline, complexant, reducing and oxidizing). High extraction yields of Ge (up to 90 %) and Ga (up to 82 %) were made using weak oxalic acid and sulfuric acid solution, respectively within 1–2 h extraction period.

Gutiérrez et al. (1997) discuss the recovery of Ga from CFA by 6 M hydrochloric acid in contact with Amberlite LA-2. Iron is precipitated with sodium hydroxide, while 83 % of Ga remains in solution, which is extracted selectively with LIX 54. Fang and Gesser (1996) studied the recovery of Ga from CFA involving a of two stage acid leaching with hydrochloric acid, followed by removal of silica and calcium, reduction of iron, extraction of Ga by foam floatation (polyurethane foam) and finally purification. This method showed that better extraction of Ga occurs at higher acid concentrations, low temperatures, short extraction times and smaller particle size. However, acid leaching is costly, requires better acid resistant equipment, and better environmental protection. Arroyo et al. (2009) discussed a water leaching process for extracting Ge from CFA, followed by solvent extraction. In this process, Ge complexation achieved with catechol, then extraction of the Ge-catechol complex with an organic reagent such as trioctylamine diluted in an organic solvent, and finally, stripping Ge from the organic extract by NaOH. This process extracted 95 % of the Ge from the leachate. In another study, Font et al. (2005) invented a low cost and environment friendly process for the recovery of Ge, where the CFA is leached in water at different ash/water ratios and temperatures. Results showed that temperature is the most important parameter in the extraction process, maximum content (up to 86 %) of the Ge extracted at 90 °C. For the recovery of Ge from solution, Hernandez-Exposito et al. (2006) developed an ion flotation method using different complexing agents (pyrogallol, catechol, hydroquinone and resorcin) at pH range of 4–7. It was found that using dodecyl amine as a surfactant, and pyrogallol or catechol as a complexing agent at pH values of 4–7, 100 % of the Ge could be recovered in 30 min. Torralvo and Fernández-pereira (2011) also reported that catechol as complexing agent for the recovery of Ge from leachates.

All the above methods were tested on bulk CFA; however, Ge and Ga are mainly concentrated in the finer fractions. This was examined in US Patent No. 4686031 (1987), where the process is mainly subjected to size classified CFA, and then extraction by various treatments. It was found that fine fraction, isolating 30 % of the finest particles while avoiding substantial rupturing of cenospheres and plerosphere, provided the best recovery.

4.3 Molybdenum

Molybdenum (Mo) is used to produce high tensile strength steel. It is also used as a biocatalyst for plants and animals and it is widely used in the manufacture of super alloys, lubricants, and electronic products. Coal fly ash can provide an alternative source for Mo recovery since some coals are enriched with Mo (Dai et al. 2015a, b). Mok et al. (1984) extracted Mo from CFA using the steps, digestion of CFA using concentrated HCl and HNO_3, diluting the solution using HCl, and selectively extracting Mo from the solution using dithiocarbamate. Ogata et al. (2011) disclosed a method for the recovery of Mo from CFA where the CFA is leached at pH levels 2, 7 and 11, and then adjusted with hydrochloric acid or sodium hydroxide solution. Subsequently, gibbsite (GB), an amorphous aluminium oxide, is used for the adsorption of Mo from elutes from the solutions. A greater concentration of Mo was eluted under acidic conditions onto GB400 (gibbsite calcinated at 400 °C) at pH 2 than from neutral or basic eluates. Recovery of Mo from eluates was 90 % using GB400. The main adsorption mechanism was ion exchange with a number of the hydroxyl groups in GB.

4.4 Lithium

There are a number of applications used for lithium carbonate, including as an additive in aluminum smelting, production of glasses, enamels, and ceramics, and production of electronic grade crystals. Its pure form is used in lithium ion batteries. Lithium has been found enriched in coal deposits (Qin et al. 2015b). US Patent No. 3857920 (1974) describes a method for the recovery of lithium carbonate from slurries resulting from the dispersion of CFA in water. In this method, CFA slurry is treated with carbon dioxide to convert the lithium carbonate to the more soluble lithium bicarbonate. The resulting bicarbonate solution is then filtered to remove insoluble material. The filtrate containing lithium bicarbonate is decarbonated either by vacuum treatment, heat, or preferably by aeration of the solution to precipitate lithium carbonate in the aqueous medium. The precipitated lithium carbonate is then filtered from the aqueous solution. This method is inexpensive and it recovered 80 %–90 % of the lithium carbonate.

4.5 **Selenium**

Selenium (Se) is mainly used in glassmaking and for pigments. It is also used for making semiconductors and in photocells. Selenium is commercially extracted as a byproduct in the refining of metal sulfide ores, most often during copper production. It can also be extracted from CFA since some coals are enriched in Se (Dai et al. 2014a, b; 2015b, c). Wang et al. (2007) studied the leaching of selenium from bituminous and sub-bituminous coal ash with respect to pH. Very low concentrations of selenium were released at a pH range from 3 to 4, while the amount increased with increasing pH above 4, and the maximum amount of Se (55 %–69 % of the total) was leached at pH 12. Iwashita et al. (2005) also reported that leaching of Se increased when pH was raised. However, recovery of this element is not attempted in these literatures.

4.6 **Radioactive elements**

Radioactive elements have many industrial and scientific applications, despite the obvious dangers if they are not handled properly. Radioactive elements are used as tracers in the medical field, the generation of electricity in nuclear power plants and in the production of nuclear weapons. Coal contains significant amounts of radioactive elements such as uranium (U) (Dai et al. 2015a, c). Thus, CFA can be alternative source of these elements. Sparton Resources Inc. (2007) reported the extraction of U from CFA. This process included acid leaching (H_2SO_4) and formation of yellowcake by passing the filtered leachate solution through a standard ion exchange resin, stripping the resin and precipitating a U compound. This process is essentially similar to the U extraction method used by the primary U ore processing plants. However, this work is preliminary and requires further work to improve recovery.

Maslov et al. (2010) discussed a method for recovering U from brown coal ash of Mangolia where the ash is leached with an acid mixture (8 M HNO_3 + 10 % HF). The acid mixture leached U and Ra as high as 99 % and 97 %, respectively. The extraction of U from the solution and its purification for the removal of accompanying elements were carried out using an anion exchanger (Dowex 2X8) with a 2 M HCl solution as the eluate. After leaching, the CFA contained neither natural radioactive elements nor their decay products, thereby allowing for the use of the ash as a building material.

4.7 Gold, silver and platinum group of elements

Precious metals such as gold (Au) and silver (Ag) have many application in industry, jewelry and medicine (Guerra and Calligaro 2003; Drake and Hazelwood 2005; Eisler 2003). Some Au compounds are used as anti-inflammatory drugs in the treatment of rheumatoid arthritis and Ag compounds are used in the treatment of infections. Gold is also found in electronic apparatus and Ag is used in photography, industry and for hard solder.

US Patent Nos. 6827837B2 (2004) and 0056548A1 (2005) released a method for the recovering of gold and one and more platinum group elements, such as platinum, rhodium, ruthenium, osmium and iridium, from CFA. In this method an ash formed by the combustion of a preselected coal is mixed with the ash contains gold and one more platinum group of elements, an inquart containing gold and one or more platinum group elements, and fluxing agent to form a 'charge'. The 'charge' is then heated in a furnace to a preselected temperature and forming a molten charge including a molten inquart; absorbing a portion of the gold and more platinum group elements of the ash into the molten inquart. Casting a dare bar (which contains gold and one or more platinum group elements) by discharging the molten inquart. Digesting the dore bar in an elctrowinning process where the dore bar is introduced as an anode in an acidic bath and an electrical potential is maintained between the cathode and anode, then precipitating and dissolving trace elements in the electrolytic solution producing an anode slime. The digestion of anode slimes is done using 30 % solution of HNO_3. The digestion is filtered and rinsed then the liquid rinsed portion containing Ag in solution is washed with HCl and filtered. The residue is mixed with aqua regia, forming a solution containing gold, platinum and palladium. Gold may then be removed by solvent extraction. Rhodium and other metals may be removed by precipitation by the treatment of the solution with potassium hydroxide and sodium borohydrite. Palladium in solution may be removed by precipitation by treatment with ammonium hydroxide, NH_4OH, and HCl as the complex $PdCl_2(NH_3)_2$. Alternatively, Pd in solution is removed by precipitation with DiMethyl Glyxime, followed by filtration and reduction and ashing of the residue. Platinum in solution is removed by any of the accepted methods including precipitation as impure $(NH_4)_2PtCl_6$ after treatment with NH_4Cl, or leaching H_2PdCl_4 from solution. An organic solvent such as DiMethyl Glyoxime is used to extract gold, after treatment with NH_4Cl, the solution should also be treated with ethanol and hydrogen peroxide. The solution is filtered and the residue, including $(NH_4)_2PtCl_6$, subjected to reduction and burning leaving an impure platinum sponge. The platinum sponge can be purified by redissolving in aqua-regia. This method is very complex, and a recovery and purity level is not mentioned.

4.8 Rare earth elements (REEs)

Rare earth elements include light REEs (Sc, La, Ce, Pr, Nd, and Pm), medium REEs (Sm, Eu and Gd) and heavy REEs (Tb, Dy, Ho, Er, Tm, Yb, Lu and Y), which are mainly used is various industrial purposes such as fuel cells, green energy devices, hi-capacity batteries, magnets for wind generation (Chen 2011; Hoenderdal et al. 2013; Franus et al. 2015). REEs have been recovered from bauxite residue (Borra et al. 2015) and electronic wastes (Anonymous 2015). It has also been recognized that some coals (Seredin 1996; Dai et al. 2014a, b) and CFA (Ketris and Yudovich 2009; Blissett et al. 2014; Dai et al. 2012a, b; Dai et al. 2014a, b; Franus et al.2015) are enriched with REE; thus, recently sparking interest in evaluating of this product as a potential alternative source of REEs (Seredin et al. 2013).

There are several methods developed for recovery of rare elements from CFA, the chlorination method is favored due to selectivity and high reactivity. The necessary apparatus required for this process is very expensive. This technique can be used to recover only certain rare elements where other complex recovery processes are not suitable. From this standpoint, US Patent No. 4649031 (1987) reported a non-destructive digestion process for the recovery of rare elements, including Lanthanides and Actinides and group V and VI elements that are adsorbed onto the surfaces of coal combustion residues. This techniques includes: (1) reducing the grain size of the combustion residues from 0.1 to 1 mm, (2) digestion of the residues with water, (3) digestion of the residues with a dilute aqueous alkali solution (1.5 %–8 %), (4) digestion of the residue with a dilute aqueous acid solution (0.5 %–5 %) at a temperature below the boiling point of water, (5) separating the solid and liquid phases after each digestion, (6) and isolating the s metals from the liquid phases. This study shows that the mild digestion agents used dissolve an unexpectedly high proportion of rare metals present in the byproducts and that increasing the heat and residence time lowers the availability of the metals. This process is not appropriate for the recovery of individual elements from the mixed solution.

Recently, US Patent No. 0287653A1 (2013) demonstrated the recovery of REEs from CFA. In this method, first the CFA was treated with a mineral acid (HNO_3) at 90 °C to form a more concentrated mineral acid solution. This solution is mixed with an organic solution (tributyl phosphate and kerosene) that includes the rare earth salts. The organic solution is mixed with water to form an aqueous solution. REEs substantially recovered from the aqueous solution by using ion exchange process. In a continuation-in-part of the above patent, US Patent No. 0139871 (2015) was carried out using the same method, but here the CFA was sorted into a substantially magnetic and non-magentic fraction. This indicates magnetic treatment can maximize efficiency of extracting REEs from CFA. However, extraction efficiency of these methods is not reported. Furthermore, ion exchange method can remove other metals from the solution. This method needs to be optimized to improve the yields for REEs.

4.9 Other trace elements (Pb, Cr, Cd, and Cu)

Coal fly ash also contains considerable amounts of trace and rare elements, and some them can be hazardous (Izquierdo and Querol 2012). There are many studies which focused on the extraction of trace metals from CFA, some employing hydrometallurgical techniques using various lixiviants: inorganic acids such as sulfuric acid, hydrochloric acid, nitric acid, organic acid such as acetic acid, citric acid, oxalic acid, alkaline reagents such as sodium and ammonium hydroxide, chelating reagents such as ethylendiaminetetraacetate (EDTA) and diethylenetriaminepentaacetate (DTPA) (Hong et al. 2000; Zhang and Itoh 2006).

Narukawa et al. (2007) studied Cr extraction from an Australian CFA using 0.01 M NaOH. However, this process did not leach Cr(III) from the CFA. Huang et al. (2011) discussed the alkaline leaching by ammonium or sodium hydroxide, where only Pb and Zn are recovered while other metals remained in the solid residues. However, this process can extract other heavy metals such as Cu and Cd if the leaching process is conducted together with subsequent leaching using acidic lixiviants. Okada et al. (2007) reported that in acidic leaching using acetic acid recovered 97 % of Zn and 98 % of Pb from CFA, while the same CFA in alkaline leaching using 3 M NaOH recovered 81 % Pb and 35.5 % Zn. Consequently, acid leaching is more effective under hydrothermal conditions. This was studied by Zhang and Itoh (2006) for CFA using five types of acids and hydrochloric acid was found to be the most effective for metal extraction from the CFA. Compared to room conditions (standard temperatures and pressures), it was determined that hydrothermal treatment accelerated the dissolution of the CFA and resulted in better leaching of metals such as Cr, Cd, and Pb. Even more effective leaching can be done using chelating agents. Hong et al. (2000) studied the extraction of heavy metals from CFA using acid and chelating agents such as nitritriacetic acid (NTA) and ethylendiaminetetraacetate (EDTA), or diethylenetriaminepentacetate (DTPA). The results showed that chelating agents provide good leaching efficiency and after the treatment the CFA's were successfully detoxified to meet landfilling guidelines. Using 3.0 % EDTA or DTPA, 20 %–50 % Cr, 60 %–95 % Cu, 60 %–100 % Pb, and 50 % –100 % Zn were extracted from CFA. However, these agents are not biodegradable and it is difficult to recover the metals from the leachate due to their strong chelating properties.

Recovery of metals from CFA using bioextractants has been studied by many authors (Seidel et al. 2001; Ishigaki et al. 2005; Sierra-Alvarez 2009). This process is important at an industrial scale. In this process, chemolithoautotrophic bacteria such as Thiobacillus species and Heterotrophic fungi such as Aspergillus and Penicillium species are used to leachthe metals. The leached metals can then be recovered and the residues may be safely utilized. US Patent No. 5278069 (1994) discusses the bioextraction process using strains of Rhodococcus and*Thiobacillus thiooxidans* for the removal of Co along with other metals (Al, Ti and Fe) from CFA. After extraction, the metals were recovered in a series of precipitation steps by controlling pH. This technique is relatively inexpensive and requires low energy inputs when compared with conventional technologies (Krebs et al. 1997). However it is a slow process requires sufficient amounts of cell culture and is not selective toward specific metals (Groudev et al.1982).

5 RECOVERY OF OTHER MATERIALS

5.1 Magnetic materials

Magnetite or other magnetic material is used by the steel industry for cleaning coal prior to combustion. One potential source of magnetite is CFA which can contains economically recoverable amounts of magnetic material. After removal of magnetic material from CFA, the remaining product can be useful for refractory and ceramic manufacturing since it contains appreciable amounts of Al_2O_3 and SiO_2 (Rao et al. 1999). The characteristics of magnetic material in CFA vary depending on the coal source and operating conditions of the coal boiler (Zyryanov et al. 2011). The basic principle behind magnetic separation deals with the fact that materials having different magnetic moments experience varying forces in the presence of a magnetic field gradient. Thus, applying an external magnetic field can extract those materials with similar magnetic properties (Yavuz et al. 2009).

There are many wet or dry magnetic separation methods for the recovery of magnetic materials from CFA. The dry process uses a rotating magnetic drum and initial separation produces middling and tailings which are discarded and a crude iron oxide fraction which must be further refined by passing it through another magnetic unit. The wet process involves pumping CFA slurry into a magnetic drum where the magnetic iron oxides and associated materials are collected and then the material is passed through a similar unit for further refining. In most of the plants using wet magnetic separation, the advantage as that the slurry can be directly fed into the feed tank. Prakash et al. (2001) demonstrated the use of a selective magnetic coating for the removal of iron from Talcher (India) CFA. They used sodium hexametaphophate (NaHMP) as a dispersing solution, followed by sodium oleate for the separation of iron from. The recovery rate for Fe was around 92 %. The wet/dry magnetic separations processes are documented by numerous US Patents (Nos. 4191336 1980; 4319988 1982; 5047145 1991; 5227047 1993). However, these methods are not cost-effective if the ash contains less than 2 % magnetite. In order to improve the process or more recovery, CFA containing sufficient amount of magnetic material (>2 %) need to be selected. Groppo and Honaker (2009) studied the economic recovery of magnetite material from CFA.

US Patent No. 3769054 (1993) released another method for the recovery of magnetic materials from CFA. In this method CFA is dry processed in an air classifier and then separated the magnetic separation. This process is expensive, and the recovered portion of magnetite often contains undesirable impurities. To resolve these problems, US Patent No. 4191336 (1980) provides an improved process for the recovery of high purity magnetite from CFA. This process involves making a water slurry of the CFA and then passing the concentrated slurry through the magnetic separators; magnetite is contained in the separated material. The separated magnetite is then pulverized to a size where 95 % of the magnetite will pass through a 325 mesh sieve. The crushed magnetite is passed through a third separator followed by purification and drying.

5.2 Cenospheres

Cenospheres are composed mainly of silica, alumina and iron oxides, and have unique properties such as sphericity, inert, lower density than water, and they are waterproof making them useful in a variety of applications. Generally, cenospheres make up 1 %–3 % by weight of the total CFA produced. They have valuable industrial applications, such as oil well drilling, ceramics, plastic, paint, coatings, polymer fillers, building products, buoyaancy, and as low dielectric constant substrates (Huo et al. 2009; Wang et al. 2011). Cenospheres are used for water purification such as the removal of excess fluoride from drinking water. In this process, magnesia loaded cenospheres are produced using a relatively simple wet impregnation method with magnesium chloride (Xu et al. 2011). Cenospheres are used as a carrier because of their low weight, sphericity, non-toxicity, and high strength. They are ideal for incorporation into materials such as silicone rubber in order to increase the conductivity. In addition, they enhance the rubber's suitability as an electromagnetic wave absorbing material, which can be used in electronic and radar applications (Pang et al. 2011).

There are several methods for the separation of cenospheres from CFA. Traditionally, cenospheres were removed from CFA via float-sink in settling lagoons. This simple procedure allows potentially toxic elements to be leached and requires a large space and significant time (Hirajima et al. 2010). Alternatively, a triboelectric separation system was used (Gurupira et al.2001) but this technique requires that the spheres have a specific gravity ≤1. Some solid silica particles with specific gravity of ≥1 may contain gas bubbles which over their specific gravity so that they mimic cenospheres during this process, rendering it less effective and attractive. Other studies that have looked at the extraction of cenospheres on a wet basis. In US Patent No. 4121945 (1978), CFA is mixed with water to form a slurry then a frothing agent is added and cenospheres lighter than water are skimmed. US Patent No. 4652433 (1987) looked at another method where CFA is separated into fine and coarse fractions and the coarse fraction is then mixed with water, subsequently the cenospheres are removed by skimming. These methods have some disadvantages. Wet processing caused environmental problems and imposed space restrictions. In addition, these methods only collect cenospheres with specific gravity ≤1.0; those that float on water. Also, these processes are not suitable for CFA that contains 10 % or more lime because high Ca bearing CFA (such as ASTM C618 Class C fly ash) has cementations properties when mixed with water and when pozolanic reactions occur the material may harden. When using these wet methods many cenospheres can be trapped in agglomerated and/or hardened masses before flotation occurs. This problem was solved by a few recent US Patents (No. 0190327A1 2008, No. 8074804B2 2011, and No. 0050519A1 2012) where cenospheres were separated from CFA in a dry state by using air as the fluid media for the separation. In one of the processes, CFA is separated into narrow particle size fractions, followed by density separation using a fluidized bed of air, preferable in the absence of water by dry screening, and then an optional further particle size separation. The advantages of this technique are that it can yield nearly 100 % cenospheres and it avoids agglomeration of ash having high Ca content.

5.3 Unburned carbon

Unburned carbon in CFA can be used during activated carbon preparation or carbon black, and can be used as a supplementary fuel. It can also be useful in the manufacture of graphite (Cabielles et al. 2008; Cameán and Garcia 2011). Removal of unburned carbon from coal ash makes it a more valuable cement additive. There are numerous methods for the recovery of unburned carbon from CFA (Hwang et al. 2002). Electrostatic separation is one of them, has been used by many authors (Gray et al. 2002; Soong et al. 2002). This method operates in a controlled electrode potential condition following the principle of bipolar charging of dry particles either by particle–particle contact or by particle–wall collision under turbulent conditions. This process separates the positively charged carbon particles from the negatively charged ash particles. This technology has been successfully commercialized.

One of the most promising methods for recovering unburned carbon from CFA is froth flotation. This method relies on the ability of air bubbles to selectively adhere to the surface of a particular mineral surface. Since carbon is hydrophobic in nature, other particles that are hydrophilic attached to the air bubbles and become positively buoyant and float on the slurry (Ucurum 2009). This process has been commercialized in the US for CFA beneficiation with reported carbon grade recoveries of up to 70 % (Baltrus et al. 2001).

A similar technology for froth flotation uses oil agglomeration, which involves wetting hydrophobic particles by adding oil to the aqueous slurry. Subsequently, the vessel can be agitated which allows the oil coated particles to collide with each other and form agglomerates. The oil/carbon agglomerates being less dense than the ash suspension rise to the top of the vessel (Mehrotra et al. 1983). US Patent No. 546363 (1995) removes carbon adding floating reagent-mixture of fuel oil and petroleum sulfonate (10:1) and has recovered 44 % after 1 min of flotation while fuel oil flotation after 20 min recovered only 38 %. Gray et al. (2001) used this method with cyclohexane as the solvent, and achieved high carbon recoveries (55 %–57 %). There are other studies reporting similar carbon purities in their agglomerates using vegetable oil as their solvent (Alonso et al. 1999). US Patent No. 6068131 (2000) describes this type of removal using a conditioning agent such as biodegradable soil.

6 CONCLUSIONS

Coal fly ash has been explored as an alternative source of various valuable elements and minerals. Recovery of these components from CFA not only provides alternative resources, but also reduces disposal cost and protects the environment. This paper has reviewed a number of hydrometallurgical methods developed for the recovery of those products from CFA and highlighted suitability and effective technique for their extraction based on US Patents and journal literature. By applying those techniques, it was found that the recovery of major elements such as Fe, Al and Si and trace elements such as Ge and Ga from CFA are satisfactory. However, for other trace and rare earth elements further research is needed to develop adequate results in terms of selective recovery and purification. Mineral matter such as cenospheres, unburned carbon and magnetic materials also significantly extracted from CFA.

Most of these methods were studied at lab-scale, their commercialization utilization is largely limited, and therefore further development in large scale is required to scale up to commercial production levels. In addition, economic viability and environmental impact evaluation of each method should be focused in the future work. This would be helpful to choosing correct method and to ensure that any operations don't result in a greater environmental footprint than mining of primary resources. Quality of raw ash obtained in parallel with the metal recovery processes needs to be assessed.

REFERENCES

1. Ahmaruzzaman M (2010) A review on the utilization of fly ash. Prog Energy Combust Sci 36:327–363

2. Alam J, Akhtar MN (2011) Fly ash utilization in different sectors in Indian scenario. Int J Emerg Trends Eng Dev 1:1–13

3. Alonso MI, Martínez-Tarazona RM, García AB (1999) Coal recovery from coal fines clearing wastes by agglomeration with vegetable oils: effect of oil type and concentrations. Fuel 78:753–759

4. Aldrich RG (1982) Separation of high grade magnetite from fly ash. US Patent No. 4319988

5. Anonymous (2015) Recovery of rare earths from electronic wastes; an opportunity for High-Tech SMEs. Directorate General for Internal Policies, Policy Department A: Economic and Scientific Policy, European Parliament.http://www.europarl.europa.eu/RegData/etudes/STUD/2015/518777/IPOL_STU(2015)518777_EN.pdf)

6. Arroyo F, Fernandez-Pereira C, Olivares J, Coca P (2009) Hydrometallurgical recovery of germanium from coal gasification fly ash. Ind Eng Chem Res 48:3573–3579

7. Arroyo F, Font O, Chimenos JM, Pereira CF, Querol X, Coca P (2014) IGCC fly ash valorisation. Optimisation of Ge and Ga recovery for an industrial application. Fuel Process Technol 124:222–227

8. Ashworth RA et al. (1987) Method for the recovery of minerals and production of by-products from coal ash. US Patent No. 4652433

9. Bai G, Teng W, Wang X, Qin J, Xu J, Li P (2010) Alkali desilicated coal fly ash as substitute of bauxite in lime-soda sintering process for aluminium production. Trans Nonferr Met Soc China 20:169–175

10. Baltrus JP, Wells AW, Fauth DJ, Diehl JR, White CM (2001) Characterization of carbon concentrates from coal-combustion fly ash. Energy Fuels 15:455–462

11. Basu M, Pande M, Bhadoria PBS, Mahapatra SC (2009) Potential fly-ash utilization in agriculture in agriculture: a global review. Prog Nat Sci 19:1173–1186

12. Barclay KM (1983) Recovery of vanadium from carbonaceous materials. US Patent No. 4420464

13. Benson S, Laumb M (2007) Removal and recovery of deposits from coal gasification systems. US Patent No. 20070274886A1

14. Blander M et al. (1984) Extraction of trace metals from fly ash. US Patent No. 4475993

15. Blissett RS, Rowson NA (2012) A review of the multi-component utilization of coal fly ash. Fuel.

16. Blissett RS, Smalley N, Rowson NA (2014) An investigation into six coal fly ashes from the United Kingdom and Poland to evaluate rare earth element content. Fuel 119:236–239

17. Borra CR, Pontikes Y, Binemans K, Gerven TV (2015) Leaching of rare earths from bauxite residue (red mud). Miner Eng 76:20–27

18. Burnet O, Murtha MJ, Wijatno H (1977) Recovery of alumina from fly ash by high temperature chlorination. In: Rose JG (ed) Proceeding 3rd Kentucky coal refuse disposal and utilization seminar, May 11–12. University of Kentucky, Lexington, pp 83–88

19. Butterman WC, Jogerson JD (2005) Mineral commodity profiles: germanium. U.S. Geological Survey, U.S. Department of the Interior, Washington, DC

20. Brown JW (1980) Process for recovering magnetite from fly ash. US Patent No. 4191336

21. Cabielles M, Montes-Morán MA, Garcia AB (2008) Structural study of graphite materials prepared by HTT of unburned carbon concentrates from coal combustion fly ashes. Energy Fuels 22:1239–1243

22. Cameán I, Garcia AB (2011) Graphite materials prepared by HTT of unburned carbon from coal combustion fly ashes: performance as anodes in lithium-ion batteries. J Power Sources 196:4816–4820

23. Cao D, Selic E, Herbell JD (2008) Utilization of fly ash from coal-fired power plants in China. J Zhejiang Univ 9:681–687

24. Chandrashekhar S (2010) Process for obtaining treated coal and silica from coal containing fly ash. US Patent No. 2010/0287827A1

25. Chen Z (2011) Global rare earth resources and scenarios of future rare earth industry. J Rare Earths 29:1–6

26. Clark LB, Sloss LL (1992) Trace elements—emissions from coal combustion and gasification. IEA Coal Research Rep. No. IEACR/49

27. Corigliano F et al. (1989) Process for the high-yield recovery of vanadium from petroleum combustion residues. US Patent No. 4814150

28. Dai S, Jiang Y, Ward C, Gu L, Seredin VV, Liu H, Zhou D, Wang X, Sun Y, Zou J, Ren D (2012a) Mineralogical and geochemical compositions of the coal in the Guanbanwusu Mine, Inner Mongolia, china: further evidence for the existence of an Al (Ga and REE) ore deposit in the Juangar Coalfield. Int J Coal Geol 98:10–40

29. Dai S, Wang X, Seredin VV, Hower JC, Ward CR, O'Keefe JMK, HuangW Li T, Li X, Liu H, Xue W, Zhao L (2012b) Petrology, mineralogy, and geochemistry of the Ge-rich coal from the Wulantuga Ge ore deposit, Inner Mongolia, China: new data and genetic implications. Int J Coal Geol 90–91:72–99

30. Dai S, Luo Y, Seredin VV, Ward CR, Hower JC, Zhao L, Liu S, Zhao C, Tian H, Zou J (2014a) Revisiting the late Permian coal from the Huayingshan, Sichuan, southwestern China: enrichment and occurrence modes of minerals and trace elements. Int J Coal Geol 122:110–128

31. Dai S, Seredin VV, Ward C, Jiang J, Hower J, Song X, Jiang Y, Wang X, Gornostaeva T, Li X, Liu H, Zhao L, Zhao C (2014b) Composition and modes of occurrence of minerals and elements in coal combustion products derived from high-Ge coals. Int J Coal Geol 121:79–97

32. Dai S, Seredin VV, Ward CR, Hower JC, Xing Y, Zhang W, Song W, Wang P (2015a) Enrichment of U–Se–Mo–Re–V in coals preserved within marine carbonate successions: geochemical and mineralogical data from the late Permian Guiding Coalfield, Guizhou, China. Miner Depos 50:159–186

33. Dai S, Liu J, Ward CR, Hower JC, Xie P, Jiang Y, Hood MM, O'keefe JMK, Song H (2015b) Petrological, geochemical, and mineralogical compositions of the low-Ge coals from the Shengil Coalfield, China: a comparative study with Ge-rish coals and a formation model for coal-hosted Ge ore deposit. Ore Geol Rev 71:318–349

34. Dai S, Yang J, Ward CR, Hower JC, Liu H, Garrison TM, French D, O'Keefe JMK (2015c) Geochemical and mineralogical evidence for a coal-hosted uranium deposit in the Yili Basin, Xinjiang, northwestern China. Ore Geol Rev 70:1–30

35. Dhadse S, Kumari P, Bhagia LJ (2008) Fly ash characterization and utilization and government initiatives in India: a review. J Sci Ind Res 67:11–18

36. Drake PL, Hazelwood KJ (2005) Exposure-related health effects of silver and silver compounds: a review. Ann Occup Hyg 49:575–585

37. Eisler R (2003) Chrysotherapy: a synoptic review. Inflamm Res 52:487–501

38. Fang Z, Gesser HD (1996) Recovery of gallium from coal fly ash. Hydrometallurgy 41:187–200

39. Fass R et al. (1994). Bioleaching method for the extraction of metals from coal fly ash using thiobacillus. US Patent No. 5278069

40. Font O, Querol X, Lopez-Soler A, Chimenos JM, Fernandez AI, Burgos S, Garcia PF (2005) Germanium extraction from gasification fly ash. Fuel 84:1384–1392

41. Fontana D, Kulkarni P, Pietrelli L (2005) Extraction of titanium (IV) from acidic media by 2-ethylhexyphosphonic acid mono-2-ethylhexyl ester. Hydrometallurgy 77:219–225

42. Franus W, Wiatros-Motyka MM, Wdowin M (2015) Coal fly ash as a resource for rare Earth elements. Environ Sci Pollut Int 22:9464–9474

43. Gardner HE (1989) Recovery of vanadium and nickel from petroleum residues. US Patent No. 4816236

44. Gieré R, Carleton LE, Lumpkin GR (2003) Micro- and nanochemistry of fly ash from a coal-fired power plant. Am Miner 88:853–1865

45. Gluskoter HJ, Ruth RR, Miller WG, Cahill RA, Dreher GB, Kuhn JK (1977) Trace elements in coal: occurrence and distribution. Illinois State Geological Survey, Circular 499

46. Grantham L et al (1974) Recovery of lithium carbonate. US Patent No. 3857920

47. Gray ML, Champagne KJ, Soong Y, Finseth DH (2001) Parametric study of the column oil agglomeration of fly ash. Fuel 80:867–871

48. Gray ML, Champagne KJ, Soong Y, Killmeyer RP, Maroto-Valer MM, Anderson JM, Ciocco MV, Zandhuis PH (2002) Physical cleaning of high carbon fly ash. Fuel Process Technol 76:11–21

49. Gropp JG et al. (1995) Method of removing carbon from fly ash. US Patent No. 546363

50. Groppo J, Honaker R (2009) Economic recovery of fly ash derived magnetics and evaluation for coal cleaning. In: World of coal ash conference, Lexington

51. Groudev SM, Gencev FN, Groudeva VI (1982) Use of microorganisms for recovery of aluminum from alumino silicate. Achiev Prospect Travaux 12:203–212

52. Guerra MF, Calligaro T (2003) Gold cultural heritage objects: a review of studies of provenance and manufacturing technologies. Meas Sci Technol 14:1527–1537

53. Gurupira T, Jones CL, Howard A, Lockert C, Wandell T, Stencel JM (2001) New products from coal combustion ash: selective extraction of particles with density less than 2. In: International ash utilization symposium, Lexington, Kentucky

54. Gutiérrez B, Pazos C, Coca J (1997) Recovery of gallium from coal fly ash by a dual reactive extraction process. Waste Manag Res 15:371–382

55. Halina M, Ramesha S, Yarmob MA, Kamarudin RA (2007) Non-hydrothermal synthesis of mesoporous materials using sodium silicate from coal fly ash. Mater Chem Phys 101:344–351

56. Harvey RD, Cahill RA, Chou CL, Steele JD (1983) Mineral matter and trace elements in the Herrin and springfield coals, Illinois basin coal field. ISGS Contract/Grant Report 1983-4

57. Heidrich C, Feuerborn HJ, Weir A (2013) Coal combustion products: a global perspective. In: World of coal ash (WOCA) conference—April 22–25, 2013, Lexington

58. Hernandez-Exposito A, Chimenos JM, Fernandez AI, Font O, Querol X, Coca P, Garcia PF (2006) Ion flotation of germanium from fly ash aqueous leachates. Chem Eng J 118:69–75

59. Hill RO et al. (1981) Process of extraction of aluminium values from coal ash. US Patent No. 4243640

60. Hirajima T, Petrus HTBM, Oosako Y, Nonaka M, Sasaki K, Ando T (2010) Recovery of cenospheres from coal fly ash using a dry separation process: separation estimation and potential application. Int J Miner Process 95(1–4):18–24

61. Hoenderdal S, Espinoza LT, Marscheider-Weidemann F, Grus W (2013) Can a dysprosium shortage threaten green energy technologies? Energy 49:344–355

62. Hong K, Tokunaga S, Kajiuchi T (2000) Extraction of heavy metals from MSW incinerator fly ashes by chelating agents. J Hazard Mater B75:57–73

63. Huang K, Inoue K, Harada H, Kawakita H, Ohto K (2011) Leaching behavior of heavy metals with hydrochloric acid from fly ash generated in municipal waste incineration plants. Trans Nonferr Met Soc China 21:1422–1427

64. Huo P, Yan Y, Li S, Li H, Huang W (2009) Preparation and characterization of cobalt sulfophthalocyanine/TiO2/fly-ash

cenospheres photocatalyst and study on degradation activity under visible light. Appl Surf Sci 255:6914–6917

65. Hurst SL et al. (1978) Fly ash beneficiation process. US Patent No. 4121945

66. Hwang JY (1991) Wet Process for fly ash beneficiation. US Patent No. 5047145

67. Hwang JY (1993) Wet process for fly ash beneficiation. US Patent No. 5227047

68. Hwang JY, Sun X, Li Z (2002) Unburned carbon from fly ash for mercury adsorption: I. Separation and characterization of unburned carbon. J Miner Mater Charact Eng 1:39–60

69. Ilic M, Cheeseman C, Sollars C, Knight J (2003) Mineralogy of microstructure of sintered lignite coal fly ash. Fuel 82:331–336

70. Ishigaki T, Nakanishi A, Tateda M, Ike M, Fujita M (2005) Bioleaching of metal from municipal waste incineration fly ash using a mixed culture of sulfur-oxidizing and iron-oxidizing bacteria. Chemosphere 60:1087–1094

71. Iwashita A, Sakaguchi Y, Nakajima T, Takanashi H, Ohki A, Kambara S (2005) Leaching characteristics of boron and selenium for various coal fly ashes. Fuel 84:479–485

72. Izquierdo M, Querol Z (2012) Leaching behaviour of elements from coal combustion fly ash: an overview. Int J Coal Geol 94:54–66

73. Jianguo Q et al. (2010) Process for recovery of silica followed by alumina from coal fly ash. US Patent No. 0119426A1

74. Jianguo Q et al. (2011) Process for recovery of silica followed by alumina from coal fly ash. US Patent No. 7871583B2

75. Jha VK, Matsuda M, Miyake M (2008) Resource recovery from coal fly ash waste: an overview study. J Ceram Soc Jpn 116:167–175

76. Joshi PB et al. (2015) Recovery of rare earth elements and compounds from coal ash. US patent No. 8968688B2

77. Joshi PB et al. (2013) Recovery of rare earth elements and compounds from coal ash. US patent No. 0287653A1

78. Karayigit AI, Onacak T, Gayer RA, Goldsmith S (2001) Mineralogy and geochemistry of feed coals and their combustion residues from the Cayirhan Power Plant, Ankara, Turkey. Appl Geochem 16:911–919

79. Ketris M, Yudovich Y (2009) Estimations of Clarkes for carbonaceous biolithes: world averages for trace element contents in black shales and coal. Int J Coal Geol 78:135–148

80. Krebs W, Brombacher C, Bosshard PP, Bachofen R, Brandl H (1997) Microbial recovery of metals from solids. FEMS Microbiol Rev 20:605–617

81. Kruger RA (1997) Fly ash beneficiation in South Africa: creating new opportunities in the market place. Fuel 76:777–778

82. Lisowyi B, Hitchcoc DC (1987) Beneficiation of gallium in fly ash. US Patent No. 4686031

83. Li L, Wu Y, Liu Y, Zhai Y (2011) Extraction of alumina from coal fly ash with sulfuric acid leaching method. Chin J Process Eng 11(2):254–258

84. Lisowyj B (1986) Method for extracting of iron, aluminium and titanium from coal ash. US Patent No. 4567026

85. Lisowyi B et al. (1986) Process for the recovery of gallium and germanium from coal fly ash. US Patent No. 4678647

86. Lisowyi B et al. (1987) Direct fuel-fired furnace arrangement for the recovery of gallium and germanium from coal fly ash. US Patent No. 4643110

87. Lisowyi B et al. (1987) Beneficiation of gallium in fly ash. US Patent No. 4686031

88. Lisowyi B et al. (1988) Condensation sand collection system. US Patent No. 4757770

89. Lisowyj B et al. (1988) Sand collection system. US Patent No. 4757772

90. Lakshman et al. (1990) Process for treatment of fly ash. US Patent No. 4966761

91. Loya MIM, Rawani AM (2014) A review: promising applications for utilization of fly ash. Int J Adv Technol Eng Sci 2(7):143–149

92. Lu S, Fang RL, Zhao H (2003) Study of recovery of highly pure super-fine powdered aluminum oxide from fly ash by way of lime sintered self powdering. Sci Study 1:15–17

93. Lukkassen D, Meidell A (2007) Advanced materials and structures and fabrication process. Book manuscript, Narvik University College, HiN

94. Masaaki N, Yoshihiro K, Setsuo T, Toru E, Iwao N (1996) Method for recovering vanadium from petroleum incineration ash. Mitsubishi Heavy Industries, Ltd. Eu Pat No. JP-A-H08325651

95. Maslov OD, Tserenpil S, Norov N, Gustova MV, Filippov MF, Belov AG, Altangerel M, Enhbat N (2010) Uranium recovery from coal ash dumps of Mogolia. Solid Fuel Chem 44:433–438

96. Matjie RH, Bunt JR, Heerden JHP (2005) Extraction of alumina from coal fly ash generated from a selected low rank bituminous South African coal. Miner Eng 18:299–310

97. Matyas B et al. (1987) Process for recovering rare metals from the combustion residue of coal by digestion. US Patent No. 4649031

98. Mayfield D, Lewis AS (2013) Environmental review of coal ash as a resource for rare earth and strategic elements. In: World of coal ash (WOCA) conference—April 22–25, 2013 in Lexington

99. McDowell WJ et al. (1981) Salt-soda sintering process for recovering aluminum from fly ash. US Patent No. 4254088

100. Meawad AS, Bojinova DY, Pelovski YG (2010) An overview of metals recovery from thermal power plant solid wastes. Waste Manag 30:2548–2559

101. Mehrotra AK, Behie LA, Bishnoi PR, Svrcek WY (1982a) High-temperature chlorination of coal ash in a fluidized bed. 1.

Recovery of aluminium. Ind Eng Chem Process Des Dev 21:37–44

102. Mehrotra AK, Behie LA, Bishnoi PR, Svrcek WY (1982b) High-temperature chlorination of coal ash in a fluidized bed. 2. Recovery of iron, silicon, and titanium. Ind Eng Chem Process Des Dev 21:44–50

103. Mehrotra VP, Sastry KVS, Morey BW (1983) Review of oil agglomeration techniques for processing of fine coals. Int J Miner Process 11(3):175–201

104. Minter BE (2004) Method for recovering of trace elements from coal. US Patent No. 6827837B2

105. Mitner B (2005) Method for recovering trace elements from coal. US Patent No. 0056548A1

106. Mok WM, Willmes H, Wai CM (1984) Solvent extraction of molybdenum from biological samples and from coal fly ash for neutron activation analysis. Anal Chem 56:2623–2624

107. Moskaly RR (2004) Review of germanium processing worldwide. Miner Eng 17:393–402

108. Moskalyk RR, Alfantazi AM (2003) Processing of vanadium: a review. Miner Eng 16:793–805

109. Murtha MJ (1983) Process for the recovery of alumina from fly ash. US Patent No. 4397822A

110. Narukawa T, Riley K, French D, Chiba K (2007) Speciation of chromium in Australian fly ash. Talanta 73:178–184

111. National Research Council (US) (2006) Managing coal combustion residues in mines. National Academies, p 256

112. Navarro R, Guzman J, Saucedo I, Revilla J, Guibal E (2007) Vanadium recovery from oil fly ash by leaching, precipitation and solvent extraction processes. Waste Manag 27(3):425–438

113. Nayak N, Panda CR (2010) Aluminium extraction and leaching characteristics of Talcher thermal power station fly ash with sulphuric acid. Fuel 89(1):53–58

114. Ogata F, Tominaga H, Yabutani H, Taga A, Kawasaki N (2011) Recovery of molybdenum from coal by gibbsite. Toxicol Environ Chem 93:635–642

115. Okada T, Tojo Y, Tanaka N, Matsuto T (2007) Recovery of zinc and lead from fly ash from ash-melting and gasification-melting processes of MSW-comparison and applicability of chemical leaching methods. Waste Manag 27:69–80

116. Padil R, Sohn HY (1985) Sintering kinetics and alumina yield in lime-soda sinter process for alumina from coal wastes. Metallurg Mater Trans B 16:385–395

117. Pang J, LiQ Wang W, Xu X, Zhai J (2011) Preparation and characterization of electroless Ni–Co–P ternary alloy on fly ash cenospheres. Surf Coat Technol 205:4237–4242

118. Pennachetti J et al. (1973) Process for the treatment of fly ash. US Patent No. 3769054

119. Prakash S, Mohanty JK, Das B, Venugopal R (2001) Characterisation and removal of iron from fly ash of Talcher Area, Orissa, India. Miner Eng 14:123–126

120. Prasad B, Mondal KK (2009) Environmental impact of manganese due to its leaching from coal fly ash. J Environ Sci Eng 51(1):27–32

121. Qin S, Sun Y, Li Y, Wang J, Zhao C, Gao K (2015a) Coal deposits as promising alternative sources for gallium. Earth Sci Rev 150:95–101

122. Qin S, Zhao C, Li Y, Zhang Y (2015b) Review of coal as a promising source of lithium. Int J Oil Gas Coal Technol 9(2):215–229

123. Querol X, Fernández-Turiel J, López-Soler A (1995) Trace elements in coal and their behavior during combustion in a large power station. Fuel 74:331–343

124. Ramme BW et al. (2008) Separation of cenospheres from fly ash. US Patent No. 0190327A1

125. Ramme BW et al. (2011) Separation of cenospheres from fly ash. US Patent No. 8074804B2

126. Ramme BW et al. (2012) Separation of cenospheres from fly ash. US Patent No. 0050519A1

127. Ramya SS, Deshmukh VU, Khanderkar VJ, Padmakar C, SuriNaidu L, Mahore PK, Pujari PR, Panaskar D et al (2013) Assessment of impact of ash ponds on groundwater quality: a case study from Koradi in Central India. Environ Earth Sci 69:2437–2450

128. Rao RB, Chattopadhyay P, Banerjee GN (1999) Removal of iron from fly ash for ceramic and refractory applications. Magn Electr Sep 10:21–27

129. Reynolds RE et al. (1979) Process for Recovering aluminium and other metal values from fly ash. US Patent No. 4159310

130. Russ JJ et al. (1978) Process for recovering mineral values from fly ash. US Patent No. 4130627

131. Russ JJ et al. (1985) Method for separating and recovering silica, alumina and iron from fly ash, minerals ores and mine tailings. US Patent No. 4539187

132. Scharifker B et al. (2009) Process to recovery vanadium contained in acid solutions. US Patent No. 7498007B2

133. Seidel A, Zimmels Y, Armon R (2001) Mechanism of bioleaching of coal fly ash by Thiobacillus thiooxidans. Chem Eng J 83:123–130

134. Seredin VV (1996) Rare earth element-bearing coals from Russian Far east deposits. Int J Coal Geol 30:101–129

135. Seredin VV (2012) From coal science to metal production and environmental protection: a new story of success. Int J Coal Geol 90–91:1–3

136. Seredin V, Dai S, Sun Y, Chekryzhov IY (2013) Coal deposits as promising sources of rare metals for alternative power and energy-efficient technologies. Appl Geochem 31:1–11

137. Shabtai J et al. (1993) Extraction of metal oxides from coal fly ash by microorganisms and a new microorganism useful therefor. US Patent No. 5231018

138. Shiroto Y et al. (2003) Production of high purity vanadium compound from vanadium-containing carbonaceous residues. US Patent No. 6652819B2

139. Sierra-Alvarez R (2009) Removal of copper, chromium and arsenic from preservative-treated wood by chemical extraction-fungal bioleaching. Waste Manag 29:1885–1891

140. Sommerville R, Blissett R, Rowson N, Blackburn S (2013) Producing a synthetic zeolite from improved fly ash residue. Int J Miner Process 124:20–25

141. Soong Y, Schoffstall MR, Gray ML, Knoer JP, Champagne KJ, Jones RJ et al (2002) Dry beneficiation of high loss-on-ignition fly ash. Sep Purif Technol 26(2–3):177–184

142. Sparton Res Inc. (2007) First yellowcake produced from Xiaolongtang coal ash, China.http://www.spartonres.ca/pressreleases/PR2007Oct15.htm. Accessed 10 April 2014

143. Styron RW et al. (2000) Method of removing carbon from fly ash. US Patent No. 6068131

144. Tolhurst L (2015) Commercial recovery of metals from coal ash. In: World of coal ash (WOCA) conference in Nashville, May 5–7. http://www.flyash.info/

145. Torralvo FA, Fernández-pereira C (2011) Recovery of germanium from real fly ash leachates by ion-exchange extraction. Miner Eng 24:35–41

146. Toshiaki A, Akira S, Masutaka F (1998) Production of ammonium metavanadate. Eu Pat. No. JPH10251025

147. Ucurum M (2009) A study of removal of Pb heavy metal ions from aqueous solution using lignite and a new cheap adsorbent (lignite washing plant tailings). Fuel 88(8):1460–1465

148. US EPA (2005) Using coal fly ash in highway construction, a guide to benefits and impacts, EPA-530, K-05-002

149. US Environmental Protection Agency (1999) Technical background document for the report to congress on remaining wastes from fossil fuel combustion: waste characterization. EPA 530-R-99-010. Science Applications International Corporation, San Diego, CA

150. Vassilev SV, Vassileva CG, Karayigit AI, Bulut Y, Alastuey A, Querol X (2005) Phase-mineral and chemical composition of feed coals, bottom ashes and fly ashes at the Soma power station, Turkey. Int J Coal Geol 61:35–63

151. Vitolo S, Seggiani M, Filippi S, Brocchini C (2000) Recovery of vanadium from heavy oil and orimulsion fly ashes. Hydrometallurgy 57:141–149

152. Wang T, Wang J, Burken JG, Ban H, Ladwig K (2007) The leaching characteristics of selenium from coal fly ashes. J Environ Qual 36:1784–1792

153. Wang B, Li Q, Wang W, Li Y, Zhai J (2011) Preparation and characterization of Fe3+ doped TiO2 on fly ash cenospheres for photocatalytic application. Appl Surf Sci 257(8):3473–3479

154. Wu C, Yu H, Zhang H (2012) Extraction of aluminium by pressure acid-leaching method from fly ash. Trans Nonferr Met Soc China 22:2282–2288

155. Xiang-yang C, Xin-zhe L, Qui-li Z, Hong-zhou MA, Hin Z (2010) Leaching vanadium by high concentration sulfuric acid from stone coal. Trans Nonferr Met Soc China 20:123–126

156. Xu M, Yan R, Zheng C, Qiao Y, Han J, Sheng C (2003) Status of trace element emission in a coal combustion process: a review. Fuel Process Technol 85:215–237

157. Xu X, Li Q, Cui H, Pang J, Sun L, An H et al (2011) Adsorption of fluoride from aqueous solution on magnesia-loaded fly ash cenospheres. Desalination 272(1–3):233–239

158. Yao ZT, Xia MS, Sarkar PK, Chen T (2014) A review of the alumina recovery from coal fly ash, with a focus in China. Fuel 120:74–85

159. Yao ZT, Ji XS, Sarker PK, Tang JH, Ge LQ, Xia MS, Xi YQ (2015) A comprehensive review on the applications of coal fly ash. Earth Sci Rev 141:105–121

160. Yavuz CT, Prakash A, Mayo JT, Colvin VL (2009) Magnetic separations: from steel plants to biotechnology. Chem Eng Sci 64(10):2510–2521

161. Zhang FS, Itoh H (2006) Extraction of metals from municipal solid waste incinerator fly ash by hydrothermal process. J Hazard Mater 136:663–670

162. Zhang Y, Talbott JL, Wiedenmann L, DeBarr J, Demir I (1999) Determination of germanium content in coal ash by wavelength-dispersive X-ray florescence spectroscopy and inductively coupled—mass spectrometry. JCPDS-International Center for diffraction data

163. Zyryanov VV, Petrov SA, Matvienko AA (2011) Characterization of spinel and magnetospheres of coal fly ashes collected in power plants in the former USSR. Fuel 90(2):486–492

Index